SALT TECTONICS

SALT TECTONICS

MALCOLM K. JENYON

Manager, Interpretation,
Seismograph Service (England) Ltd,
Holwood, Keston, Kent, UK

ELSEVIER APPLIED SCIENCE PUBLISHERS
LONDON and NEW YORK

ELSEVIER APPLIED SCIENCE PUBLISHERS LTD
Crown House, Linton Road, Barking, Essex IG11 8JU, England

Sole Distributor in the USA and Canada
ELSEVIER SCIENCE PUBLISHING CO., INC.
52 Vanderbilt Avenue, New York, NY 10017, USA

WITH 185 ILLUSTRATIONS

British Library Cataloguing in Publication Data

Jenyon, Malcolm K.
Salt tectonics.
1. Evaporites 2. Geology, Structural
I. Title
553.6'32 QE471.15.E8

Library of Congress Cataloging-in-Publication Data

Jenyon, Malcolm K.
Salt tectonics.

Bibliography: p.
Includes index.
1. Salt. 2. Salt deposits. 3. Geology, Structural.
4. Evaporites. 5. Diapirs. I. Title.
QE389.4.J46 1986 553.6'3 86-11638

ISBN 1-85166-015-1

Printed in Great Britain by Galliard (Printers) Ltd, Great Yarmouth

To my wife, Ann,
and daughters
Amanda, Elizabeth and Joanne

Preface

This monograph is addressed to all earth scientists, whether of geological or geophysical background, who are interested in earth processes in general, and evaporite geology in particular.

The 'tools of the trade' used by those engaged in the earth sciences vary. Some use hammer and clinometer, others the microscope or geophysical instruments. Those used in this work are seismic sections, structure maps derived therefrom, borehole information, and other displays of geological and geophysical data. From these, observations are made and hypotheses formulated which whenever possible have been tested—in Popperian spirit—to destruction.

Since the seismic examples shown are intended to argue geological cases, it is not believed that any profound knowledge of the seismic method is required on the reader's part, although as an aid for those less familiar with seismic usages, some of the basic aspects as they relate to geological interpretation are discussed briefly in the second chapter. Because of its brevity and modest intentions, this treatment may for some require augmentation, and for those wishing to pursue the seismic aspects in greater depth, several excellent books are included in the Bibliography.

MALCOLM K. JENYON
Petts Wood, Kent, UK

Contents

Acknowledgements

My thanks go first to the Directors of Seismograph Service (England) Limited for permission to use many examples of high quality seismic data proprietary to the Company as illustrative material.

Sincere gratitude is also owed to the following: to Dr A. A. Fitch, who kindly undertook the considerable task of reading the first draft of the book, and who made innumerable valuable suggestions. I count myself as fortunate indeed to be able to draw upon his vast knowledge of matters both geological and geophysical; to Dr J. C. M. Taylor of V. C. Illing and Partners, who read a part of the draft and who gave me the benefit of his expertise in evaporite geology; to Dr G. D. Hobson, for kindly providing useful information and comment on several matters; to Thomas G. Fails of Denver, Colorado, USA, for his generosity in providing information and publications—some not easily obtainable in England—on US Gulf Coast salt tectonics; to the National Bureau of Standards, US Department of Commerce, for kind permission to make full use of material from their excellent Monograph No. 167, *Physical Properties Data for Rock Salt* (1981); to Mrs Bernice Farrell for her work on illustrations and, last but not least, to my family, to whom this book is dedicated, for their patience whilst the work was in progress.

I am also grateful to the publishers, authors and companies who kindly gave permission for the reproduction of illustrations from other works. Acknowledgements for these appear in the individual figure captions, and full entries for the works involved appear in the Bibliography and References list at the end of the book. For conciseness, some abbreviations have been used in the caption acknowledgements, and these are as follows:

AAPG	The American Association of Petroleum Geologists
CSPG	The Canadian Society of Petroleum Geologists
GSA	The Geological Society of America
SS(E)L	Seismograph Service (England) Limited

M.K.J.

CHAPTER 1

Forethoughts

1.1 INTRODUCTION

The geology of evaporites has, since the mid-nineteenth century, engendered a vast literature. Cramer (1969), for instance, in a selected bibliography, lists over 700 references with an emphasis on sources relatively easy to obtain dealing with marine evaporites of the North American area. In the global context, this number could be multiplied many times, and a great many of these works are devoted partly or wholly to structures resulting from the deformation of mobile salt acting as a member of evaporite cycles. This being the case, is there really any need for a book of this nature?

With the increasing ability of modern applied seismology to depict a two-dimensional, and even (in a limited sense) a three-dimensional, representation of the subsurface geology in an area, much that was previously based only on inferences drawn from borehole correlations can now be seen more or less clearly on the seismic section. In an area as complex and subtle as an evaporite basin in which salt movement and/or dissolution has taken place, stratigraphic sections based on information from widely spaced boreholes are quite inadequate to resolve many of the structural situations which result. Only the much higher sampling of the subsurface (albeit with a lower degree of resolution than for the data from an individual borehole) by the reflection seismic method offers the possibility of arriving at some solutions to the many problems presented by such an area.

And there lies the rub! The vast majority of the countless thousands of line kilometres of seismic data acquired in salt areas is commissioned by organisations involved in hydrocarbon exploration, and as such is subject to stringent confidentiality. This is entirely understandable. The acquisition and processing of seismic data represents a substantial financial investment on the part of the organisations concerned, and also, in this highly competitive industry, knowledge is power. Another circumstance which militates against new geological information coming into the public domain is that the overworked interpreters of the bodies owning the seismic data normally have not the time to consider the broader, perhaps more academic framework of salt tectonics. They are under pressure to answer specific questions on hydrocarbon accumulation in what are normally quite restricted areas.

This book results from the writer having access to a large body of state-of-the-art seismic data, and having been given special permission to use some excellent examples as illustrations. It is intended in part to be a review of salt tectonics as seen in the light of excellent modern seismic data. It is also hoped to include some new ideas and suggested lines of thought on certain aspects in a subject which is not lacking in controversy and inconsistencies. An attempt is made to relate the known geological phenomena associated with salt tectonics to some characteristic features on the seismic section, and in turn to infer the geological meaning of seismic events, the significance of which is not immediately obvious.

1.2 THE USE OF REFLECTION SEISMIC DATA

There are limitations to the seismic reflection method; some of these are inherent in the method, while others stem from the stage reached currently by a developing technology. The latter will doubtless disappear—or change—with time. The former, which involve basic physical laws, will not. The geologist who may in the course of his work look at seismic data or maps generated therefrom must be fully aware of this. The use of data produced by this most successful of applied geophysical techniques is, because of the limitations mentioned, beset with traps and snares with which the interpreter must be intimately acquainted. The oft-quoted 'He that diggeth a pit shall fall into it' (Proverbs, 10:8) might almost have been written about the first man to dig a seismic shot-hole!

It is certainly true, however, even after such cautionary words, that the interpreter of modern seismic data should be able to extract from it a degree of geological detail which would have been unthinkable a few years ago. It is therefore with a modest measure of confidence that the study of salt tectonics through good seismic data may be approached.

A salt basin, in spite of the complexity often present, is perhaps one of the most satisfying of geological environments to study by seismic means. There are frequently strong, clear reflection events associated with a salt interval—at least at one or both of its bounding interfaces if not within the salt itself. Plastic deformation of the salt produces well-defined and interesting structures in the overlying sequence as well as fascinating distortions of any competent bands of differing lithology within the salt interval. In many areas, leaching of deeply buried salt layers by migrating water undersaturated with respect to the salt produces collapse features in the overburden which would be difficult or impossible to study in detail without seismic data. Fault patterns may influence, or be the result of, salt movement. Where hydrocarbons are present—and in the majority of the world's major oil and gas fields there is close association with evaporites—there is clearly much potential for structural trap formation above the salt, which may also provide a most efficient seal formation for underlying accumulations. In such an area, there is thus a great deal to engage the interest of the interpreter.

Piquancy is added to the study by the fact that many questions still remain to be answered regarding salt geology. The largest question, which involves the mode of origin of the many giant evaporite basins which have formed through geological time with global distribution, will not be addressed here, although it will be discussed briefly in review. Suggestions will, however, be made regarding some of the many remaining problems presented by salt features on both regional and local scales. Certain inconsistencies which have become apparent between laboratory and field observations related to salt deformation will also be commented on, in the light of seismic and other studies.

1.3 ABOUT THE BOOK

A few explanatory words are required about the illustrations of seismic data used. Unless otherwise stated, all the examples shown are of migrated seismic sections, since we are here concerned with the real-earth shapes of structures, which are approximated by migrated seismic sections when these are in the true dip direction for the structure under consideration. Unmigrated ('stacked') sections, which are those used by the interpreter for carrying horizons round closed grid-loops in the survey area to give tied control, depict distorted responses to subsurface shapes. In occasional instances where they are used to illustrate particular points, their nature is stated. Elsewhere, whether stated or not, the examples can be assumed to be of migrated data.

All the seismic sections shown will have a vertical scale of two-way time (TWT) in seconds. This is the standard vertical scale on all seismic sections used in interpretation. Vertical scales in depth, which are more familiar to the geologist, are seldom used on sections during the interpretation of seismic data. Any depth conversion is normally executed at a late stage directly on a map. Rather obviously, two-way time is used for the vertical scale since the seismic wave must travel down to the various reflecting horizons, and then back up to the surface, before being recorded at the detectors, and there is no necessity to convert to one-way time until depth calculations are required.

The horizontal scale, where necessary, is shown on the illustrations, and is determined by the trace spacing on the cross-section, and on the spacing of detector groups on the seismic line (on land) or in a marine streamer cable. A point to note is that there is a considerable vertical exaggeration present in seismic sections, which is about 2·5:1 in most of the examples shown; this should always be borne in mind when looking at the examples, since it has a marked effect on the realities of angles of dip, fault hade, etc.

Throughout the book, references will be made to physical model studies carried out by a number of workers. Such studies have, in the past, given many insights into the processes involved in this field of study. Many suggestions and predictions derived from model behaviour have later been confirmed to be valid for processes in the real earth by drilling, seismic and other methods. Much of value probably still remains to be learned from well-designed and correctly scaled models, both physical and computer-based. It may also be true that some incorrect inferences have been drawn from such studies, and there will be discussion of this later in the book.

As regards the overall plan of the book, it will be seen that the following three chapters are essentially reviews of basic aspects of the study. Chapter 2 discusses in a preliminary way some current concepts regarding salt geology, and their relationship to some of the typical features seen in the seismic data of a salt area. Chapter 3 is an overview of current ideas on evaporite basins, and the chemistry and physics of salt deposits. In Chapter 4, the development of ideas on salt movement and diapirism is linked to physical model study work as represented by certain key papers and studies selected from a very large field. From Chapter 5 onwards, several different aspects of salt tectonics are examined in some detail, based largely on observations derived from seismic data. On some points, these observations and ideas to which they have given rise appear to conflict more or less with current concepts, and these points will be discussed.

In any work of this nature, it is both necessary and desirable to utilise, and build upon, the work of those who have gone before, and the writer wishes to record his sincere gratitude, and acknowledge his indebtedness, to the many workers in this field whose names are cited throughout the book.

The first halokinetic area with which the writer was concerned as an interpreter of seismic data was in the northern part of Spain, north of the Ebro Valley, in the early 1960s. The fascination of the spectacular structures present stimulated an interest, fed by later work in salt areas in the Middle East and Europe, which has not diminished. If some of this enthusiasm can be transmitted to the reader, then the book will have been worthwhile.

CHAPTER 2

Preliminary Aspects of the Interpretation of Seismic Data in Salt Areas

2.1 INTRODUCTION

The most effective way of discussing the use of seismic data in the interpretation of geological features is to present examples of the seismic response to various geological structures. This, the objective of the present chapter, will be combined with the requirement to introduce some ideas basic to material in later chapters dealing with salt area phenomena in more detail.

In order to carry out this scheme, certain ideas regarding the properties and deformational characteristics of salt rock will be presented in a preliminary way; many of these ideas will be developed in more detail later in the work. They are presented here as what are believed to be concepts currently held by many—perhaps most—geologists. Some of them involve inconsistencies which will be explored at greater length in certain of the chapters which follow. They will be accepted as 'working' concepts for the present, although comments will be made on certain issues.

2.2 TERMINOLOGY

There is a degree of ambiguity in some current terminology descriptive of features associated with salt rock and salt mobility, and it is proposed at this point to define some terms as they will be used hereafter in this work.

A positive salt feature produced by a thickening of the salt, the maximum amplitude (base to top salt interval) of which is equal to or less than twice the current thickness of the salt source layer adjacent to it (i.e. a thickness increase equal to or less than the current adjacent source layer thickness), will be referred to arbitrarily as a salt *swell*. A salt thickness increase greater than this, up to the stage of actual piercement of the overburden, will result in a feature called a salt *pillow*. The term *diapir* will be restricted to a salt structure where actual piercement of the overlying sequence has taken place. (This definition is necessary since, in recent years, the term 'diapir' has sometimes been used to refer to a non-piercement structure, which is an antilogy.) The use of other terms as alternatives to diapir, such as 'dome' or 'stock', will be avoided except where they occur in literature citations. The terms salt *roof* and salt *floor* will refer to the surfaces of contact with a salt interval of the upper and lower confining sequences respectively.

Further definitions of terminology involved in salt movement will be given at relevant points in later chapters.

2.3 SOME PHYSICAL PROPERTIES OF SALT ROCK

Salt rock is formed mainly from the mineral halite (chemical formula NaCl), which in its pure form has a hardness of 2·5, and a density of approximately 2·17 g/cm^3. An important aspect of the mineral is that its density is effectively invariant with depth, i.e. it shows no effects of compaction with increasing depth of burial. Since density is probably the most important determining parameter of seismic velocity, it is not surprising to find that salt rock, made up of interlocking crystals of halite, shows constant seismic velocity independent of thickness and burial depth.

The remarkable consistency of seismic velocity through a halite interval is demonstrated by the tracing from an acoustic velocity log shown in Fig. 2.1. Three portions of the log, covering a total depth interval of over 6500 ft (2000 m), are seen, and over the whole interval the log trace gives a transit time value of about 68 μs/ft, equivalent to a seismic interval velocity of 14 625 ft/s (4500 m/s). The two minor log excursions marked A are produced by thin anhydrite bands within the halite.

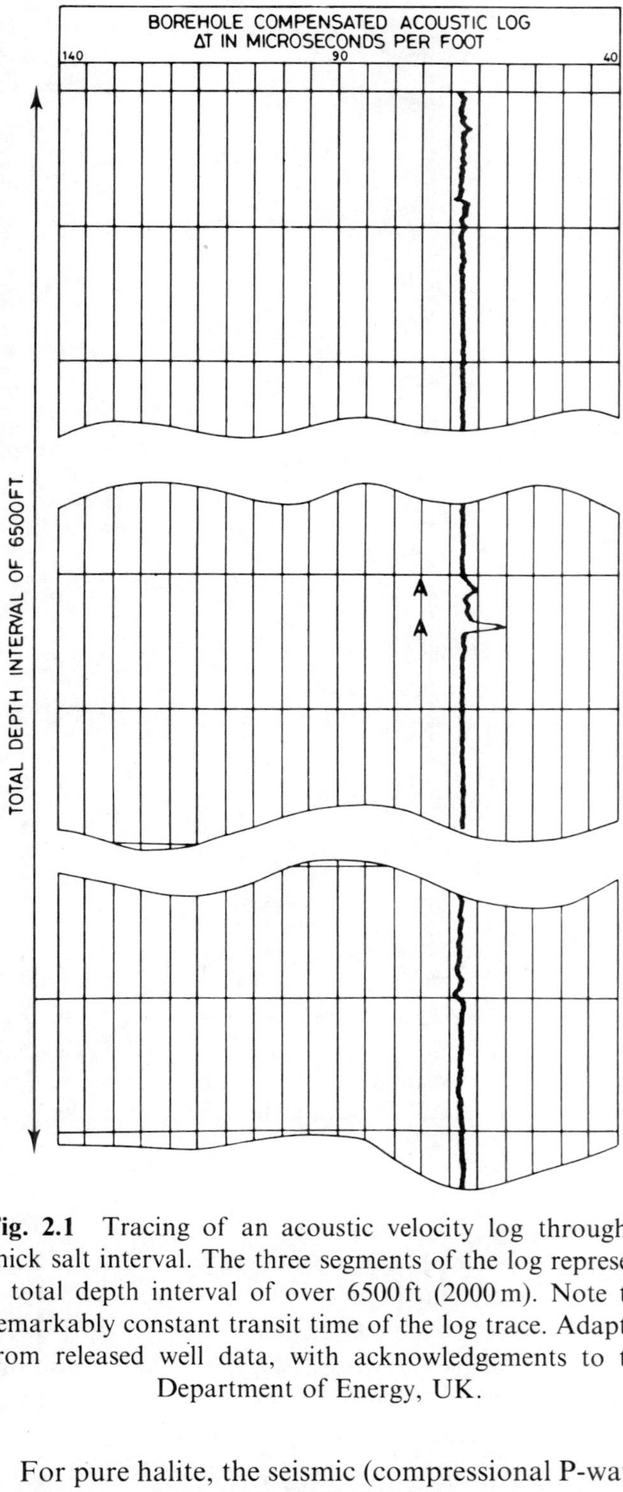

Fig. 2.1 Tracing of an acoustic velocity log through a thick salt interval. The three segments of the log represent a total depth interval of over 6500 ft (2000 m). Note the remarkably constant transit time of the log trace. Adapted from released well data, with acknowledgements to the Department of Energy, UK.

For pure halite, the seismic (compressional P-wave) velocity is of the order of 4500 m/s, as seen in this illustration. Impurities are commonly present, however, and the most common of these is anhydrite ($CaSO_4$), which in its pure form has a density of about 2·98 g/cm^3 and a velocity of the order of 19 500 ft/s (6000 m/s). The velocity variations in a binary system with halite and anhydrite as the end members are related in a linear way approximately (either by weight or volume) to the densities of various mixtures of the two components between the end values quoted. It is not very unusual to find salt rock containing an appreciable proportion of disseminated anhydrite, with a seismic velocity of 15 600 ft/s (4800 m/s) or higher. Also, the presence of thick bedded anhydrite and carbonate units within a total salt rock interval may have an appreciable effect on the average seismic velocity through the salt interval.

The admixture of anhydrite and other impurities with halite in salt rock appears to have no sensible effect on the *invariance* of density and velocity of the rock.

Since, unlike salt rock, clastic sediments are compactible to varying degrees, and thus undergo density increase with increasing burial depth, it follows that in a section containing a salt interval there will be some critical depth below which there will be a density inversion at the top of a salt layer. Typically, with normal clastic sequences and depending on the lithologies involved, the average overburden density would equal that of salt rock somewhere within the thickness range 2000 to 3250 ft (600 to 1000 m). The frequent mention in the literature of about 1000 m of overburden as being the minimum loading required to initiate salt movement would thus seem to be related to the average overburden density.

If a salt rock interval occurs in the stratigraphic column in any area at a depth greater than that for the 'critical loading' mentioned, a gravitationally unstable situation will exist, with a buoyant salt layer beneath a heavier overburden. The subsequent history of such an area will depend upon whether or not the salt remains in an effectively rigid state. Should the salt become mobile, for reasons to be discussed, it will respond to any imbalance in overburden loading by flowing plastically from regions of high load stress, laterally, towards regions of lower load stress, e.g. from a basin centre towards the margins, or away from beneath a linear edge of thickening of the overburden. Apart from lateral flow, vertical movement into swells, pillows and diapirs may occur in response to irregularities in the salt roof or floor (due for instance to faulting), or in response to purely gravitational (halokinetic) effects of variation in density of the overburden, or in the viscosity of the salt rock itself. A belief held by some geologists is that, even where no load imbalance exists, salt features including diapirs may initiate quite spontaneously due purely to a situation of gravitational instability. This is believed to be a controversial concept, and will be discussed at some length in the chapters on model studies and diapirism.

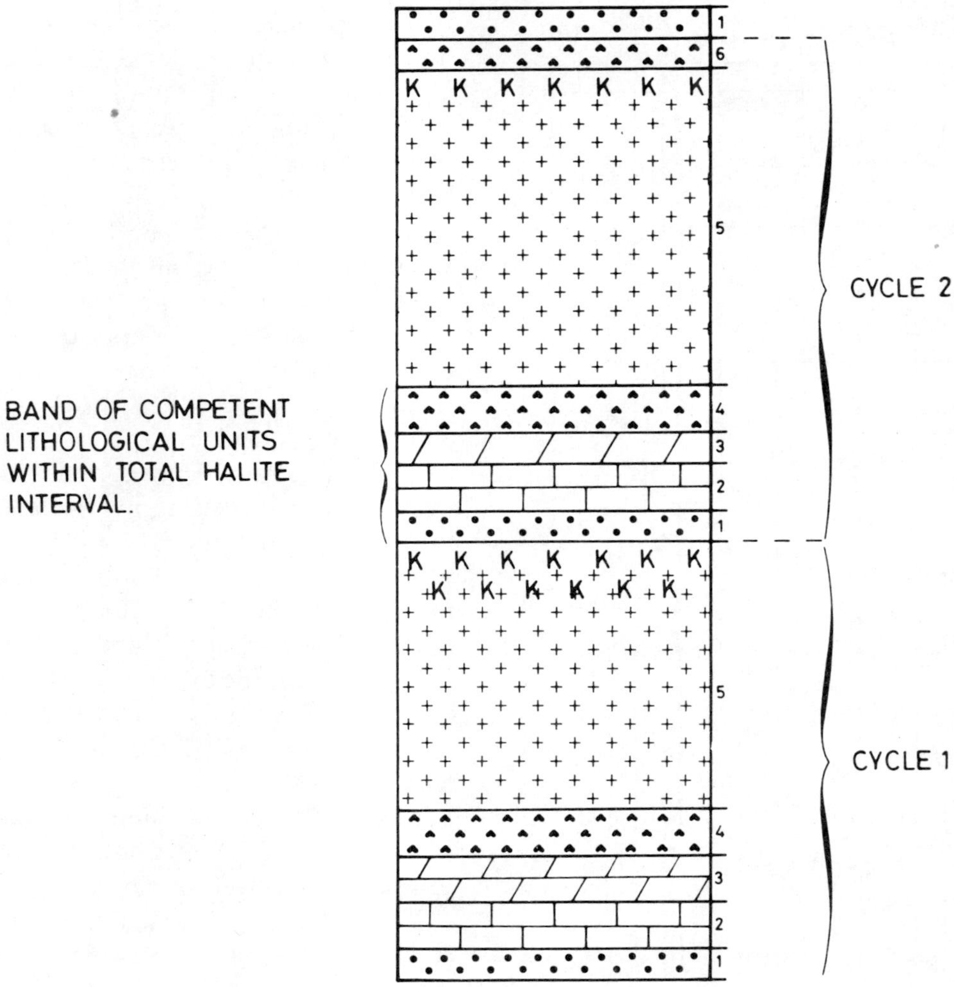

Fig. 2.2 Sketch of two idealised evaporite cycles. No scale is involved, and there is no significance in the relative thicknesses shown. Key: 1, thin clastic member; 2, limestone; 3, dolomite; 4, anhydrite; 5, halite, with potash (K) and other bittern salts interbedded in the late stages of the cycle. Cycle 2 also has a 'regressive' anhydrite (6) in the last stage, which is not present in cycle 1. Note that members 1 to 4 of cycle 2 form a band of competent rocks within the total salt interval of cycles 1 and 2.

2.4 MOBILITY IN SALT ROCK

Early ideas on mobility in salt included the belief that it had been melted by igneous intrusion, and had flowed in molten form in order to produce the observed features. Later work in the laboratory indicated, however, that halite passes from the elastic to the plastic state at temperatures far lower than its melting point (under atmospheric pressure) of about 800°C.

For halite in its crystalline form, it is found that the elastic limit varies with temperature slightly differently when stress is applied in different crystallographic directions; it begins to become plastic at 200°C, and is completely plastic at 350°C. At 300°C it is plastic in all but one direction, and will flow readily if the required stress is applied (Gussow, 1968).

As regards deformation under pressure, Van Tuyl (1930) showed that at normal temperatures the rate of creep of salt becomes 'excessive' at a pressure of 12 000 psi (equivalent to a geostatic load of about 11 000 ft (1 ft = 0·3048 m) of overburden of average clastic type. However, it has long been known from observations in mine galleries at quite shallow depths that salt creeps under much lower differential pressures than this.

At this point it must be remarked that there appear to be considerable discrepancies between ideas on deformation parameters in salt rock as based on laboratory determinations and those based on field observations by some workers. One of the most obvious cases is that of the emergent salt diapirs ('salt glaciers') of Hormuz in Iran. Kent (1979), quoting the

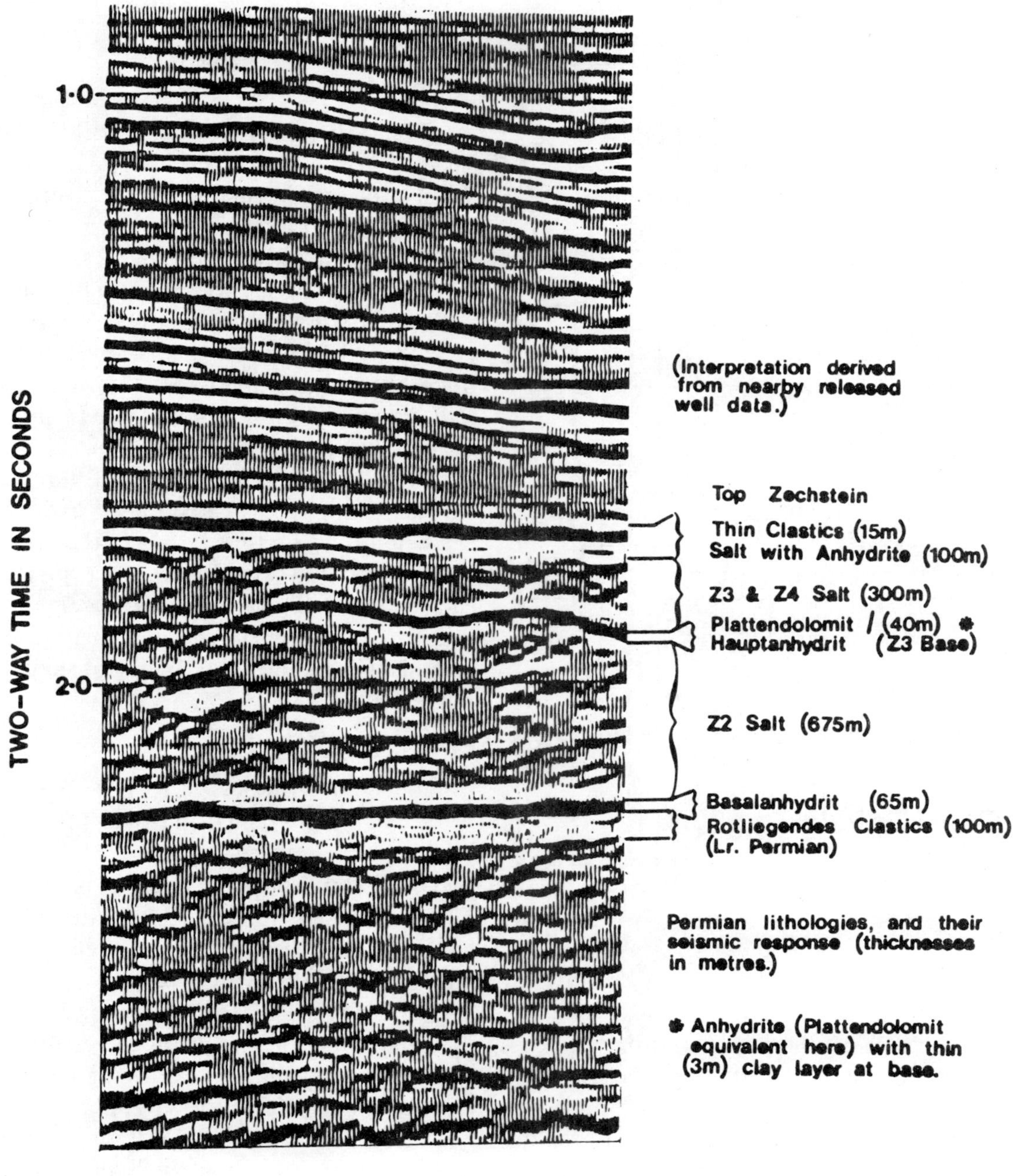

(a)

Fig. 2.3 (a) The response of an evaporite sequence to the seismic reflection technique. Note particularly the strong event produced by the competent band of the Plattendolomit equivalent at the base of the Zechstein Z3 cycle, which separates the Z2 salt below from the Z3 and Z4 salts above. Some movement of the salt is indicated by the deformation of the Plattendolomit band. (Courtesy of Seismograph Service (England) Ltd—SS(E)L.) (b) Tracing of an acoustic velocity log, showing the appearance of the Plattendolomit band (P–P) on the log trace (not in the same location as the example in (a)). The 50 μs/ft transit time through the P–P interval indicates an interval velocity in excess of 6000 m/s, and this suggests, as confirmed by the lithology log, that the Plattendolomit equivalent in this area consists mainly of anhydrite. The halite is showing a transit time of 68 μs/ft, which is typical for the salt velocity in this area. The very irregular segment of log trace below the P–P band is thought to be due partly to faulting passing through the well, although within it can be seen a short fragment of log trace showing the typical salt transit time. Adapted from released well data, with acknowledgements to the Department of Energy, UK.

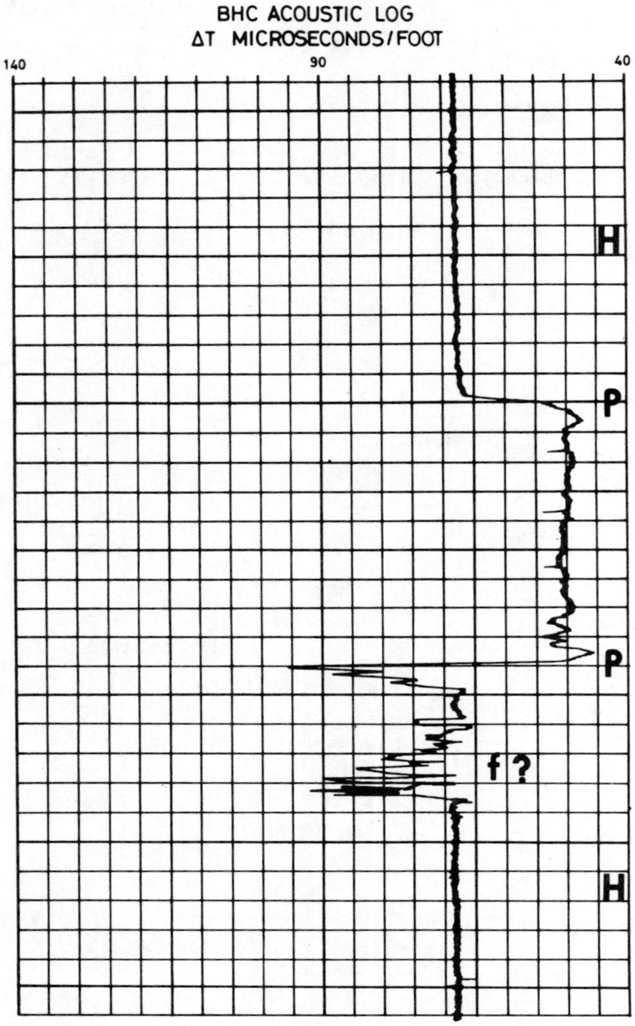

(b)

Fig. 2.3—*contd.*

results of his own and other workers' observations, avers that seasonal, and even diurnal, variations in flow rate of the salt at the surface have been measured in recent years at ambient surface temperatures not greater than 46°C; laboratory-based analysis would insist that the ambient temperature for such flowage must be in the range of 200°C to 300°C. Note also that this is taking place at atmospheric pressure, with only minor local gravitational gradients operating as stress differentials.

The above is given as one example of a number of discrepancies which exist in the matter of salt deformation. These will be pursued at length in the chapters on halite petrophysics and diapirism. For the purposes of this present chapter, and on commonsense grounds, it seems acceptable to state that there is probably a wide range of temperature/differential pressure scenarios which when applied with other factors will bring about the change of state of halite from the elastic to the plastic, and facilitate deformation.

2.5 MECHANISMS OF DEFORMATION

Briefly, of the normally accepted modes of plastic deformation, intergranular movement will not be of importance in an interlocking mass of halite crystals. The principal modes involve intragranular movement.

Plastic deformation occurs within individual halite crystals by various forms of dislocation and gliding in dodecahedral (and perhaps also cubic) planes within the lattice. This is accompanied—probably at all stages—by pressure solution, the Riecke principle, which involves solution of parts of the crystal in planes perpendicular to the direction of greater stress, with reprecipitation on those parts of the crystal subject to least stress (Gussow, 1968). In a given stress field, each crystal will deform in the same way, and the integrated effect of the shape changes of the individual crystals will result in the change of shape of the whole polycrystalline body.

2.6 SEISMIC EVIDENCE OF SALT DEFORMATION

Having established a minimal background of the physical aspects of salt rock and its deformation, this can now be related to effects seen in actual seismic data from areas of salt geology.

Certain types of evidence indicate the effects of deformation of salt rock on the seismic section, and these may be grouped roughly under three headings: (a) indications within the salt interval, (b) indications outside, and at the roof and floor of the salt interval, and (c) other indications in sequences contiguous to the salt.

2.6.1 Indications of Deformation within the Salt Interval

Indications within the salt interval stem from the usual mode of occurrence of the salt as a member of an evaporite cycle. Where several such cycles are superimposed, as in the sketch of Fig. 2.2, the salt in the total evaporite interval has within it bands of other lithologies—typically carbonates, anhydrites, and bittern lithologies—which represent the lower (and sometimes the upper) part of individual cycles. Thus

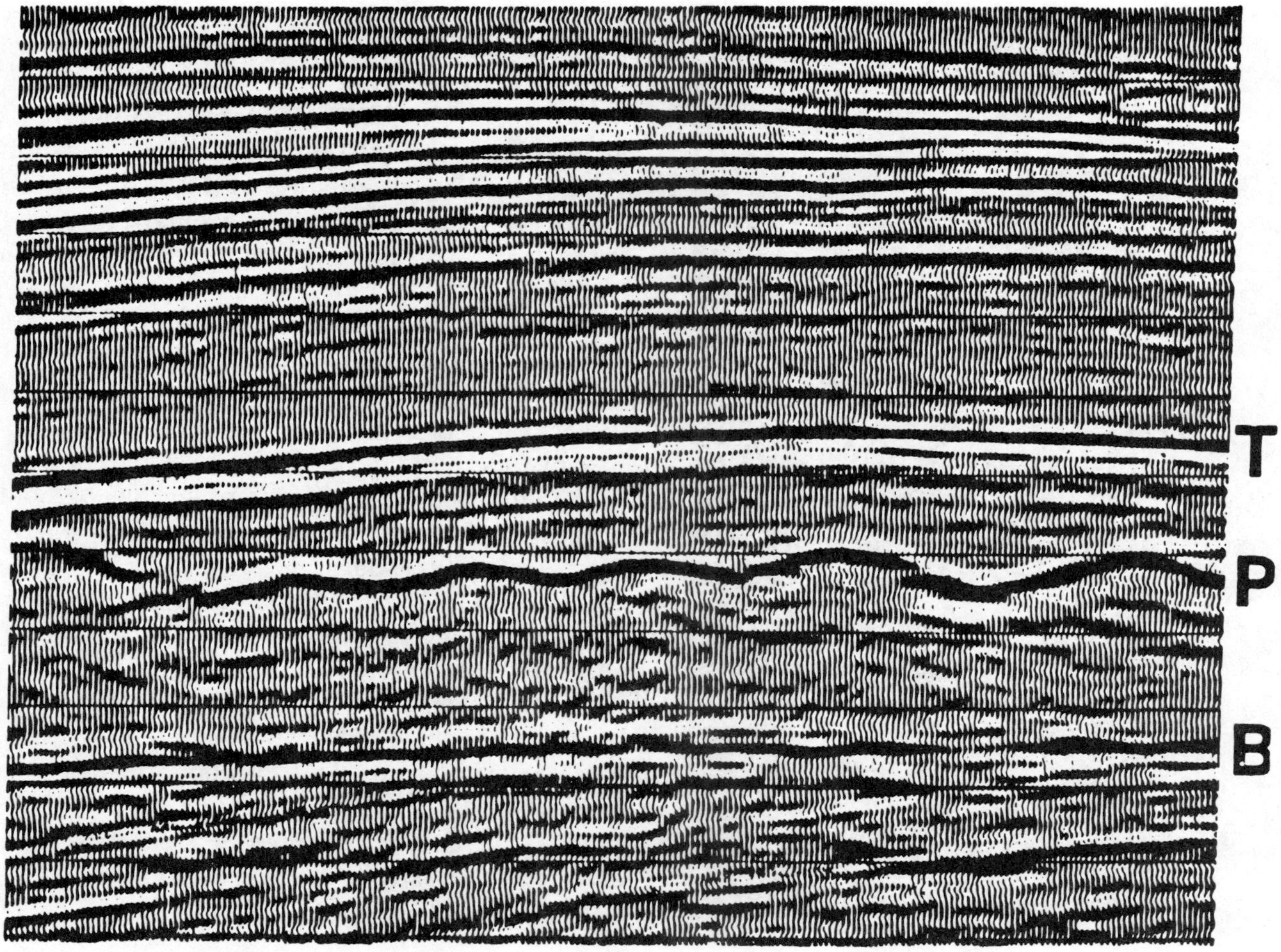

Fig. 2.4 Gentle deformation of a competent band, P, within the salt interval B–T. (Courtesy of SS(E)L.)

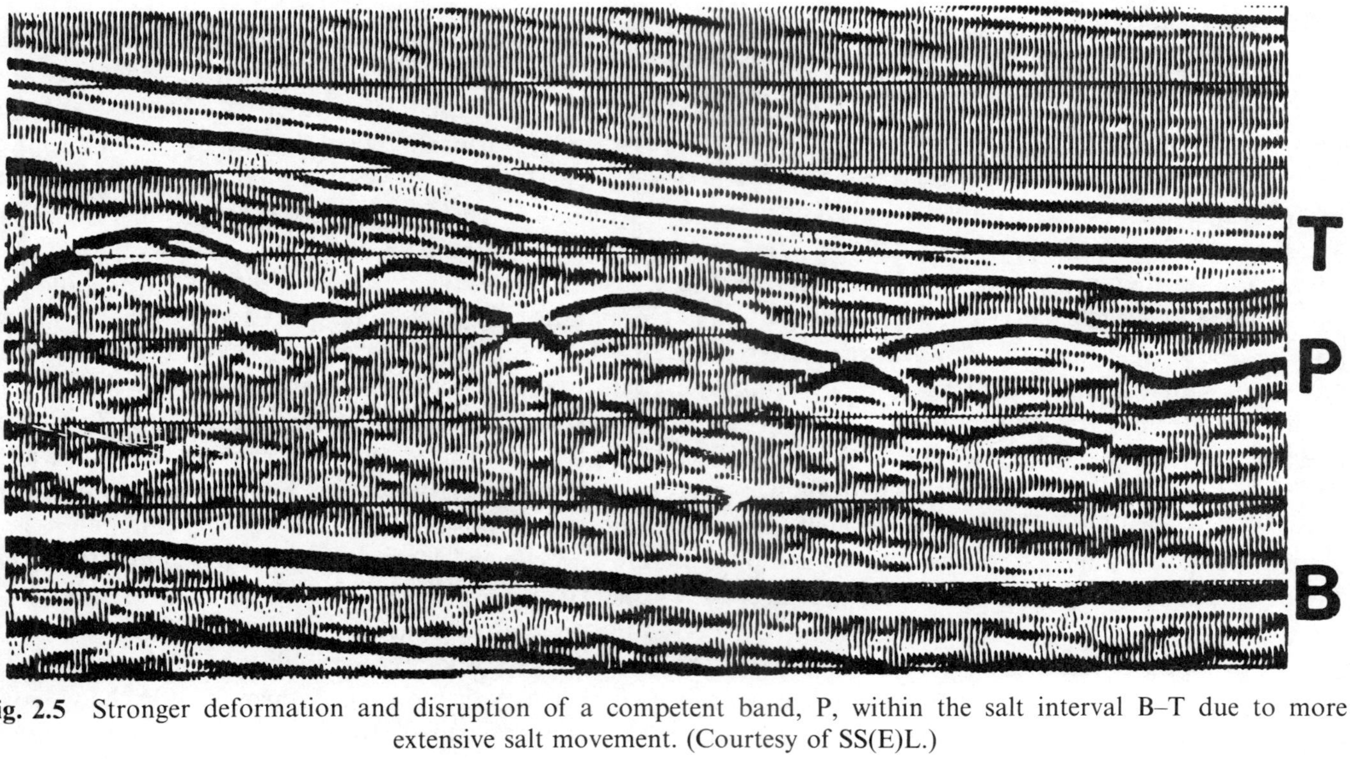

Fig. 2.5 Stronger deformation and disruption of a competent band, P, within the salt interval B–T due to more extensive salt movement. (Courtesy of SS(E)L.)

Fig. 2.6 Evaporite core from borehole in NE England, showing deformation of thin anhydrite bands in halite which has been mobile. The 1p coin gives scale. (Photograph by Dr D. B. Smith, reproduced by courtesy of the National Coal Board, UK.)

in the total evaporite sequence there may be one or more bands of competent rock types, beneath and above which are often much thicker salt intervals. As these competent lithologies are usually much denser than halite, if they are sufficiently thick compared to the dominant seismic wavelength at the particular level on the seismic section, they will produce strong and persistent seismic reflection events within the normally reflection-free salt interval.

Such a situation is seen in Fig. 2.3(a), where a competent band of Zechstein Z3 carbonates and anhydrites—the Plattendolomit equivalent—is seen separating the Z2 salt below from the Z3 salt above. Figure 2.3(b) shows the acoustic log trace of such a band.

Such competent bands form excellent indicators of the early stages of salt movement, since when the salt deforms plastically, the deformation of the competent bands, whether by lateral or vertical movement, is faithfully recorded by the reflection events to which they give rise. Figure 2.4 shows the quite gentle deformation produced in the same Zechstein Z3 unit in an example from the British sector of the Southern North Sea. The deformation in this case is probably caused by rather restricted horizontal movement of the Z2 and Z3 halites towards the basin edge, which is off section to the right.

More extensive movement of the salt produces much stronger deformation in these bands, as seen in Fig. 2.5. On a smaller scale, this type of deformation is seen in the core in Fig. 2.6.

Sometimes, by a process which appears to be akin to magmatic stoping during igneous batholithic intrusion, large sheets of the salt roof, or the uppermost competent bands of the last evaporite cycle, are peeled away due to forceful injection of the salt, and project inwards into the salt mass, as in Fig. 2.7. This effect tends to be seen very frequently at a late stage in the immediate vicinity of a diapir where the overlying layers have been shattered by piercement so that the roof sequence is in places unsupported. It is also probably more common where the roof sequence is made up of relatively incompetent rock types, rather than, say, compact indurated rocks such as some carbonates.

2.6.2 Indications Outside, and at Roof and Floor, of Salt

Thickening and thinning of a salt interval produces structures in the overlying sequence. The overlying formations adapt themselves to the shapes taken up by the salt, usually by gentle folding initially, but at a later stage by more pronounced folding and by stress-zone faulting (discussed later). An example of the gentle, open type of folding is shown in Fig. 2.8.

Since stress relief in a salt interval in response to gravitational or tectonic forces must be by lateral or upward movement, it would be unusual to find deformation of the salt floor sequences due to salt movement. However, apparent highs and lows may appear beneath the salt layer; these are known to the seismic interpreter as 'pull-up' and 'pull-down' features, or positive/negative velocity anomalies, and are the effects of the velocity differentials upon the underlying sequences caused by the variations in thickness of the constant-velocity salt layer. In which sense these velocity anomalies occur—or, indeed, whether they occur at all—depends upon the value of the seismic velocity in the salt compared to the velocity

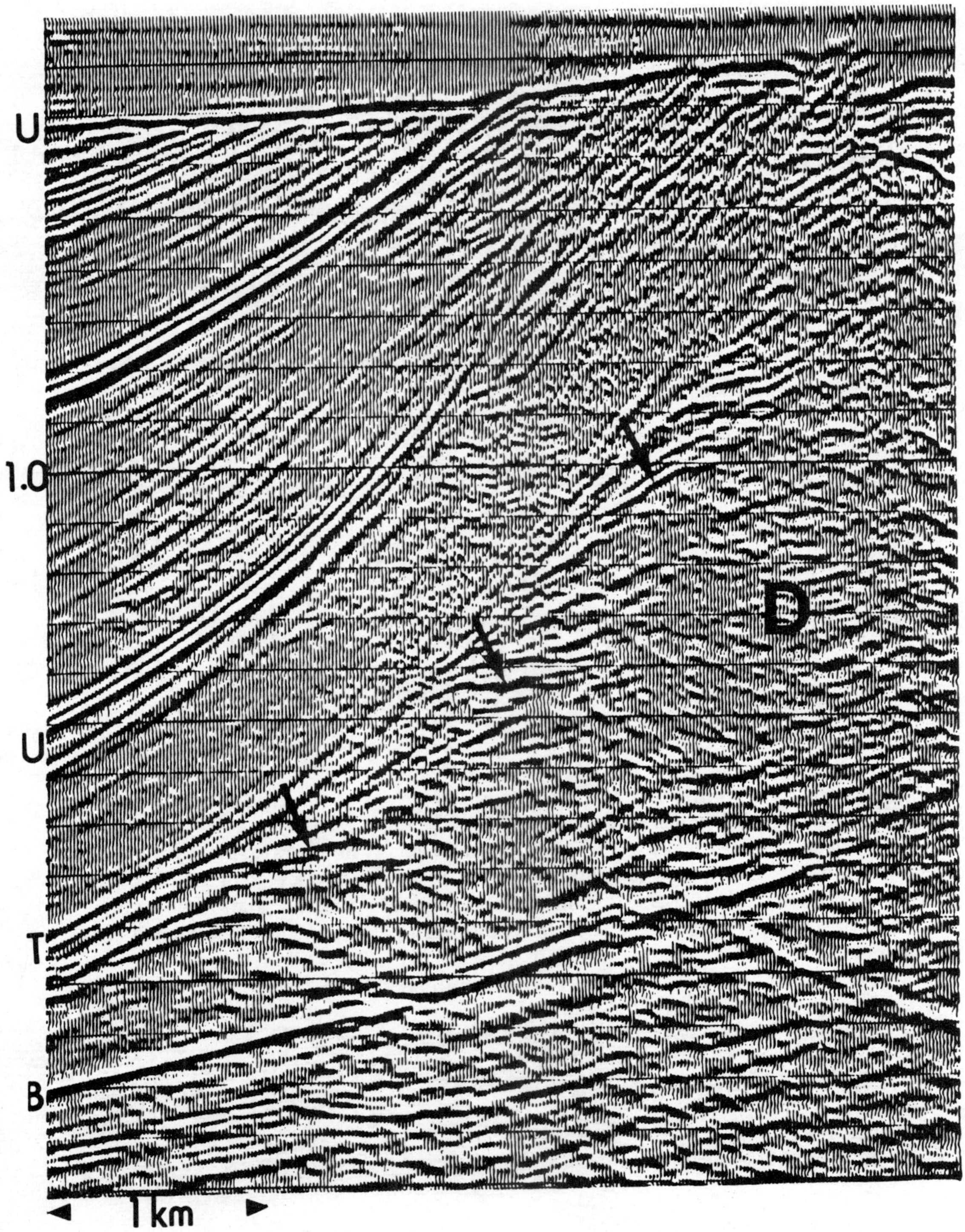

Fig. 2.7 Shards of salt roof rocks producing events, some of which are arrowed, projecting into the body of the salt of a diapir (D). This is probably an effect of shattering of the roof sequence due to piercement by the rising salt mass. (Courtesy of SS(E)L.)

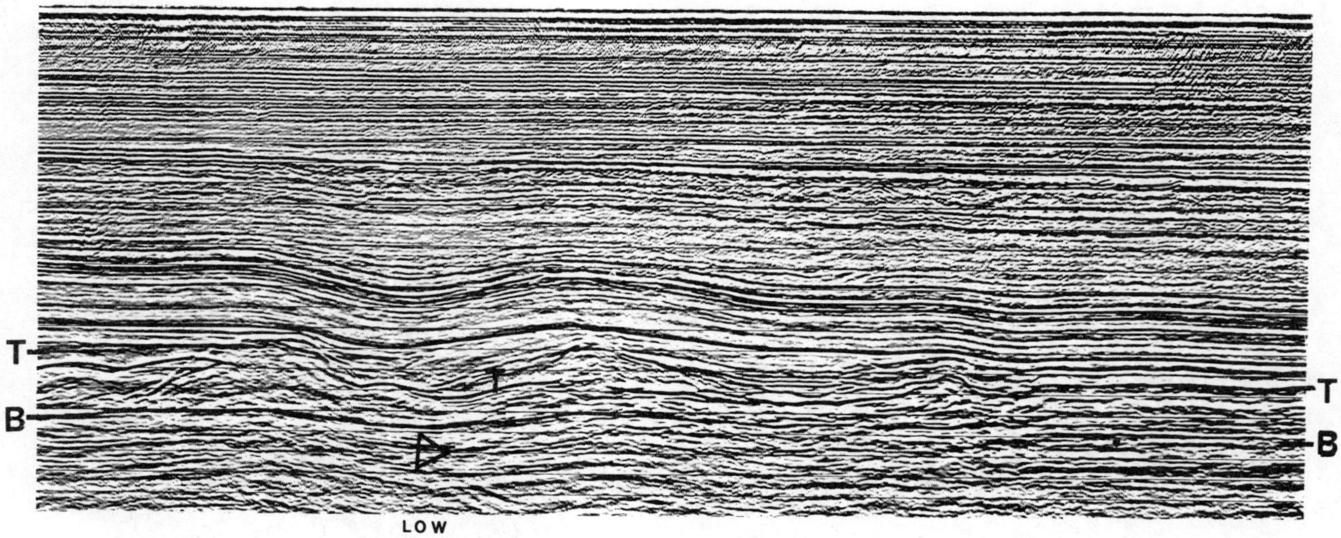

Fig. 2.8 Gentle, open folding in the overlying sequence due to salt movement in the interval B–T. Where the salt has thinned, a localised apparent low has appeared at the base of the salt interval, as indicated by the open arrow, and the shape of the salt top T above the low. See also Fig. 2.9. (Courtesy of SS(E)L.)

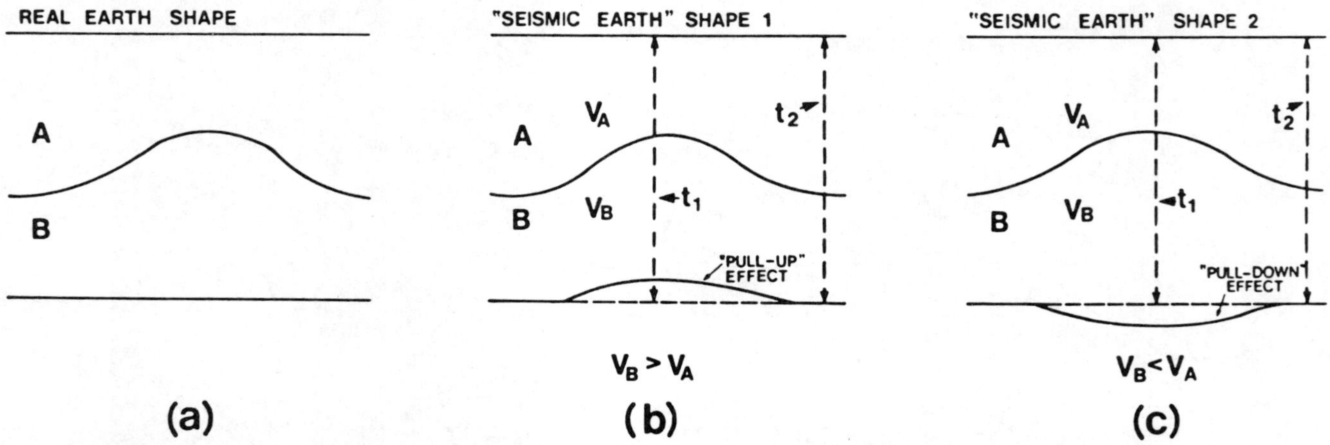

Fig. 2.9 Sketches showing how the thickening of a layer B in the real earth leads to 'pull-up' and 'pull-down' anomalies beneath, depending upon whether the interval velocity in layer B is greater or less than that in the overlying layer A. In each case, it is the time difference between the travel paths through the thick and thin locations of layer B which determines the shape of the underlying anomaly. In (b) and (c), the dashed line at the base of B is the real-earth shape, and the solid line is the shape of the resulting anomaly. In a salt area, B would be the constant-velocity salt layer, and A the overburden.

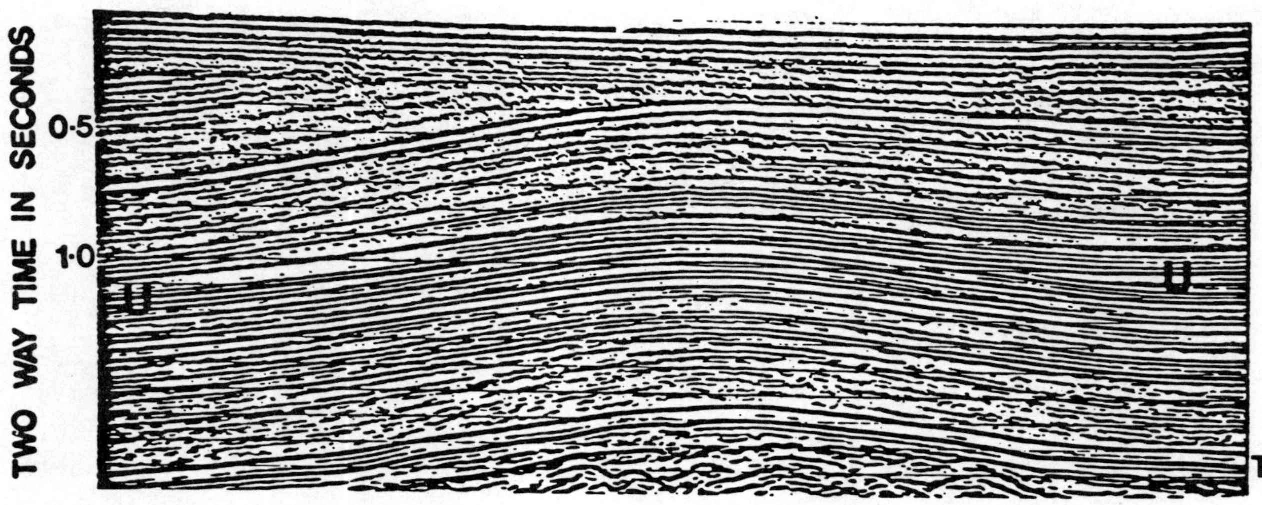

Fig. 2.10 Subtle thinning and wedging in the overburden due to salt movement both before and after the erosional episode marked by the unconformity U. T is the top of the salt interval. (Courtesy of SS(E)L.)

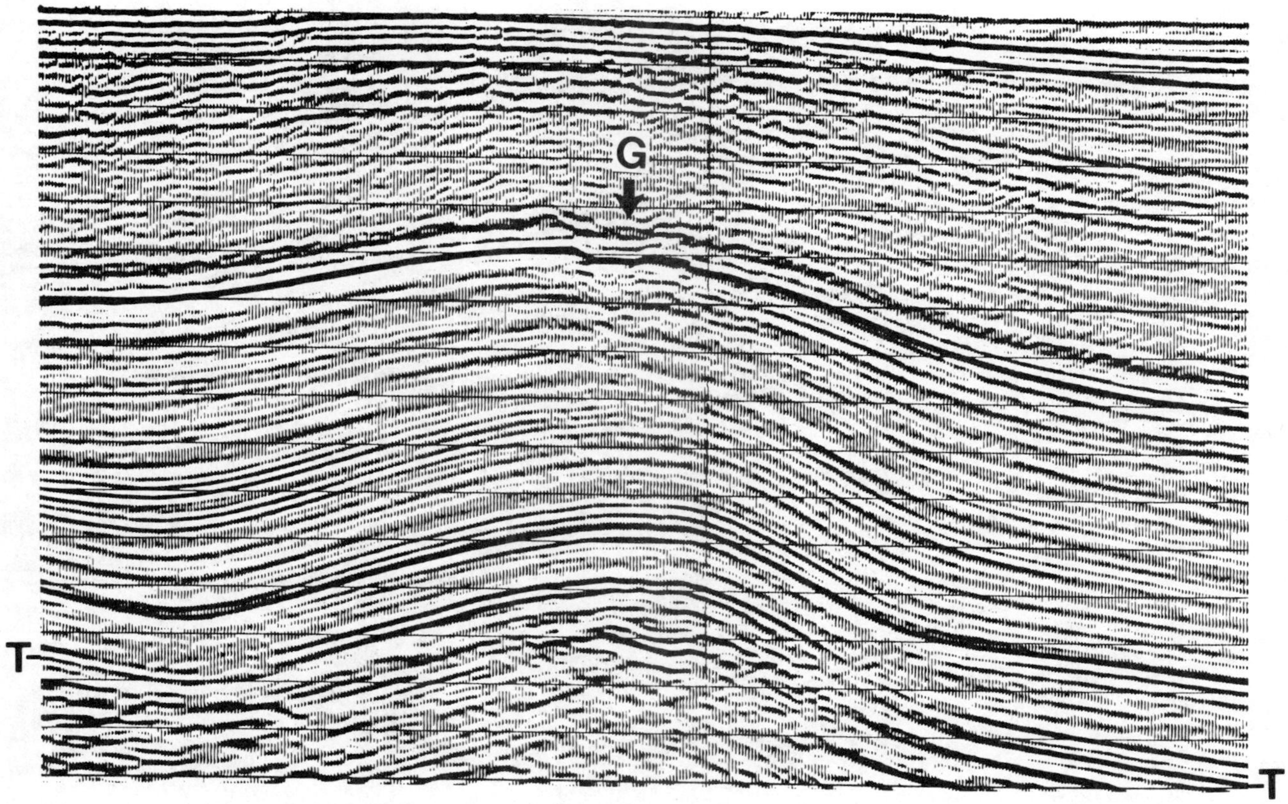

Fig. 2.11 Shallow structural graben, G, produced by stress-zone fracturing and faulting resulting from a salt uplift, the salt interval being below T. (Courtesy of SS(E)L.)

Fig. 2.12 The much larger type of collapse graben produced when stress-zone fractures extend down to the top of a rising salt mass, leading to dissolution and removal of salt from the top of the salt feature, followed by overburden collapse. Salt top is at T. (Courtesy of SS(E)L.)

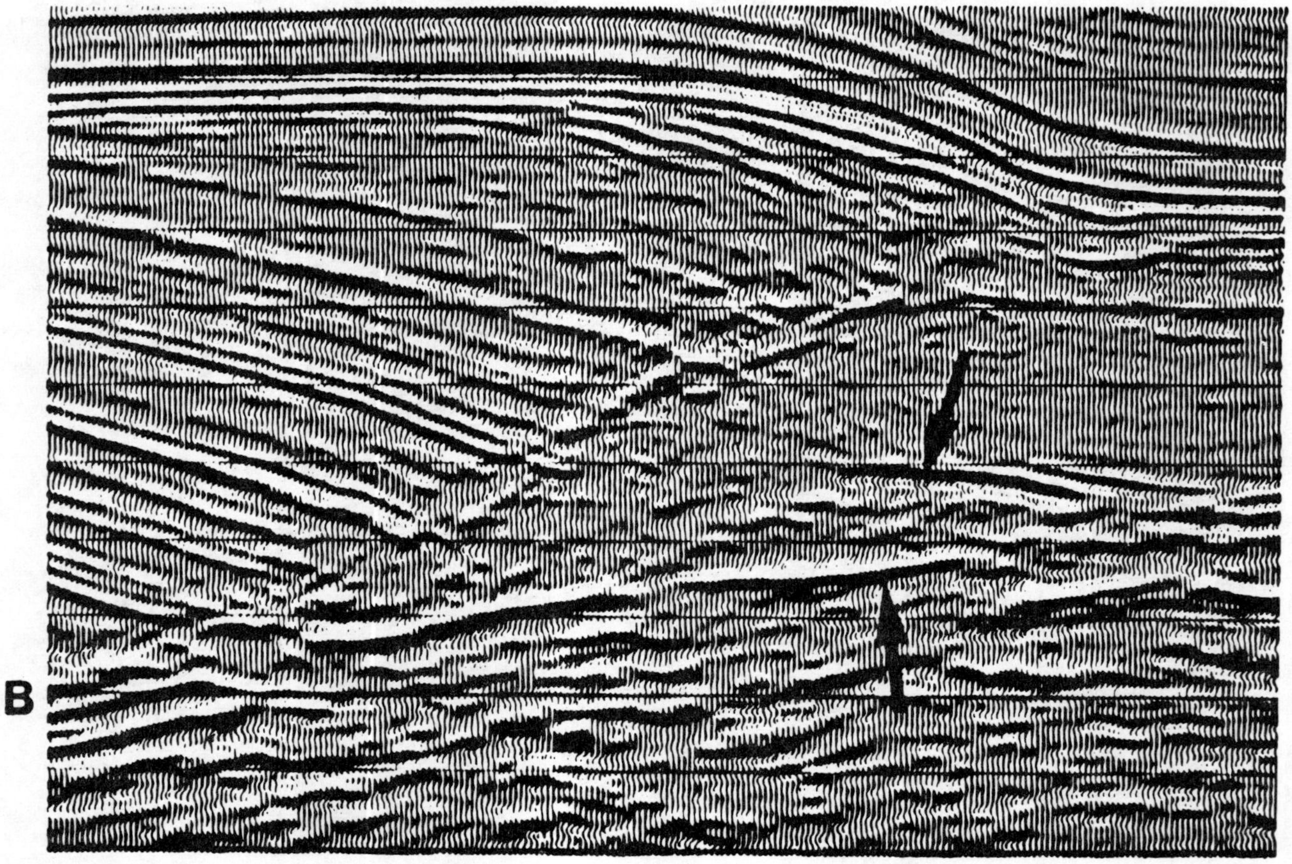

Fig. 2.13 Bedding 'splay'—the forcing apart of bedded units (arrowed) due to lateral movement of salt from left to right in this example. Basinal salt moving towards the shelf edge under high pressure, probably due to load stress differential between the basin centre (off section to the left) and the basin edge, has injected and forced apart the bedded shelf facies units seen here. The base of the salt at left is marked B. Effectively, no salt remains above B now, due to the subsidence following salt withdrawal of a massive wedge of overlying rocks—represented by the strong right-dipping events above B. (Courtesy of SS(E)L.)

through the overburden at the thicker, and thinner, salt locations. This time differential effect is shown in Fig. 2.9. It can be seen to some extent in Fig. 2.8, where the base salt horizon, B, which is probably quite flat in reality, shows a localised low just left of centre, beneath a more marked low at the level of the top of the salt, T. which indicates a localised thinning of the salt layer.

Structure above a high feature due to deformation of the overlying sequences, and pull-up or pull-down effects below a feature, are suggestive of salt movement. They are not diagnostic, however, as there are other possible explanations for such effects, as, for example, where carbonate buildup features are present in the section.

As noted, the density of salt rock is effectively invariant with depth, and therefore seismic velocity within the salt interval is constant. Thus, excepting the occurrence of competent bands of other types of rock within the salt interval, there will be no acoustic impedance contrasts and hence no reflections within a salt layer on the seismic section. There will be seismic noise—coherent noise due to seismic pulse tails and multiple reflections, together with background (incoherent) noise—but no strong, continuous primary seismic reflection events.

At the salt roof and floor, however, we may expect to see strong reflections, either due to normal acoustic impedance contrasts between salt and the confining clastics, or because at the top and/or base of the salt layer, strongly contrasting bands of the other members of the evaporite cycles, notably carbonates and anhydrites, are present. This will be seen clearly in many of the seismic examples shown in this chapter, and is again suggestive but not diagnostic of a salt interval.

If sedimentation has been completed in a thick sequence of overburden clastics before the onset of any salt movement, then simple post-depositional anticlinal and synclinal shapes, as in Fig. 2.8, will result from salt movement.

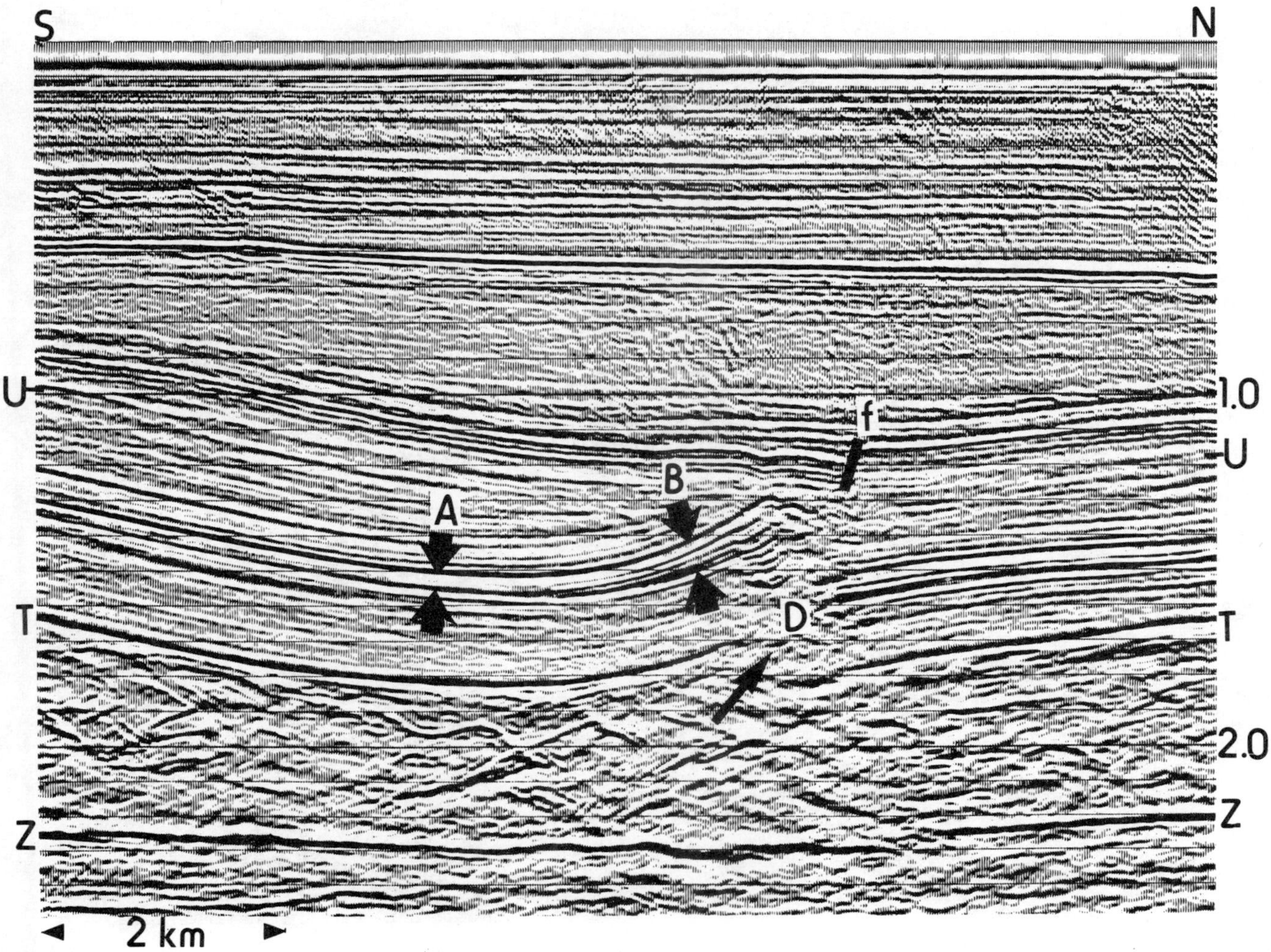

Fig. 2.14 Another example of bedding injection by salt. An oblique diapiric invasion by salt of the overlying sequence (at D), to form a 'trapdoor' diapir, has injected the bedding in a clastic interval to the south of the diapir. The arrowed interval at A is seen to be void of reflections internally. Northwards, however, at B, the same interval is gradually thickening, and increasingly strong reflection events are seen within it due to tongues of salt injected from the diapir. These have generated 'new' acoustic impedance contrasts within the interval. Source salt layer is within the interval Z–T. (Courtesy of SS(E)L.)

If, however, syndepositional movement of the salt takes place, then thinning and wedging-out patterns will develop in the overburden above the salt feature as a result of the related uplift/subsidence effects. An example of this is seen in Fig. 2.10 where pre- and post-unconformity salt movement has occurred.

With continued uplift of an overburden anticline due to the stresses set up by a rising salt structure, stress-zone faulting may occur eventually. In the part of the anticline remote from its core, there will develop a zone of extensional stress across the crest of the fold down to the traces of the inflection points on both flanks, with the maximum stress occurring in the region of the axial trace of the fold, where the maximum curvature is found. In the earth cross-section of this hinge region, radial weakness zones will develop which in many cases become actual fractures and normal faults. Commonly, two antithetic sets form across the fold crest, resulting in a shallow structural graben (see Fig. 2.11). This purely structural type of feature should not be confused with the much larger graben which may evolve from it due to dissolution of invading salt, and collapse of the overburden, as in Fig. 2.12. These larger graben are more surely diagnostic of a salt area than are the shallower structural graben just mentioned (Lohmann, 1980).

In a fold cross-section below a zone of extensional stress, a surface of null stress exists, its position depending on the nature and geometry of the applied stress field, the exact shape of the resulting fold, and to some extent the rock types present in the fold.

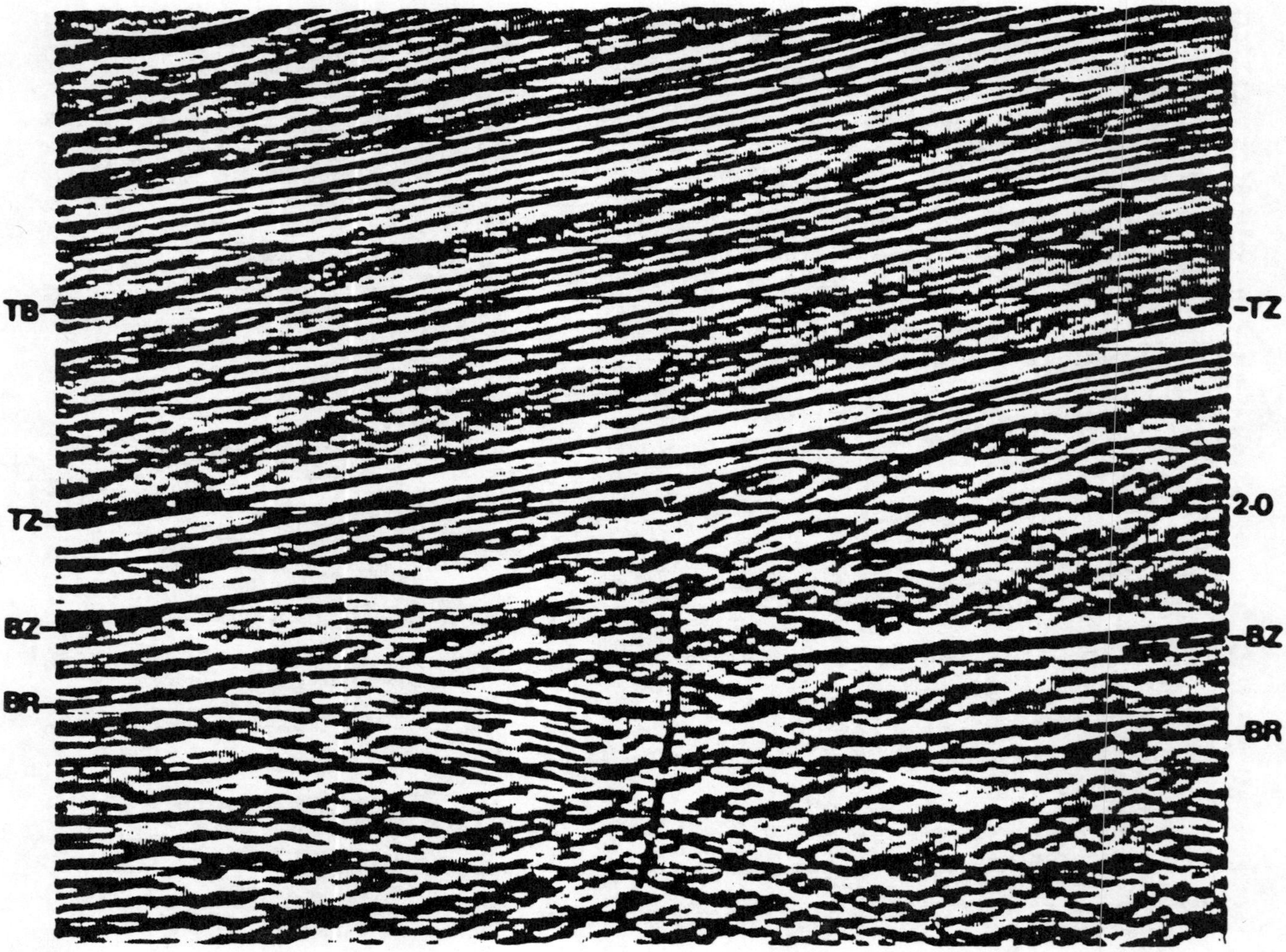

Fig. 2.15 The accommodating effect of a salt layer on underlying strike-slip fault movement. Thick and thin Lower Permian (BR–BZ) and Upper Permian (BZ–TZ) intervals in juxtaposition. This example will be seen again in Chapter 5, where it will be discussed in more detail. (Courtesy of SS(E)L.)

Below this surface, there will be a zone of increasing compression in towards the fold core where crushing, shearing and out-of-fold thrusting develops. This often chaotic zone is usually not well resolved on the seismic section, and in a pure uplift fold may be quite absent (see Chapter 5). In an overburden syncline in a salt area, the stress-zones described are also present, but in reverse order as compared with the anticlinal situation previously described. The tensional zone exists in the lower beds, and the compressed zone—if it develops—is in the upper beds. Typically, salt diapirism leads to tight anticlines and gently folded synclines in the overburden, and hence the synclines are less likely to show strong stress-zone fracturing than any positive structures present.

2.6.3 Other Indications in Sequences Contiguous to Salt

Under some circumstances, lateral movement of salt under high pressure may cause contiguous bedding planes in a clastic sequence to be forced apart, as indicated by characteristic 'splaying' of events in the seismic data. This cannot be caused, as far as is known, by gravitationally controlled vertical movement of salt, and so is a good indicator of lateral salt movement under very high pressure. It may occur at a basin margin, where basinal salt undergoes a facies change to shelf lithology—carbonates, anhydrites, and non-mobile shelf halite, for example (Jenyon, 1985a). The interdigitating of the mobile basinal salt with these rock types over a relatively short distance produces a damming effect if the basin salt is moving laterally from the basin centre towards the marginal shelf region owing to differential overburden loading. The pressure buildup causes the basinal salt to be forcefully injected into the bedding-planes of the shelf units, driving them apart, as in Fig. 2.13.

This effect is also seen further into the basin during diapiric episodes. In Fig. 2.14 is shown an example where forceful injection of salt into the bedding of the overburden on the high side of what is known as a 'trapdoor' diapir has, because of the 'new' acoustic impedance contrasts set up by the tongues of injected salt within the sequence of invaded overburden lithology, generated reflections within what was probably a previously reflection-free interval. The salt tongues thin away from the diapir, and the resulting reflections also die away as the thickness of each tongue of salt falls below the limits of ability of the seismic data to resolve the top and bottom of it.

2.7 FAULTING IN SALT AREAS

Salt movement, as was noted previously, commonly causes faulting in the overburden. On the other hand, a salt interval may under certain circumstances act as a form of insulating layer, preventing some types of faulting below it from affecting the overburden. The plastic mobility of the salt accommodates the sub-salt fault movement in such cases.

The types of faulting involved are dip-slip of small throw, and strike-slip or oblique-slip with very little dip-slip component. Some examples of strike-slip faults which have undergone large horizontal displacements beneath a salt interval without affecting the post-salt section are known from seismic data.

An apparently bizarre situation of this nature is seen in Fig. 2.15—bizarre, that is, unless the nature of the fault is understood. It appears at first sight that the Base Zechstein on the right of the fault has been downthrown by some 200 ms or more—an amount of throw at least equal to the thickness of the whole Zechstein sequence on the left of the fault—and yet the Top Zechstein has remained unaffected.

The latter fact, together with the sudden thickness change of the Zechstein (and to a lesser extent the underlying Rotliegendes interval) is diagnostic of strike-slip in this case, with the fault movement having been accommodated by the Zechstein salt. The fault has undergone left-lateral (sinistral) movement estimated, from isopach studies, at about 10 km.

The subject of the effects and significance of the fault-salt interval association is of great importance; it is believed to be fundamental to the understanding of the mechanisms of formation of many—perhaps most—salt diapirs. As such, it merits a chapter on its own, which will be found later in the book.

CHAPTER 3

Brines, Evaporite Basins and Halite Petrophysics: A Review

3.1 INTRODUCTION

Before considering the more detailed aspects of salt tectonics, it is necessary to prepare the ground by discussing some basic matters. These include the important subjects of (a) how the evaporite minerals, particularly halite, formed from brines, (b) what the environment of deposition was like in the basins in which they formed, and (c) what is known of the petrophysics of these minerals—again with particular emphasis on halite—which will explain the striking deformational and rheological behaviour observed.

This chapter will be devoted to a review of these subjects, based on the work of many people. The author acknowledges his debt to those whose works are cited, and to many others upon whose observations and experimental work our present knowledge is based.

3.2 BRINES

3.2.1 Preliminary Comments

The order of precipitation of minerals when sea water is evaporated to dryness is well established, and there is a general belief, supported by fluid inclusion and other studies, that the chemical composition of sea water has not changed significantly throughout Phanerozoic time, and probably for much longer.

In spite of this, a great variety of minerals is found in evaporite basins, including some which do not result from the total evaporation of sea water. Also, the precipitation order and the distribution of evaporite minerals within a basin is frequently—one could say usually—at variance with a model based on simple evaporation to dryness.

To understand why this should be requires a study of the chemistry of brines, this being one of the main factors controlling the observed variations. Brine chemistry is a complex subject with a large literature of its own. We will here concern ourselves with only a brief overview, the modest objective of which will be to examine some of the more important reasons for the variability of mineral species, order of precipitation, and distribution found in evaporite basins. Other reasons, of a more physical nature, involving climate, basinal configuration, and water movement, will be dealt with in the later section of the chapter which discusses evaporite basins.

The sediments termed evaporites may be defined as rocks formed by evaporation of a concentrating body of brine leading to precipitation of the minerals involved. This is a somewhat oversimplified definition which will, however, serve for working purposes.

3.2.2 Marine Brines: Some Basic Matters

It is apparent that where thick deposits of evaporite rocks occur, large quantities of brine must have been available from which the minerals were precipitated, and that the most likely source of brine in such quantities must be oceanic water. It is believed by most authorities, based on the analysis of liquid inclusions in mineral components of rocks distributed throughout geological time, that the composition of oceanic brine has altered very little during the Phanerozoic eon, and probably for an extensive period before this.

Present day ocean water contains less than 4% of dissolved solids by weight, nearly 80% of this being NaCl. Clarke (1924) gives the average content of various solids, which are shown in Table 3.1.

Usiglio (1849) was the first to investigate the evaporation to dryness of sea water under controlled laboratory conditions of temperature and humidity. He found that the various salts precipitated in inverse

Table 3.1

NaCl	77·76 (weight %)	27 213 (ppm)
$MgCl_2$	10·88	3 807
$MgSO_4$	4·74	1 658
$CaSO_4$	3·60	1 260
K_2SO_4	2·46	863
$CaCO_3$	0·35	123
$MgBr_2$	0·22	76

The figures for $CaCO_3$ include traces of other solids, including Fe_2O_3.

order of solubility, and this is shown graphically in the sketch in Fig. 3.1, down a vertical axis of decreasing water volume.

Very small amounts of Fe_2O_3 and $CaCO_3$ began to precipitate when the water had concentrated to about 50% of its original volume. At about 20% of the original volume, $CaSO_4.2H_2O$ (gypsum composition) began to precipitate, and this continued down to about

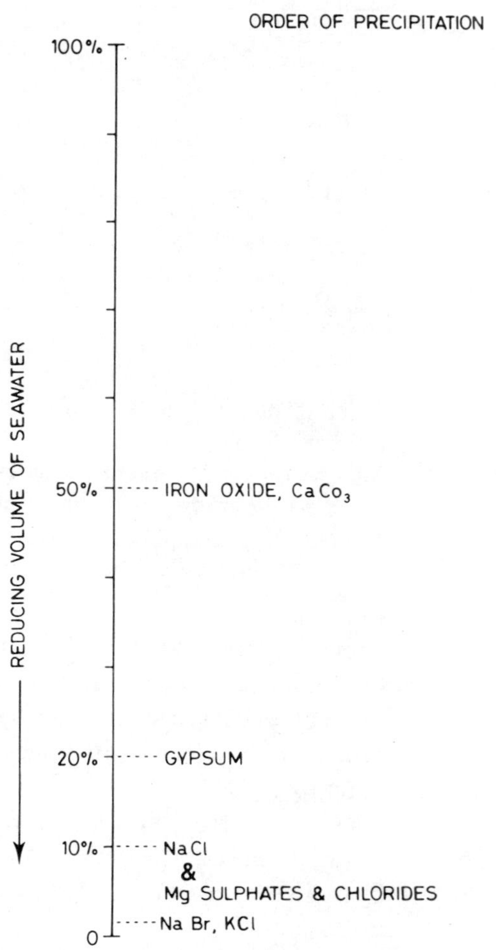

Fig. 3.1 Diagram showing the order of precipitation of minerals from sea water which is evaporated to dryness. As the sea water volume reduces, the minerals precipitate in order of increasing solubility. (After Usiglio, 1849.)

10%, when NaCl, $MgSO_4$ and $MgCl_2$ first began to come out of solution. These minerals continued to precipitate to dryness except for the $CaSO_4.2H_2O$, which ceased precipitating at about 3% of original volume, while NaBr began to appear at about 4% of original volume. In the last bitterns, NaCl, $MgSO_4$, $MgCl_2$, NaBr and KCl were all present. It is noteworthy that the NaCl did not begin to exsolve until 90% of the water had evaporated. It is also of interest to note that, in spite of the large number of evaporite minerals found in nature (only a few of which are volumetrically important), very few cations (Na^+, Mg^{2+}, K^+, Ca^{2+}) and anions (Cl^-, SO_4^{2-}, CO_3^{2-}) are in fact involved in the composition of the great majority of them, as remarked by Sonnenfeld (1984), who also noted that, although the aggregate volume of chlorides in the evaporites is much greater than sulphates, a much greater variety of the latter is found.

Ideally, an evaporite cycle, as exemplified by a composite model based on the Zechstein of the North Sea, is the result of increasing salinity due to evaporative concentration. It consists of a thin clastic member at the base—the result of the initial marine incursion—followed upwards by limestone ($CaCO_3$), dolomite ($CaCO_3.MgCO_3$), anhydrite ($CaSO_4$), halite (NaCl) and various Mg and K salts of highest solubility from the last bitterns (see, for example, Richter-Bernburg, 1959).

Whilst the order of precipitation as determined by Usiglio bears a general resemblance to this idealised cycle, there are certain important differences. The experimental work did not result in the precipitation of various minerals frequently found in evaporite deposits, the most notable being anhydrite and dolomite, but including others which often occur in the late stages of an evaporite cycle by primary crystallisation from the bitterns. Also, no *pure* halite (NaCl) was precipitated during the experiments. Since Usiglio evaporated the brine to dryness, and we can assume that nothing 'escaped', how then can other minerals result from the evaporation of natural brines? Equally, why was the precipitated NaCl contaminated when halite of greater than 99% purity sometimes occurs naturally?

On the other hand, the 'idealised' evaporite cycle is seldom found in real earth deposits. There is a large number of variables involved in the deposition of evaporites, such as diagenetic alteration of original minerals, climatic (particularly temperature and humidity) changes, run-off rate, oceanic water movements, water depth changes due to eustatic or tectonic causes, sedimentation rate, and subsidence.

All of these factors can have a strong influence on the quantity and type of minerals precipitated, and on their preservation unchanged through time.

Crystallisation of a mineral is affected by the relative concentration of the salts in solution, which may be modified by influx or reflux water movements to or from the basin of deposition; it may also be affected by the amount of clastic sediment present in the brine. Temperature increase frequently increases the solubility of minerals, delaying crystallisation; rarely, an increase in solubility of a mineral occurs with *decreasing* temperature, as with kainite. The principal zone of precipitation in a brine is where evaporation is highest, i.e. at the air–water surface, and changes in atmospheric humidity and temperature have marked effects upon the evaporation rate.

In view of these and other factors, it is hardly surprising that in naturally occurring evaporite deposits, omissions and reversals of various stages are found. Complete reversal of the order of precipitation may even occur, due to increasing dilution of a brine by an incursion of fresher water.

3.2.3 The Precipitation of Halite

The high solubility of halite in water was noted previously from the fact that a body of brine must be reduced by evaporation to one-tenth of its original volume before NaCl begins to come out of solution.

Parenthetically, it is mentioned that deposits of salt rock in evaporite sequences are often loosely referred to as 'halite'. Strictly speaking, this is incorrect, since halite is only one component of salt rock, although by far the most important quantitatively. The US Bureau of Standards Monograph No. 167 (1981) gives the chemical compositions of salt rock samples from five different areas in the USA. The weight percentage of NaCl varies between 97·27 and 99·01; the rest of each sample is made up of small, variable percentages of $CaSO_4$, $CaCO_3$, $MgCO_3$, SiO_2, Fe_2O_3 and H_2O, the majority of these being present in all the samples. Also, in many apparently massive salt rock beds there are to be found many thin (often only a few inches) partings of shale and anhydrite, and occasionally alternations of bands of pure halite with bands of anhydritic halite of a darker colour.

From the same source, another series of analyses involved 46 different samples of rock salt from 18 localities in the USA, which included samples from both layered and diapiric deposits. Figure 3.2 is a bar chart presenting the results in terms of weight % of NaCl in numbers of samples. It is noted that rock salt of a remarkably high degree of purity with respect to NaCl is of frequent occurrence. Apart from a very few samples which show rather high content of water insolubles, the majority of the samples show very high weight % of NaCl, with the small balances to 100% being made up largely of other water-soluble salts and frequently only a few tenths of one percent of water insolubles being present.

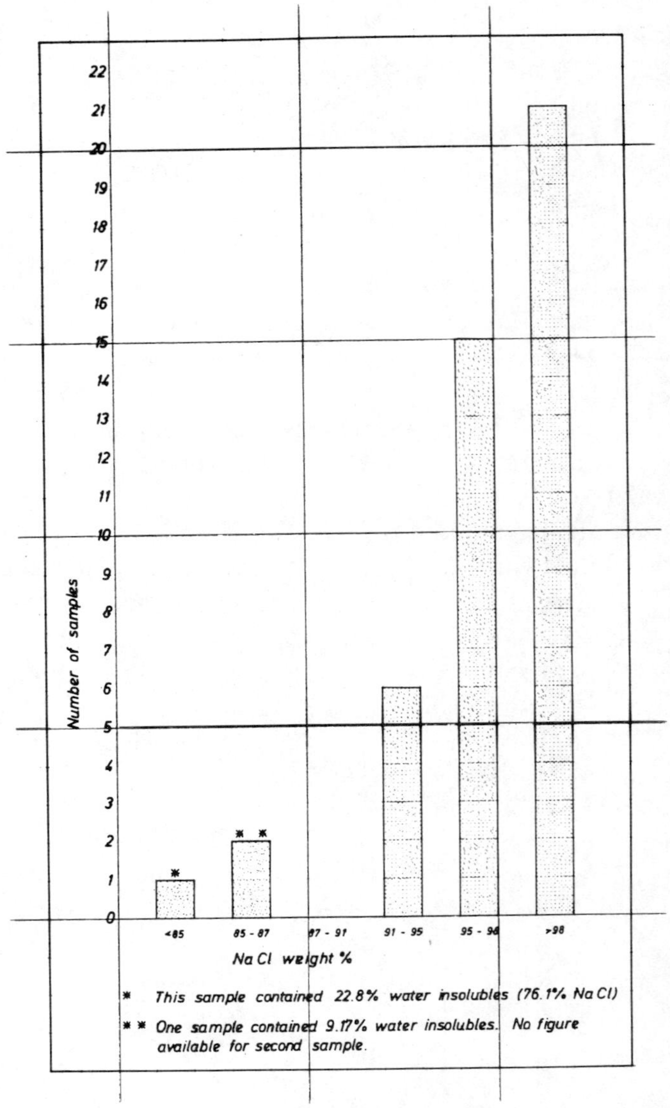

Fig. 3.2 Weight % of halite (NaCl) in 46 samples of salt rock from 18 different localities in the USA. The largest number of samples shows a content of over 98% of halite. (From NBS Monograph 167 (1981), courtesy of the National Bureau of Standards, US Department of Commerce.)

Such results are of interest when regarded from the viewpoint of palaeogeography and palaeoclimatology. Superficially, it would seem that an evaporite basin, situated in an arid or semi-arid setting, should receive quite large amounts of wind-borne water insoluble material. Recent studies of past continental climates by the examination of deep sea sediments (see, for example, Prospero, 1985) appear to indicate, however, that the most abundant sources of fine aeolian-transported sediments are those areas which are *changing* from moist to arid. Currently arid areas are, relatively speaking, already deflated with respect to finer particles; the rocky gravel plains and dune fields remaining are comprised mainly or solely of coarser materials not easily transported any distance by winds.

From experimental work, and observations of naturally occurring halite in layered deposits, it appears that there are two principal zones in the original brine where nucleation of precipitating halite crystals normally takes place: at the brine/air surface, and on the floor of the brine layer.

At the surface, evaporation produces a thin film of concentrated brine; this seems to lead to precipitation of crystals of initially platy habit which adhere to the surface tension film while continuing to grow in a downward direction. Some of these crystals develop as the hollow-faced 'hopper' type, since the margins of the flat plate attached to the surface film undergo more growth than the inner face of the plate away from the margins. When the weight of the growing crystal exceeds the strength of the surface tension, it is released, and sinks to the bottom.

There is therefore, under certain circumstances of brine and air temperature, movement and air humidity, probably a continuous shower of halite crystals sinking to the floor of the brine layer. At the floor, as evaporation of the brine body continues, a layer of higher density brine forms due to the sinking of the dense surface layer of water undergoing evaporation, and in this layer halite crystals form (a) by nucleation, and (b) as overgrowths on detrital crystals which have sunk from the surface.

Arthurton (1973) investigated the formation of halite crystals by evaporation in shallow glass vessels using artificial NaCl brines, and found that under experimental conditions surface nucleation tended to occur mainly during the day, while bottom nucleation occurred mainly at night when ambient temperatures were low compared to the day temperatures, leading to a solubility decrease. Continued growth of the crystals on the floor of the brine layer led eventually to competitive upward growth of the crystals, the form of which was elongate perpendicular to the floor of the container. Similar growth perpendicular to bedding surfaces has been noted in natural occurrences in several places, including Saskatchewan, in the Prairie Salt, by Wardlaw and Schwerdtner (1966) who gave the term 'chevron halite' to the vertically elongated crystals because they contained many brine-filled inclusions disposed in upward-directed chevron patterns. These inclusions produce an opaque crystal which is believed to be the norm in primary halite, whereas clear, transparent halite is considered to be a recrystallised variety (see, for example, Tucker, 1981) by some, although primary transparent halite certainly also seems to occur as overgrowths on chevron halite, or possibly where crystal growth has been slower.

Arthurton noted that the surface-nucleated crystals tended to aggregate, eventually forming continuous mats which sank, intact, to the floor of the container. Mat-like fabrics similar to this have been observed in the field, and in other laboratory experiments. Such fabrics may be difficult to identify, however, under natural conditions, due to later overgrowths or re-solution.

Two other points mentioned by Arthurton in the paper cited above are worthy of note, and concern the effects of terrigenous sediment on crystal growth, and the presence of planar solution or deflation surfaces at the tops of layered units in the Lower Keuper Saliferous Beds of Cheshire. In the experiments, the deposition of a thin layer of clastic sediment on growing crystals terminated their growth, and a new crop of bottom-nucleated crystals appeared on the surface of the clastic sediment. However, if the sediment layer was very thin, and barely covered the halite crystals, they continued to grow upwards partially by displacement, and partially by inclusion of the clastic particles.

The Cheshire salt beds were probably formed in a shallow lake, one or two metres deep. This conclusion results from comparing the salt-rock fabric with those produced experimentally. The presence of bedding disruptions and planar solution or deflation surfaces at the tops of layered units suggest frequent and possibly extended periods of emergence during the formation of the deposits.

Formation of halite crystals at the surface of the brine and the floor of the basin is controlled by temperature, air humidity, water and air movements, and other factors. There is no evidence that the processes are dependent on water depth, which, in any case, is difficult to establish for many evaporite basins.

So far, the precipitation of halite (and, indeed, other evaporites) due to evaporative concentration at a brine/air surface, and due to temperature-related solubility changes at the bottom of a brine layer, has been discussed. Another mechanism for halite precipitation,

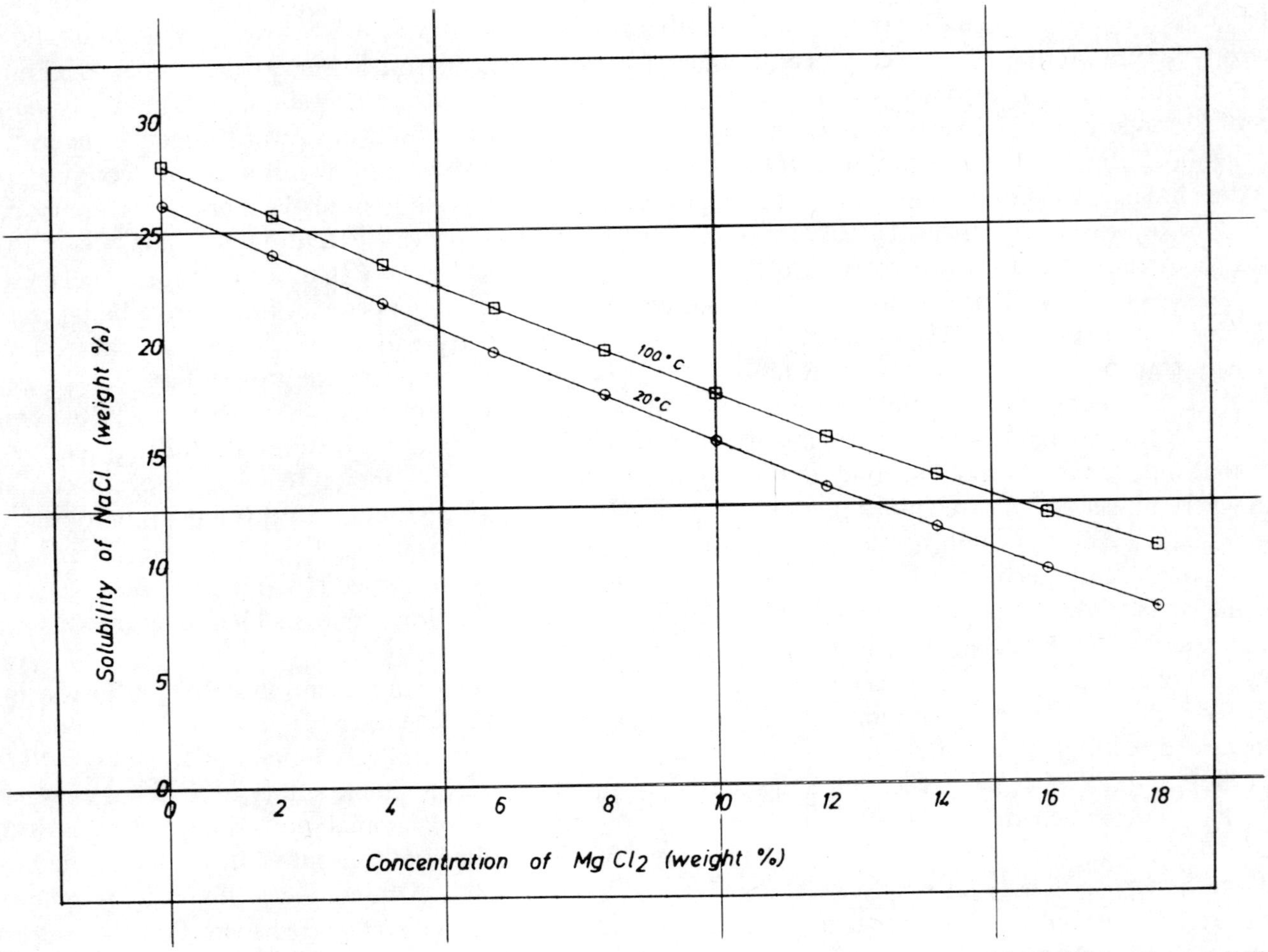

Fig. 3.3 Curves at two temperatures showing the marked effect of $MgCl_2$ in solution with NaCl on the solubility of the latter. Note also the rather minor effect on NaCl solubility of an 80°C difference in temperature. (Derived from figures in NBS Monograph 167 (1981), courtesy of the National Bureau of Standards, US Department of Commerce.)

related to the mixing of brines of differing composition and specific gravity, has been suggested by Raup (1970).

This depends upon the significant lowering of the solubility of NaCl and KCl in the presence of $MgCl_2$, and the fact that the progressive evaporation of sea water leads to an increasing concentration of $MgCl_2$ in the remaining brine. In experiments, Raup was able to precipitate halite by diffusive mixing at the boundary between a normal sea water brine and a denser $MgCl_2$-rich artificial brine. In the most favoured barred basin model for evaporite formation, evaporation and concentration of the sea water leads to sinking of the higher-density concentrated brines relatively rich in $MgCl_2$, resulting in density stratification, the bottom layer of water being a dense, $MgCl_2$-rich brine. Raup suggests that precipitation of halite (and sylvite, KCl) could occur at the boundary of the lower and higher density layers without any associated water loss by evaporation, and (from his experimental results) even from brines which were undersaturated before mixing.

Figure 3.3 shows the reduction of solubility of NaCl (weight %) in the presence of $MgCl_2$ of increasing concentration (weight %) in solution at two different temperatures: 20°C and 100°C. Note the very significant effect on NaCl solubility of quite small concentrations of $MgCl_2$, and also the relatively insignificant effect on NaCl solubility of an increase in brine temperature of 80°C. Due to the high solubility of $MgCl_2$ in the brine, it is reasonable to suppose that increasing concentration of this order may occur at later stages in evaporation of the brine.

It should be clear that the suggested mechanism requires the barred basin model for evaporite formation in order to operate, since the basin configuration would need to favour the retention of the denser brines in the deeper part of the basin, with no removal by reflux (see Section 3.3).

Areas in which such dense, hypersaline brines

occur are known at the present day. One of the most recent discoveries has been made in the eastern Mediterranean, where a bottom stratum of highly concentrated brine in which gypsum crystals appear to be forming (in this instance by floor nucleation) at a depth below 3000 m has been found, and will be discussed later in the chapter, as will more recent work by Raup suggesting that gypsum may also be formed by the brine mixing mechanism outlined above.

3.2.4 Calcium Sulphate Precipitation

Volumetrically, the two most important evaporite types after halite are forms of calcium sulphate minerals—gypsum ($CaSO_4.2H_2O$) and anhydrite ($CaSO_4$). In some evaporite basins, gypsum and/or anhydrite occur without any associated halite or other late-stage evaporites, which seems to indicate the operation of some reflux mechanism in the original basinal brines, whereby the more concentrated brine (remaining after the deposition of the sulphates) was removed, or there was an influx of fresher water which prevented the precipitation of halite and other minerals.

The dehydration of gypsum to anhydrite, represented by the reaction

$$\underset{\text{(gypsum)}}{CaSO_4.2H_2O} \leftrightharpoons \underset{\text{(anhydrite)}}{CaSO_4} + \underset{\text{(water)}}{2H_2O}$$

occurs under the influence of solar heating in the uppermost 1 m or less of sediment in supratidal areas (Kinsman, 1974). On burial, the appropriate temperatures are again reached at depths of 2000–3000 feet, depending on the geothermal gradient (Murray, 1964). The effect of the resulting volume change would be that a 10 m thick gypsum bed would reduce to a 6·2 m thick bed of anhydrite. The water/brine released is probably the active agent in much solution metamorphism, and may explain the largely secondary nature of minerals in the bittern salts where these are present in the later stages of an evaporite cycle. Some dissolution effects may also be ascribed to the same agency.

Under laboratory conditions, it has been found that the precipitation of sulphates as either gypsum or anhydrite is dependent on temperature, and also on the concentration of other salts in solution. Above 42°C, in a pure calcium sulphate solution, anhydrite is precipitated, while below this temperature the precipitate is gypsum (Posnjak, 1940). If NaCl is present in the solution, the critical temperature drops significantly to 30°C or less, depending on the NaCl concentration (Van t'Hoff and Weigert, 1901). These temperatures fall within the range of possible ambient temperatures in arid and semi-arid areas, and thus (theoretically) either gypsum or anhydrite could be the primary precipitate. Were gypsum to be precipitated and subsequently buried, it would dehydrate to anhydrite, while anhydrite, exposed at the surface, or occurring at shallow depth, would hydrate to gypsum. It is generally believed, however, that only gypsum is normally precipitated in nature from sea water, and that anhydrite forms secondarily by the dehydration process. The complex problems of gypsum/anhydrite stability and metastability, and changes of one mineral to the other, have been reviewed authoritatively by Kinsman (1974), who concluded that gypsum will probably always, under today's conditions, be the phase precipitated from standing bodies of brine, but that both primary anhydrite and penecontemporaneous anhydrite after gypsum are very common products of the sabkha environment. Also, he considers that if isolated bodies of marine brine existed well below sea level, as suggested for the Mediterranean during the Messinean, then temperatures could be high enough for the precipitation of anhydrite as a primary phase.

It appears likely that the sabkha environment existed in the past, based on the evidence of sedimentary fabrics observed. Some laminite deposits, consisting of thin (a few millimetres or less) layers of anhydrite and/or gypsum alternating with equally thin layers of carbonates and/or organic matter, have been ascribed to this type of environment. Doubts exist, however, in many cases, because of the consistent character of the layers over very large areas. There is a substantial body of literature arguing the case for a rather deep, subaqueous origin for laminites, based on the correlation of detailed thickness patterns over long distances, as in Richter-Bernburg (1960) for evidence related to the Zechstein, Anderson and Kirkland (1966) and Anderson *et al.* (1972) for the Castile, and Davies and Ludlam (1971) for the Elk Point evaporites.

Very large thicknesses of sulphates are sometimes encountered, which poses a problem when the quite small content of sulphates in sea water is considered. Twenhofel (1950) quotes the case of a 360 m section of anhydrite found in a Texas borehole, calculating that the evaporation of a column of sea water over 300 000 m high would be required to produce this as a primary precipitate. Even if it were considered to be caused by the filling and evaporation of a basin

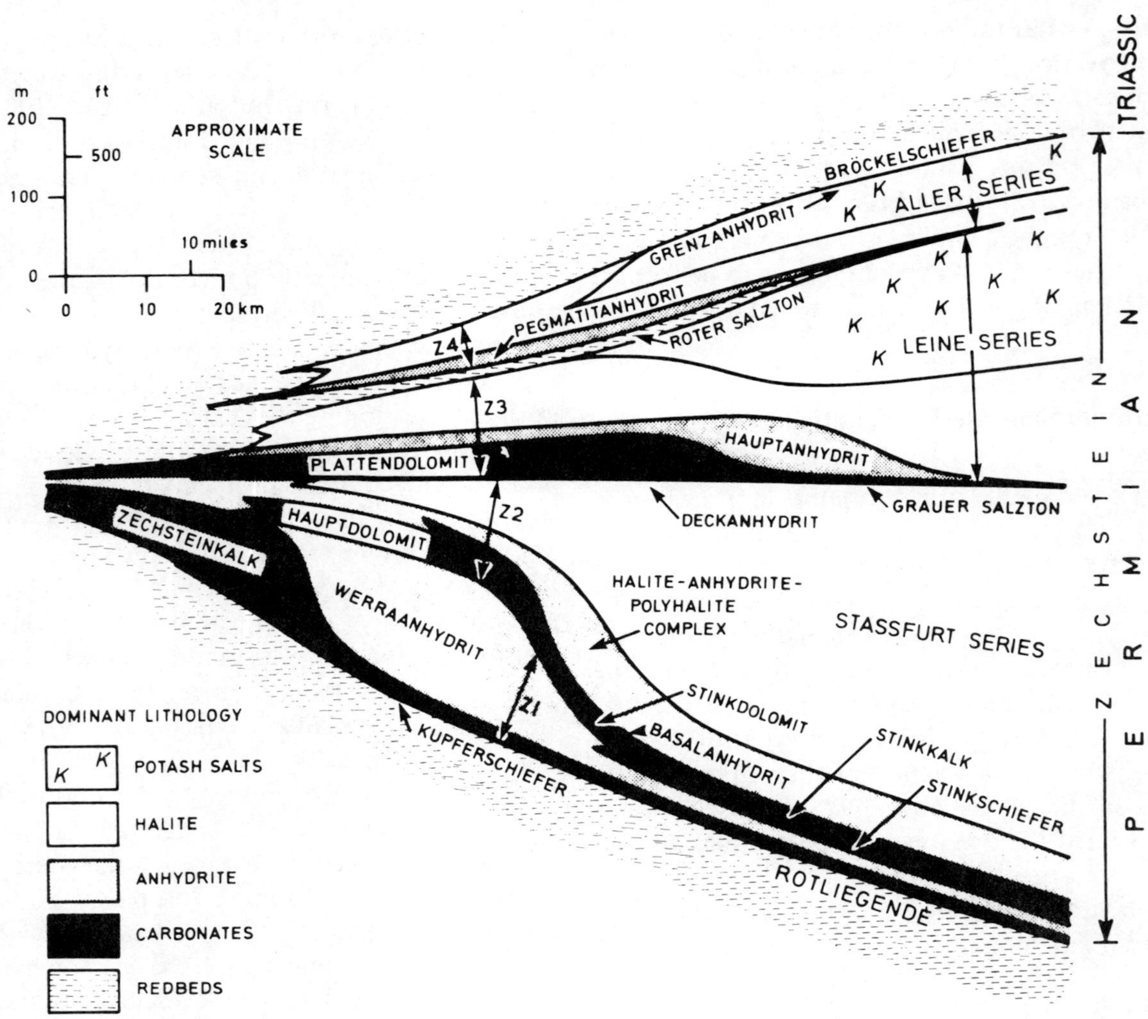

Fig. 3.4 Generalised shelf-to-basin section of the Zechstein evaporite cycles in the Southern North Sea. Note the three major carbonate units (Zechsteinkalk, Hauptdolomit, Plattendolomit), each overlain by a massive ramp of anhydrite prograding into the basin. This situation is difficult to reconcile with, particularly, the deep desiccating basin model in its basic form (Fig. 3.5c). (From Taylor, 1984, courtesy of the author and Blackwell Scientific Publications.)

repeatedly, the number of times this would have to happen stretches credibility too far. Twenhofel suggested that the anhydrite had been washed down at a later stage from the peripheral parts to the central part of the basin, thus representing evaporation over a very much larger area than it currently occupies. However, sulphate deposits of considerable thicknesses are known which cover substantial areas of basin margins, and Borchert and Muir (1964) offered the suggestion that where there is a steady influx of sea water into a restricted basin, and the physical conditions controlling evaporation remain stable, the brine may be held within the normal concentration range for sulphate precipitation over long periods, with continuously renewed supplies of sulphates leading to thick and widespread deposition. In this situation, the concentration stage for halite precipitation may never be reached, or may be delayed. In the latter case, as in the Zechstein cycles of the Southern North Sea sub-basin, thick ramps of anhydrite may prograde out over generally thin to very thin carbonates that had been deposited beneath water 300–600 feet (or more) deep, which form the earlier 'pre-evaporite' stage of the cycle (Fig. 3.4).

A difficulty raised by Borchert and Muir's suggestion is the delicate balance required to stabilise the small inflow of sea water which would be permissible if sulphate minerals were to be precipitated for geologically significant periods. Attention was drawn to this point by Lucia (1972), who calculated that the inflow channel would need to have a cross-sectional area of less than some one-millionth of the surface area of the basin. This subject has also been addressed by Kinsman (1974). Alternative suggestions which have been made to circumvent this difficulty will be discussed in Section 3.3.

Raup (1982) conducted experiments which show the possibility of gypsum being precipitated within a body of brine by the mixing of brines of different composition and specific gravity. The mixing took place at boundaries between stratified brines of different densities, as was the case in the precipitation of halite under similar circumstances in experiments carried out previously by the same author (Raup, 1970). This suggests the possibility of the operation of this mechanism where basinal brines of intermediate concentration react by diffusive mixing at a boundary with denser, more concentrated brines rich in magnesium chloride and magnesium sulphate. Raup noted that gypsum was the only form of calcium sulphate to be precipitated during the experiments. Again, such a mechanism requires a barred basin type of environment, with no removal of denser brines by reflux. It was also noted that carbonate minerals may be precipitated by a similar mechanism, involving brines of fairly low concentration, the carbonates being the least soluble of the salts in solution in sea water brines.

Dense, hypersaline brines occurring in the depths of a restricted basin, as suggested in the above discussion, could be expected to be highly hygroscopic, and it has been suggested that any gypsum crystals sinking into such a brine layer would be dehydrated to anhydrite. It is interesting in this connection to note the recent discovery (Scientific Staff, Cruise *Bannock* 1984–12, 1985) of gypsum precipitation from cold brines in an anoxic basin in the eastern Mediterranean. Normal salinity was recorded to a water depth of 3000 m, but below this, at depths between 3228 and 3428 m, locally high temperature and salinity gradients were logged, and black muds stinking of H_2S were recovered from the bottom, suggesting the presence of anoxic hypersaline bottom water. Large masses of interlocking gypsum euhedra were dredged up from the bottom, and found in sub-bottom cores. Various lines of evidence made it most likely that these had formed *in situ* from the hypersaline brine. It is believed that the source of the latter is not evaporative concentration, but leaching from Late Miocene (Messinian) evaporites at or close to the sea floor. However, the growth by bottom nucleation of the gypsum crystals in the dense brine, and the lack of recovery, even from cores, of any crystals other than gypsum, suggest that if the hygroscopic dehydration of gypsum to form anhydrite in such brines is a reality, then the time frame for the transformation must be a long one (some of the cores having penetrated to (probable) middle Pleistocene sediments).

3.2.5 The Carbonates

Marine basins in which evaporites are deposited are basically basins of carbonate deposition where, for various reasons, water exchange has become restricted enough for evaporites to form. In many basins, the evaporitic stage is never reached. Where it is, before evaporites begin to be deposited, extraction of calcium carbonate from the marine brine is dominated by biological processes, particularly towards the margins of a basin in warm, shallow and well-lighted waters. This is the case even though the brine may be supersaturated with respect to calcite and aragonite.

In the deeper parts of a marine evaporite basin, the pre-evaporitic carbonates found are usually lime-mudstones. Above these, similar lithological types formed during the sulphate phase of evaporite formation may occur, grading up into thin limestones and dolomitic limestones. The pre-evaporite carbonates and at least the lower part of the evaporitic stage carbonates tend to be dark, foetid, aphanitic, euxinic types without infauna, and laminated. The earlier lime-mudstones derive mainly from calcium carbonate carried in suspension from the shelves and slopes, while in the evaporite-stage rocks, calcite may be provided largely by the action of sulphate-reducing bacteria, and reaction with organic matter of which, together with argillaceous material, there is usually an appreciable presence.

On the slopes of the basin, the argillaceous content decreases, and more shelf-derived carbonates are included, such as turbiditic sediments, slump material, and fine sediment winnowed from the shelf area during storms and, on the upper parts, benthonic organisms.

At the basin margins, in the favourable conditions mentioned previously, the production of carbonates tends to be most rapid, with the life processes and remains of biota being responsible for a significant part of the sediment build-up on broad shelves, with barriers, lagoons, intertidal and sabkha zones. Reef building is common in these conditions, with sediment production and binding functions by algae being of particular significance. Dolomitisation of the carbonates may occur, due to contact with more concentrated brines rich in magnesium produced contemporaneously in marginal sabkhas, and more generally later during the sulphate phase of evaporite deposition. Some are of the opinion that even if dolomitic types are not present, this is due to dedolomitisation (see, for example, Clark, 1980), although this is not generally agreed.

The question of whether any primary carbonate minerals are chemically precipitated from sea water is a difficult one. Generally, the phase relationships and stability fields of $CaCO_3$ in the presence of other salts in a brine are very complex. Aragonite is the form of $CaCO_3$ normally precipitated from sea water, but it is metastable and alters rather readily to other minerals. One of the factors affecting the precipitation of both aragonite and calcite from brines is that the solubility of both forms increases with increasing concentration of both NaCl and Mg ions. This is an inhibiting factor for carbonate precipitation in increasingly concentrated brine, and Sonnenfeld (1984) remarks that this may be a contributory factor in the preferred distribution of carbonates around the margins and towards the entrance of a restricted brine body.

Also, there is thus an increasing degree of oversaturation with increasing evaporation beyond the stage at which $CaCO_3$ would normally precipitate. This has led to a belief that localised precipitation may occur if brine in this condition is suddenly diluted by, say, rainfall or the localised influx of fresh water from some other source. The localised 'precipitation' of aragonite in the form of myriads of needle-like crystals in sea water (the 'whiting' effect) has been observed (e.g. Wells and Illing, 1964). Black (1933) and Cloud (1962) favour the idea of precipitation to explain the phenomenon. However, de Groot (1965) and Ellis and Milliman (1985) believe that it is due to the material in suspension having been 'stirred up' from the sea bottom. The latter authors state that in the southern Red Sea sediments, in spite of the oversaturation of the sea water with respect to aragonite and magnesium calcite, there is no good evidence for the presence of 'chemogenic' carbonates in the water, suspended grains being of biogenic or detrital origin. They note, however, that aragonite ooliths do grow interstitially in disturbed sediments, and magnesium calcite *cements* can form in permeable undisturbed sediments on the sea bottom.

In carbonate chemistry, thermodynamic predictions are seldom reliable. Reasons for this include the complex nature of natural brines; the spectrum of solubilities of the solid phases (dolomite, magnesian calcite, calcite, and aragonite) which are not even constant for a mineral of a particular composition but vary according to the state of the surface and with particle size; and kinetic effects, including the influence of living and inert organic matter, complexing of ions, etc.

If there is a consensus, it is probably that the carbonate mineral thermodynamically in equilibrium with normal sea water is dolomite, which is actually the most difficult of the carbonate minerals to precipitate. Seas which are warm and shallow are in practice oversaturated with respect to aragonite, magnesian calcites, and even more highly supersaturated with respect to ordinary calcite and dolomite. There appears to be little evidence that any of these precipitate spontaneously in the water column although, as mentioned above, aragonite and, less commonly, magnesian calcite are precipitated as coatings on sand-sized grains, and as interstitial cements on the sea bed.

By far the bulk of lime sediments are the result of organic processes—mainly from skeletal remains or their breakdown products. Corals, many molluscs and green calcareous algae provide the material for aragonite, whilst red calcareous algae, foraminifera and echinoderms provide magnesian calcite. Organic and inorganic processes combined result in a sediment consisting mainly of the metastable minerals aragonite and high magnesian calcite, with some ordinary calcite. Ancient carbonates, in contrast, consist of the stable minerals calcite and/or dolomite.

As suggested earlier, the most important single factor controlling the type of carbonate minerals precipitated from sea water is probably the Mg ion, which strongly inhibits the growth of calcite in natural brine in spite of the mineral's low solubility. Magnesian calcites form so slowly that at the high supersaturations of tropical sea water they are usually swamped by the growth of aragonite. During later diagenesis, lowering of Mg ion concentrations by, for example, flushing with meteoric or compaction waters allows calcite to form. Conversion of aragonite and magnesian calcite to normal calcite may then take place by a wet dissolution and reprecipitation process. Solid state inversion is negligible: aragonite inverts to calcite at a temperature of 100°C, which requires a burial depth of some 10 500 feet with a geothermal gradient of 25°C/km. If on the other hand the Mg/Ca ratio of the brine is raised—by, for example, sulphate mineral precipitation—existing aragonite sediments may be dolomitised. This can occur quite rapidly in hot supratidal aragonite muds, and leads to microdolomite formation, or more slowly in coarser and/or deeper sediments, producing coarser dolomites. The model suggested by Clark (1980) is adequate for this purpose, and the dolomitisation of the bulk of carbonates associated with basinal evaporites appears almost inevitable. Paradoxically, the *dilution* of sea water, it has been suggested, may also result in dolomitisation (Badiozamani, 1973).

Important works dealing with the solubility and formation of carbonates are Garrels *et al.* (1961), de Groot (1966), and Weyl (1967). There is also an important paper by Bathurst (1971).

Once formed, dolomites may later be affected by the process of 'dedolomitisation', or calcitisation (Shearman *et al.*, 1961). This seems to occur where there exists an abundance of Ca ions, made available frequently by dissolution of the sulphate minerals gypsum and anhydrite. Instability may also be produced by the presence of CO_2 which may be present either in depth due to the decarboxylation of organic matter (often particularly well preserved in the basal parts of evaporite cycles), or when, as a result of uplift, surface water charged with CO_2 comes into contact with dolomites (Clark, 1980). This process can create porosity under certain circumstances, and reservoir-quality units of this type are suspected in Zechstein slope carbonates close to source rocks in basinal plain sediments (Clark, in press). The various diagenetic processes affecting pre-evaporite carbonates and evaporite rocks at different burial depths is an interesting and important field of study in petroleum geology; an excellent concise account is given by Clark (1985).

It should be mentioned, following the discussion of the previous paragraph, that the presence of an excess amount of CO_2 can actually have an adverse effect on porosity by causing the congruent dissolution of dolomite. This was indicated by Yanat'Eva (1955), and supported by experiments carried out by de Groot (1967). Dedolomitisation *per se*, in the absence of leaching by Ca-rich water, should cause a volume increase, and a reduction of porosity.

3.2.6 Post-Halite Stage Precipitates: The Bittern Salts

In some basins, where evaporation to dryness eventually occurs at the end of an evaporite cycle, the most highly soluble salts are precipitated from the bitterns. These salts, often referred to generally as the 'potash salts', include primary precipitates, secondary minerals which are alteration products of less stable primary minerals, and exotic mineral species resulting from entry of solutions into the potash deposits after these have been buried. In all, some 40 different salts have been found in various deposits but, of these, relatively few are present in more than minute amounts.

Baar (1977) tabulates the following minerals in descending order of abundance as the main constituents of potash deposits, based on a list by Braitsch (1971): sylvite (KCl), carnallite ($KMgCl_3.6H_2O$), bischofite ($MgCl_2.6H_2O$), tachyhydrite ($CaMg_2Cl_6.12H_2O$), which are chlorides, and kieserite ($MgSO_4.H_2O$), langbeinite ($K_2SO_4.MgSO_4$), epsomite ($MgSO_4.7H_2O$), polyhalite ($K_2MgCa_2(SO_4)_4.2H_2O$), which are sulphates, plus kainite ($4(KClMgSO_4).11H_2O$), which is a chloride–sulphate.

It is not proposed here to discuss the complex phase relationships and physical chemistry of the potash minerals. For this the reader is referred to such standard works as Borchert and Muir (1964) and Braitsch (1971), in which the lists of references will open up a large literature. The chemical and physical factors determining the presence or absence of any particular potash mineral are many and varied. Most of the minerals in the above list may occur primarily or secondarily in a deposit. Some, including tachyhydrite, langbeinite, epsomite and polyhalite, are almost invariably secondary after other minerals, whilst kieserite is unstable and readily alters to other forms of hydrated $MgSO_4$ salts.

It has been said that the most remarkable thing about the potash minerals is that they have survived at all. Many are so soluble that they deliquesce in damp air, and undoubtedly in most cases they have been precipitated and re-dissolved many times. Frequently they form intergrowths with other minerals such as halite, and are more rarely found as pure deposits.

An interesting characteristic of these minerals is that, in a series of increasing solubility (and decreasing density) from halite onwards, they show a decreasing resistance to deformation due to progressively lower yield stresses in the series as densities decrease. This can have important effects when an evaporite cycle containing potash salts in its late stages is subjected to applied stress, and this will be discussed further in Chapter 8.

It cannot be regarded as surprising that potash deposits are by no means commonly found in evaporite basins. Even if the original brine has concentrated to the bittern stage, brine reflux movements leading to removal of the minerals in solution or their wide dissemination, together with later solution removal or metamorphism, will often lead to the bittern salts being absent. Where they are preserved, they will frequently bear little resemblance to the primary mineral assemblage which was precipitated.

3.2.7 The Significance of Bromine

The bromide content of evaporites has been shown to provide an important indicator of the degree of brine concentration at which the evaporites were formed, and also acts as a useful diagnostic to distinguish between evaporites of marine and non-marine origin.

The bromide content of primary halite in evaporites of normal marine origin has been shown to fall within the range of 50 to 100 ppm. Solution and reprecipitation causes this concentration to fall to 20 ppm or less on the first cycle, and decreases for any further cycles (Holser, 1979), especially if the water involved in the re-solution were not of marine origin.

Work by Schulze (1960) on evaporites of the Zechstein Z2, Z3 and Z4 cycles in Germany shows increases of the bromide content from the early evaporitic stage up to the late potash precipitation stage of the Z2 cycle, and then a sharp fall indicating dilution of the brine at the beginning of the Z3 cycle. This is followed by a range of bromide values controlled by the degree of recrystallisation.

Halites of the Williston, Alliance and Delaware Basins in the USA have recently been studied from the same viewpoint (Wilgus and Holser, 1984). The bromide content of halites of the Alliance Basin falls within the range 76–98 ppm, indicating precipitation from normal marine brines. Those for the Delaware Basin showed, at some levels, concentrations higher than 150 ppm which suggests brine concentration towards the potash stage. The Dunham Salt of the Williston Basin showed no measurable bromide content, supporting the interpretation of this as a recycled halite, while samples of the Opeche Salt from the same basin showed considerable variations from 2 to 70 ppm of bromide, suggesting strong influence of non-marine waters, and reprecipitation in a close continental basin—a suggestion supported by the presence in core samples of much terrigenous silt.

3.2.8 The Association with Hydrocarbons

Important hydrocarbon accumulations in many parts of the world are associated with evaporite basins. In many instances, the most obvious role of the evaporites is the relatively 'passive' one of providing traps in the overburden resulting from salt deformation, and seals above underlying oil or gas traps. There can, however, be a closer connection than this. Because of the very nature of the processes which occur in an evaporite basin, hydrocarbon source rocks may be included as members of evaporite cycles in the form of organic-rich and bituminous shales, marls and limestones.

In the early stages of evaporative drawdown in a basin, although the body of the brine may be slowly increasing in NaCl concentration, the surface layer of water may still be much fresher, approximating normal sea water in composition. In such warm, light, well aerated water, a rich assemblage of biota will thrive, their remains will sink into the lowest layer of denser, more concentrated brine, and under anoxic conditions will be at least partly preserved. Together with argillaceous material, the organic matter will be included in lime muds to become marls and marly limestones.

At a later stage, heat and pressure due to increasing burial depth convert the sapropel to kerogen, and ultimately to gaseous and liquid hydrocarbons.

Studies in recent years suggest that organisms can survive in brine which has reached a more advanced stage of concentration than was previously thought possible. Kirkland and Evans (1981) put forward the view that in a 'mesosaline' environment, i.e. that part of a basin with brine of salinity between 4% and 12%, where only carbonates can precipitate, some organisms may be present in great numbers. Present-day examples of saline environments such as the Rann of Kutch in the west of India are known to support algae, phytoplankton and other life forms, and although the number of different species is limited in comparison with that in other environments, there is a compensatory proliferation of individuals in the species which *do* exist. There appears to be a maximum development at about 2 to 3 times the concentration of sea water; with increasing concentration beyond this, there is a decline to zero at concentrations at which the potash and magnesium bittern salts begin to precipitate. At this stage, dehydration by osmosis has reached a level insupportable by any life form.

Kirkland and Evans suggest that dark, organic-rich basinal carbonates in many areas may be source rocks produced in such a mesosaline environment. They point out that the depositional history of prolific hydrocarbon areas such as the Middle East is frequently dominated by thick sequences of carbonates, often of 'basinal limestone' type, and evaporites.

The rate of deposition of basinal carbonates is usually very low. In some areas, limestones of this type may exist and be of source rock quality, but may be too thin to be viable. Basinal Zechstein carbonates in the Southern North Sea may be of this type (Taylor, 1981).

3.3 EVAPORITE BASINS

3.3.1 Introduction

The problem of the origin of major evaporite deposits has preoccupied geologists and geochemists for a century or more, and still no universally acceptable solution has been arrived at.

With dry humour, Hsü (1972) states that the origin of evaporites '. . . should constitute no particular problem: evaporites will form where evaporation exceeds influx, and every little pond, lagoon, or man-made depression in an arid environment is an actual or potential salt pan'. He then goes on to highlight the true nature of the problem, which lies in the sheer volume and areal extent of many of the large evaporite deposits known in various parts of the world.

The volumetric aspect can be appreciated by considering the amount of dissolved salts in normal ocean brine. Pettijohn (1957) points out that the results of evaporating to dryness a 1000-ft column of sea water would be deposits of about 5 inches of $CaSO_4$. $11\frac{1}{2}$ ft of halite, and 3 ft of potassium/magnesium bittern salts, or in other words a total of only some 15 ft of precipitated solids. Based on simple, one-stage evaporation to dryness, the 4500 ft of Zechstein halite believed to have been precipitated in the Southern North Sea Basin would have required a water depth of some 400 000 ft. Clearly, this is an absurdity.

An alternative to such patently impossible water depths would be to consider that a basin had been filled, emptied by evaporation, and refilled, in a number of cycles. With the thickness of evaporite deposits in many basins, however, this would require the complete process to be repeated anything from scores to hundreds of times. The problem becomes even more acute when it is considered that in some basins, halite may not even be present, but several thousands of feet of sulphate minerals—anhydrite or gypsum—may exist. When the low $CaSO_4$/NaCl ratio in normal brine is borne in mind, the single- or multi-stage process of evaporation to dryness is even less credible.

Another factor which adds to the difficulties is the presence in parts of some thick evaporite sequences of what appear to be clear signs of shallow-water deposition—interbedded silts of terrigenous origin showing cross-lamination, nodular anhydrite, desiccation cracks, and dissolution surfaces associated with halite. As regards fossil evidence, occurrences of marine biota are extremely rare, but material of terrigenous origin is not uncommon. On the other hand, the presence in some thick deposits of sequences of undisturbed laminites covering very large areas seems to indicate deposition in water of moderate to substantial depth.

It is proposed to discuss briefly the most important hypotheses which have been put forward to explain the origin and distribution of basinal evaporites, and to mention what may be an emerging consensus. Comment will also be made on the principal physical factors which are known to govern evaporation of brine in the basinal context.

3.3.2 Brine Movement

It is unfortunate that the principle of uniformitarianism popularised by Lyell, with the 'present being the key to the past', cannot be applied to the problems under discussion here, since no large evaporite basins on the scale of examples known in the geological column are currently present at the earth's surface. The Gulf of Kara Boghaz, a lagoon-like body of water connected to the Caspian Sea, has often been cited as a modern example which has progressed over a relatively short period through the various stages of precipitation of evaporites up to and including that of halite. As Hsü points out, however, it cannot be regarded as a realistic model of major marine basinal precipitation, being a relatively small, shallow lake off what is in effect a larger inland lake. Present-day marine lagoons in arid climates exhibit only normal marine sedimentation.

Evaporites are certainly being precipitated at the present surface, not only in inland lakes, but in other environments in arid areas. Anhydrite, for instance, is known to be forming in supratidal zones and sabkhas of the Persian Gulf area (see, for example, Kinsman, 1966, 1969; Kendall, 1978a). Nodular 'chicken-wire' anhydrite similar in form to this is present in the peripheral parts of various palaeo-evaporite basins.

Such features can do little to resolve the basic problems mentioned in the Introduction, however. It may be that these are related to others noted earlier in the chapter regarding the wide variation in amount and type of different minerals in various basins. In particular, the mineralogy of evaporites which form in inland lakes and salt pans is different in several respects to that of evaporites of marine origin, and there are certain quite fundamental differences, such as the absence of bromine in 'terrigenous' evaporites, which suggest a different mode of origin.

Of considerable significance to the question of the mode of origin of marine evaporites is the matter of mineral assemblages. In some basins, massive thicknesses of sulphate minerals (sometimes several thousands of feet) are present, but with no halite or bittern salts. In other areas, thick halites exist with only minor developments of sulphate minerals, and either with or without bittern salts. Such variations suggest that (a) the source brines have in some cases been 'held' within the limits of precipitation of some minerals for long periods, or that (b) brines of increasing concentration have been removed from the basin at some stage.

The first of these possibilities implies that incoming fresher brine has maintained the basinal brine within certain limits of concentration, and the second that the more concentrated brines have been carried away. Both of these possibilities require that movement of the brines has taken place during the evaporation process, either as an inflow continuously replenishing the basinal brine as evaporation proceeds, or as an outflow. Possibly both these processes may operate together in some instances. Basic ideas of this nature have led to the formulation of various theories regarding the origin of major evaporite deposits in the past.

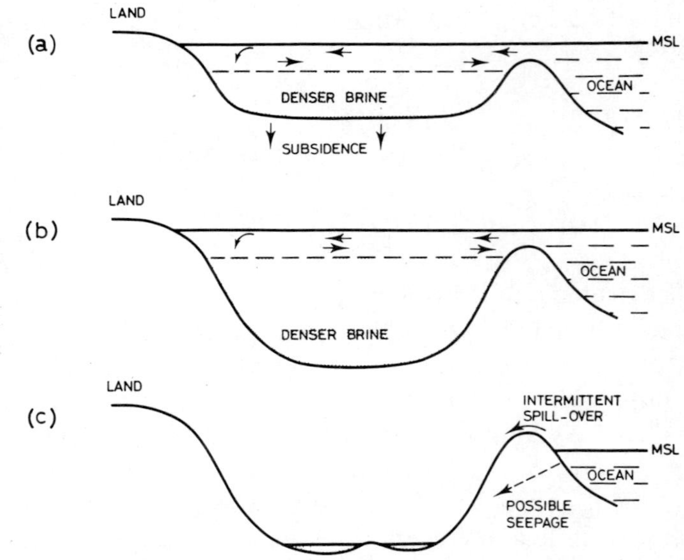

Fig. 3.5 The three principal models normally invoked to explain the origin of major evaporite basins. (a) The shallow basin–shallow water model, which requires basinal subsidence to account for thick evaporite deposits; (b) the deep basin–deep water model, with reflux taking place in a less dense brine layer above a deep, denser brine layer; (c) the deep basin–shallow water model (or deep desiccating basin), with shallow evaporites forming far below normal sea level. (See, for example, Kendall, 1978b.)

3.3.3 Theories of Marine Evaporite Deposition

The three principal models put forward to explain the development of large deposits of marine evaporites derive from the basic concept of a barred or restricted basin—that is, a body of brine connected to the open ocean by a narrow channel of limited depth which acts as a sill, allowing the influx of sufficient sea water to make up a deficit resulting from evaporation in the basin. This is necessary to enable continuous precipitation of evaporites over long periods (Kendall, 1978b, gives a good overview).

The first model, sketched simply in Fig. 3.5(a), is that of a shallow basin, shallow water situation, with a layer of denser brine sinking to the bottom of the shallow water layer. Syndepositional subsidence of the basin is required in order to allow very thick deposits of evaporites to develop by surface and bottom nucleation of crystals. This model offers an explanation for the evidence in many evaporite deposits of shallow water formation. It appears to have developed from observations of evaporite formation in some sabkhas and playa lake deposits. As noted earlier in the chapter, Arthurton (1973) concluded that, from the evidence present, layered Triassic halites in Cheshire, England, were probably deposited in a shallow lake with a water depth of only one or two metres.

The second model, sketched in Fig. 3.5(b), is that of a deep basin with deep water. As proposed by Schmalz (1969) and others, this model has the advantage over the previous one of requiring no basinal subsidence for the build up of thick deposits. This avoids objections regarding rates of subsidence in specific areas when considering the thickness of evaporites present, or the stratigraphic evidence of adjacent areas. By postulating the eventual development of a considerable depth of higher density brine up to the level of the restricting sill, with reflux (see later) occurring only in the less dense, shallower brine layer above, it does not require the operation of a continuous dense brine outflow along the basin floor. It also explains the presence of apparently euxinic sediments on the floor of many evaporite basins, and (in its late stages only) how evidence of shallow water deposition may be present. Schmalz considered that the Salina and Castile evaporites of the USA, the Zechstein evaporites, and those of part of the Elk

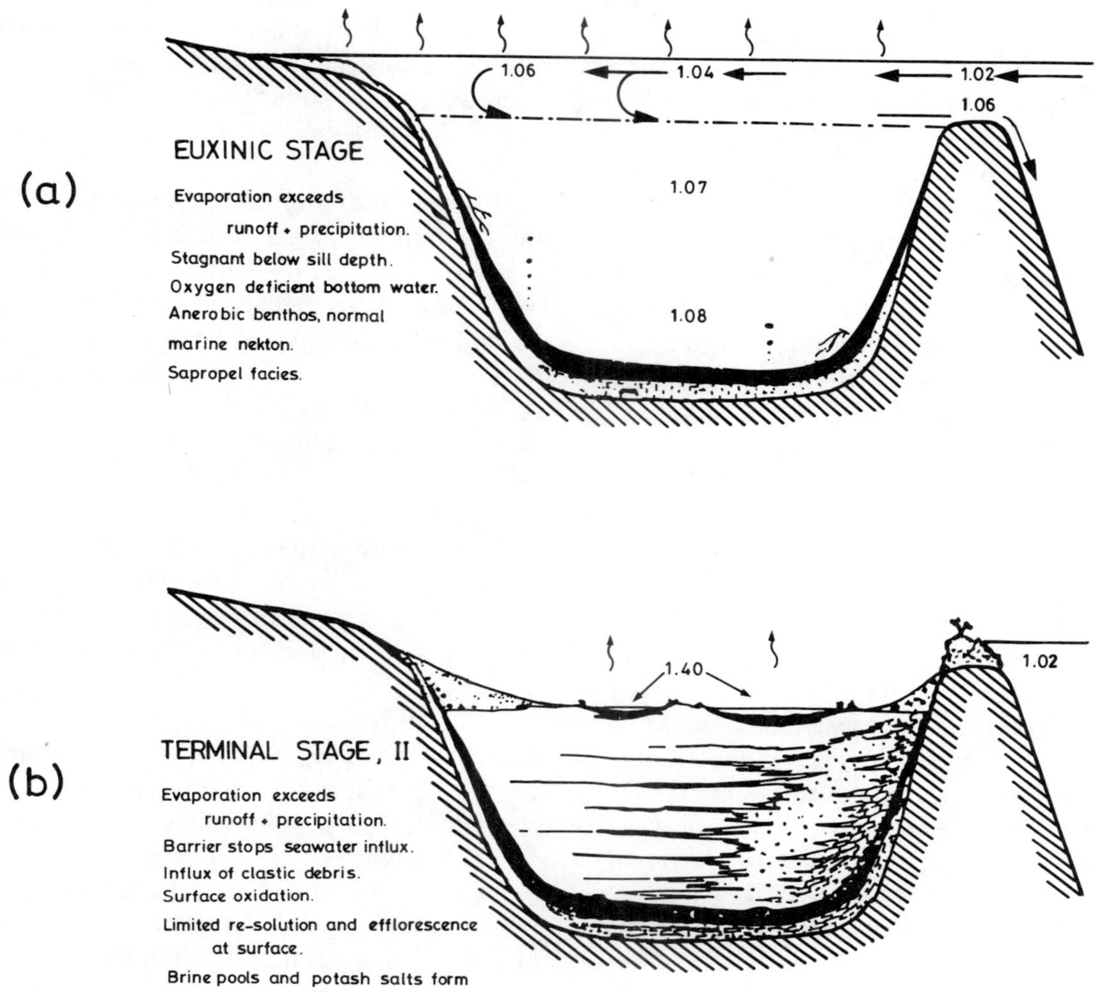

Fig. 3.6 Sketches showing an early stage (a) with free water circulation and euxinic sediment deposition, and (b) one of several alternative late stages with the basin filled and desiccating, as envisaged by Schmalz in the development of a deep basin–deep water origin for evaporites. (From Schmalz, 1969, courtesy of the AAPG.)

Point (Alberta) area show evidence of such an origin, although not everyone has agreed with his assessment of the evidence in these areas.

Figure 3.6(a) sketches an early stage in the development of this model, and Fig. 3.6(b) shows one of the alternative late stages suggested by Schmalz.

The third model is that of the deep basin–shallow water type (or desiccating deep basin). Hsü (1972) advocates this model to explain the origin of the Messinian (Tertiary) evaporites of the Mediterranean, and a similar model was invoked in studies of the Elk Point basin of Canada by Shearman and Fuller (1969). As Hsü points out, there is much evidence for the deposition of ancient evaporites in deep *basins* with floors very much below contemporary sea levels (including that already mentioned of euxinic black shales interbedded with evaporites, and the lack of evidence of major subsidence related to some thick evaporite deposits), but little or no positive evidence of deep *water* deposition. On the other hand, there is frequent and widespread evidence of shallow water deposition.

As the sketch in Fig. 3.5(c) suggests, in this model the deposition of evaporites is supposed to take place in widespread playas and desert lakes covering the floor of the basin, which receives intermittent inflows of sea water when the sea level rises above the level of the restricting sill. The situation could include a deep water phase as well as a complete desiccation phase from time to time. The observed geology of the Messinian evaporites appears to be accommodated by this model. There are other lines of evidence mentioned by Hsü such as the presence of incised canyons due to rivers which ran down into the desiccating basin.

However, the situation found in other basins, such as the existence of massive shelves of displacement anhydrite surrounding basins like the Zechstein (Taylor, 1984), is hard to fit into this model.

Hsü considers that the desiccating basin model explains the frequently-observed 'bullseye' pattern of deposition in ancient evaporite basins, with a peripheral development of 'pre-evaporite' carbonates, an intermediate development of sulphates, and a central zone of halite with or without bittern salts. Carbonates and sulphates are envisaged as being deposited on the peripheral shelves and slopes while progressive lowering of the brine level with increasing concentration leads to halite and potash salt deposition on the abyssal plain. The whole massive deposit would develop by upward accretion to a thickness of several kilometres.

Regarding the zonation of the salts, an alternative concept is that of reflux—or outflow—of denser brines from the basin. Ideas on outflow currents from evaporite basins have existed since earlier times, and Scruton (1953) advanced a hypothesis which is represented in the sketch of Fig. 3.7. This supposes a lateral salinity gradient in restricted basins in arid areas, with brine density increasing progressively from the inlet channel to the distal parts of the basin. Such gradients have been observed in some locations in recent times. Scruton goes on to postulate that movement of 'fresh' sea water from the inlet promotes the development of a bottom reflux current carrying dense brines out of the basin, thus delaying the precipitation of later stage evaporites. Concentration increases in the brine eventually lead to the precipitation of sulphates, halite, and potash minerals in a series of zones progressively

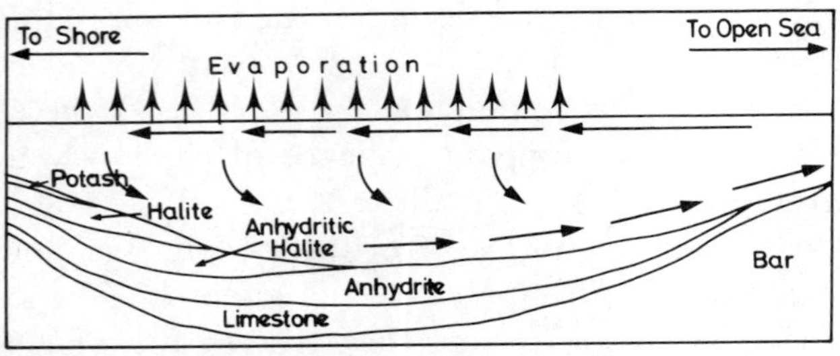

Fig. 3.7 Sketch indicating the reflux mechanism suggested by Scruton (1953) to explain zonation of evaporites in a basin. The brine density is increasing landwards, and a reflux current of dense brine on the floor of the basin is supposed to flow back out of the basin due to antiestuarine circulation. The existence of such deep basin floor currents of dense brine has not been established. (From NBS Monograph 167, 1981, after Scruton, 1953, courtesy of the National Bureau of Standards, US Department of Commerce.)

more distant from the inlet. The zonal distribution of evaporites in a basin could result from this process; alternatively, the zonation could be interpreted as a modification of the 'bullseye' distribution mentioned earlier.

However, the reflux current involving outflow across the floor of the basin of dense brines has not been established as a reality. Outflow currents not involving dense basal brine layers certainly do occur at present in similar situations. Water movement of this type, which is termed *antiestuarine* circulation, is the result of evaporation not being balanced by run-off input of water from adjacent land (Sonnenfeld, 1984); a lowering of sea level in arid areas is thus prevented. The opposite to this, which is *estuarine* circulation, occurs in more temperate, wetter areas where surface run-off surplus is instrumental in pushing surface water out of a basin or bay, resulting in an inflow of the deeper water from the open ocean to maintain sea level.

As noted previously, Schmalz supposed antiestuarine reflux in his deep basin, deep water model, although confined to the less dense brines above the level of the restricting sill, and not involving the deeper and denser brines.

Each of the models mentioned so far has both advantages and disadvantages as regards the provision of an explanation for the origin of most, if not all, giant evaporite basins of the past. Some advocates of the various concepts have, in their enthusiasm, attempted to promote each idea to the exclusion of others. It may be that there is no universally applicable model for the majority of evaporite basins. If there *is*, however, then there must be some merit in the idea of reviewing all the concepts put forward to date, and from them constructing a model which is capable of reproducing the effects which seem to be mutually exclusive as observed in various basins, together with any common factors which also exist.

Such a composite model has recently been suggested by Taylor (1980), and is represented by the sketch in Fig. 3.8. This is based on the necessity to imagine a process which balances influx of brine into the basin against evaporative loss from the basin, such that large amounts of any given evaporite mineral may be precipitated over extended periods. The model was devised to explain certain aspects of the North Sea Zechstein evaporite sequences, but may have applications beyond this.

Recourse was had to two existing concepts—that of a permeable barrier (e.g. a reef or carbonate complex; see, for instance, King, 1947; Maiklem,

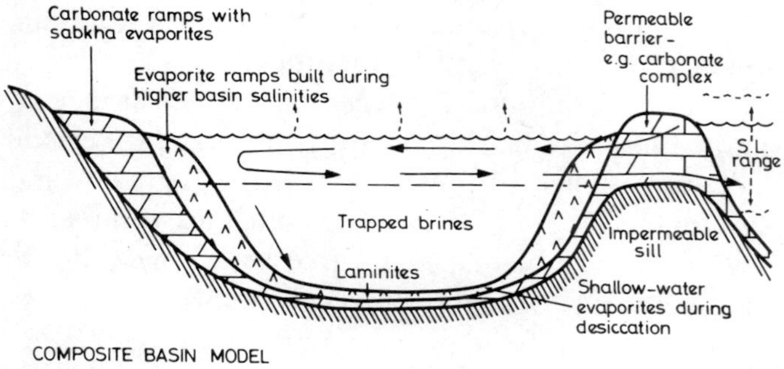

Fig. 3.8 A composite model which, as discussed in the text, could explain most of the features observed in the North Sea Zechstein evaporite basin, and possibly those in other basins. (From Taylor, 1985, courtesy of the author and JAPEC (UK).)

1971), and that of the air humidity/brine temperature controls on maximum evaporative grade (Kinsman, 1976; Walton, 1978).

In the model, an early stage of free circulation is followed by restriction due to the build-up of the carbonate barrier on the entrance sill. Relatively minor changes in ocean level within the range shown could lead to restricted flow through the permeable carbonate barrier, resulting in increased brine concentration within the basin and fall in the brine level until an evaporation/flow balance is reached. If the ocean level falls below the top of the impermeable sill, very marked evaporative drawdown within the basin could ensue and, dependent upon the air humidity and brine temperatures which develop, later stage evaporites (halite and potash salts) may be precipitated, with possible evidence of shallow water deposition as the brine level continues to fall. If ocean level is later restored to its original position, the early phase of free circulation may then recommence. For a more detailed discussion of this model, the original work by Taylor (1980) is recommended.

3.3.4 Climatic and Other Controls on Evaporite Precipitation

In the global context, it appears likely from previous work that major evaporite basins have developed in the past in semi-arid regions of the globe which have marked wet seasons, rather than in very arid areas (see, for example, Sonnenfeld, 1984).

The principal climatic factors governing the rate of evaporation from a body of brine are, according to Kinsman (1985), (i) the temperatures of the brine/air system (the higher the temperatures, the more water vapour is needed to saturate the air), (ii) humidity (as brine concentration increases, evaporation rate decreases, ceasing at $<77\%$ relative humidity of the air when the brine is saturated with NaCl), and (iii) wind speed above the brine surface, related to palaeoclimate and wind patterns.

Tectonic controls include effects leading to the raising or lowering of level of a restricting sill at the inlet to a basin, and the raising or lowering of sea level due to localised tectonism.

Other important factors are (i) eustatic changes in sea level (not related to localised tectonism), (ii) the configuration of the basin, and (iii) the presence or absence of permeable barriers affecting influx or outflow of brines, either permanently, or for shorter periods during the basin's history.

An external factor which may affect the processes of precipitation in a basin is the general one of availability and transportation of terrigenous sediments, which brings in factors of palaeogeography, drainage and wind systems, and the lithologies of any adjacent sediments involved.

3.4 BASIC ASPECTS OF HALITE DEFORMATION PETROPHYSICS

3.4.1 Introduction

Early ideas on salt mobility included the belief that salt rock which was observed to have deformed had been melted, and had flowed in liquid form, due to the presence of igneous rocks. Later, laboratory work (see, for example, Gussow, 1968, for an account of this) demonstrated that halite becomes plastic at much lower temperatures than its melting point of about 800°C. Figure 3.9 shows curves established from determinations of the elastic limit of crystalline halite where stress is applied with increasing temperature. The curves are slightly different when stress is applied in different crystallographic directions, and the limiting curves for the range, for stress applied along (100) and (111), are shown. The curve for stress applied along (110) lies within this range. The average strength of a single halite crystal, as determined by Joffé (1928), is approximately constant over the range shown by the dashed line at 44 kg/cm^2. It can be seen that the elastic limit falls rapidly with increasing temperature to 200°C (with stress applied in the (100) direction). Thereafter, it falls below the strength curve, and the halite passes into the plastic state in the (100) direction with continued application of stress. With temperatures above 350°C at the stress level of the average strength curve, the halite is plastic no matter in which direction stress is applied.

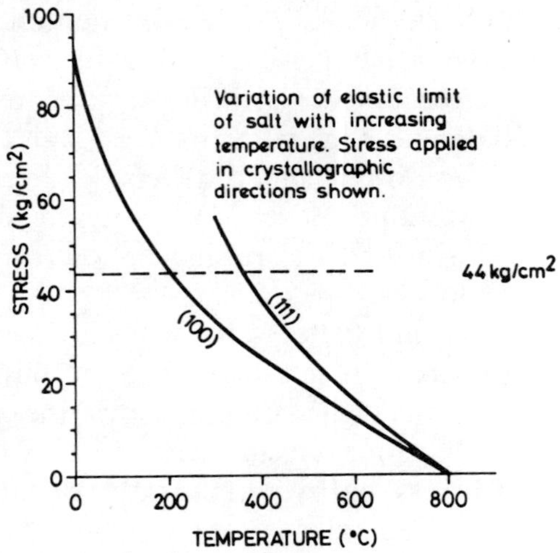

Fig. 3.9 Changes in value of the elastic limit of a single halite crystal with increasing temperature. (Redrawn, simplified, after Gussow, 1968, and Joffé, 1928.)

There is general agreement that halite is thermoplastic under natural conditions, but some disagreement about the actual temperatures involved. The laboratory determinations suggest that a temperature range of about 200–350°C to bring about a partially or wholly plastic state in halite. Many people, based on field observations, believe that these figures are too high. The emergent salt diapirs of Hormuz, Iran, are often cited as an example of a situation in which salt rock is flowing under ambient surface temperatures not greater than 46°C, and subjected to only minor gravitational gradients resulting from surface topography. It is claimed that seasonal, and even diurnal, variations in flow rate have been measured (see, for example, Kent, 1979). Richter-Bernburg (1970) has referred to flow behaviour in Zechstein salt in mine workings at much lower temperatures than the laboratory figures quoted. Baar (1977) claims that the laboratory figures are affected by work-hardening which occurs in halite crystals under laboratory conditions but not in nature.

As regards the effects of pressure, early laboratory work by Van Tuyl (1930) led to the conclusion that creep in halite becomes 'excessive' under a pressure of 12 000 psi at normal temperatures. Much work has since been done on salt creep rates in connection with mining, and it is known that salt rock creeps to varying degrees in open mine workings at quite shallow depths and under pressures much lower than that just quoted. Laboratory work suggests that whilst confining pressure plays a relatively minor role in salt deformation, differential pressure is much more important, especially at elevated temperatures.

It has been emerging from recent work that the presence of even minute quantities of free water may have pronounced effects on the deformation characteristics of halite, effectively lowering the elastic limit very markedly with increasing temperature. Although Baar (1977) states that salt rock is impermeable at depths below 300 m, free water may still be available in the form of the myriads of brine inclusions found, particularly, in primary unrecrystallised halite.

The manner in which the factors mentioned may interrelate will now be discussed.

3.4.2 Modes of Deformation

There are three generally recognised modes in which rock deformation takes place:

(i) Cataclastic fracture and intergranular movement.
(ii) Intragranular dislocation and gliding.
(iii) Diffusive mass transfer (pressure solution).

In an interlocking polycrystalline body of halite, mode (i) may be discounted, certainly at all except the shallowest depths. Plastic deformation in salt rock is understood to take place in modes (ii) and (iii).

3.4.2.1 Dislocation and gliding

The intragranular translational movements referred to as dislocation and gliding take place in the dodecahedral planes of the halite crystal lattice. Six different slip systems exist (Borchert and Muir, 1964; Shlichta, 1968) which means in practice that when a halite crystal is in a plastic state, movement in any direction is possible, the slip systems acting as shear planes. Dislocations, which are classified as edge- or screw-type, and which involve only flaw-like, small movements of a very localised nature in parts of the lattice, probably occur at lower levels of stress. Gliding, where complete sheets of the lattice move over each other along the slip planes, probably develops from dislocations as the stress builds up. In individual crystals, relatively large translational movements may occur varying, according to Borchert and Muir, from a few Ångströms up to several millimetres.

3.4.2.2 Diffusive mass transfer

Two principal mechanisms have been recognised to account for the phenomena of mode (iii)—diffusive mass transfer, or pressure solution. Although these mechanisms are still imperfectly understood, it is believed that the first, referred to as Coble creep, takes place by the transference of material from points of high intergranular stress to points of lower stress through the medium of a thin intergranular aqueous film which is capable of withstanding shear stress. The driving force of the diffusion is a stress-induced chemical potential gradient, and other gradients due to local chemical reactions (Rutter, 1983; see also McClay, 1977).

The second mechanism, the Nabarro-Herring creep effect, involves lattice diffusion, whereby material is transferred internally through the crystal from higher to lower load stress regions. It is unlikely that this mechanism operates within the context of salt rock deformation, since it is understood to take place at higher temperatures, becoming important in areas of thermal metamorphism.

3.4.2.3 Deformation and 'flow'

It is clear that the operation of both intragranular dislocation/gliding and diffusive mass transfer will lead to a polycrystalline salt rock body deforming, that the deformation will be an integration of the effects in the individual crystals, and that in any given field of applied stress, the pattern of strain which develops in all the individual crystals will tend to be the same. Thus a flat, horizontal polycrystalline slab under high vertical load stress would tend to become areally more extensive, but thinner along the vertical axis.

Although salt rock is referred to as undergoing 'flow', it should be realised that this is a non-rigorous (although convenient) use of the term, since flow implies interparticulate movement. A body of salt rock in a plastic state does not flow in the strict sense of the word; rather, it changes shape.

3.4.3 'Creep' in Halite

In engineering parlance, a material is said to exhibit 'creep' if it deforms very slowly over a long time period when subjected to constant loading. Metals and ceramics show this effect, and it describes quite well the behaviour of a salt rock body under stress.

Much work has been carried out on investigating the phenomenon in halite, both in the laboratory and in salt mine workings. In the laboratory, creep rates are normally expressed as the relationship between strain in the specimen and time over which the measurements are made. Figure. 3.10 shows curves

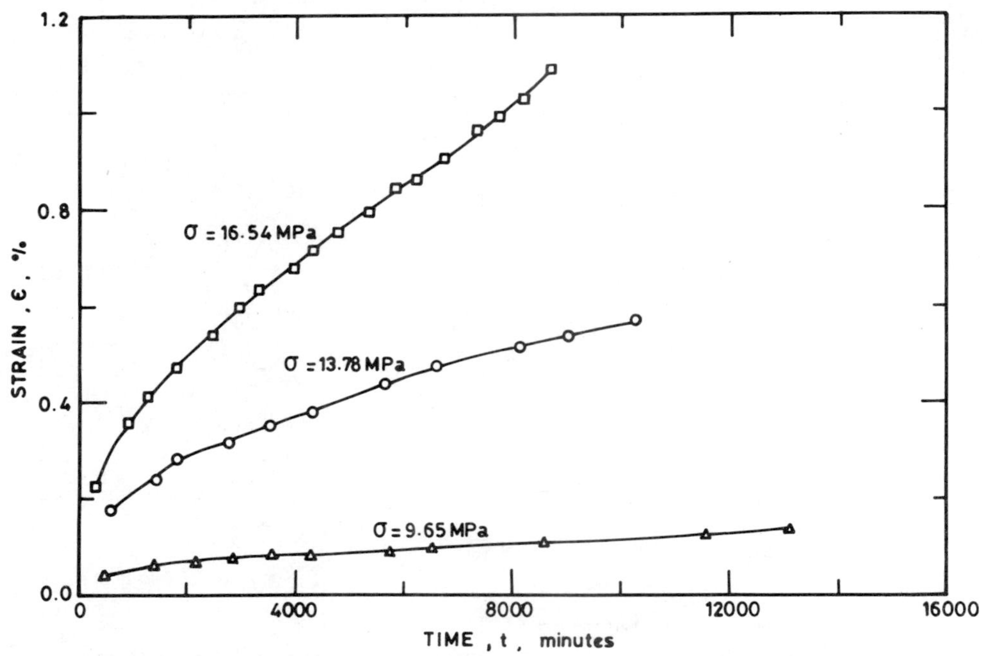

Fig. 3.10 Creep rate measurements made on four halite samples at a temperature of 306°K and with confining pressure of 11·37 MPa. Three different differential stresses were applied to the specimens, ranging from 9·65 MPa to 16·54 MPa. The creep rate is rather uniform under each differential stress. The curves show the importance of differential stress on creep rate. (From NBS Monograph 167, courtesy of the National Bureau of Standards, US Department of Commerce, after Thompson, 1965.)

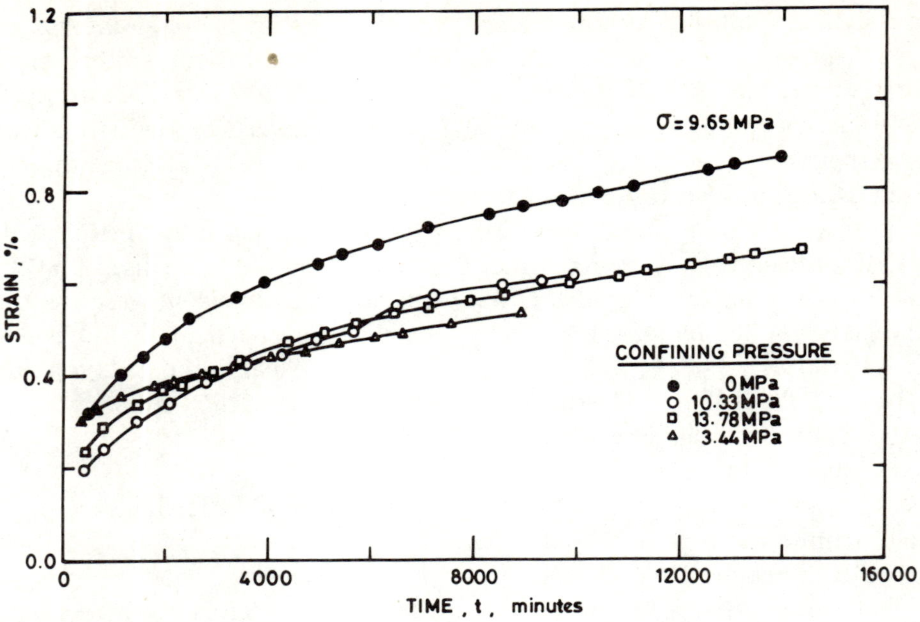

Fig. 3.11 Curves showing the negligible effect of confining pressure on creep rate in halite. Differential stress was kept constant in these tests. (From NBS Monograph 167, courtesy of the National Bureau of Standards, US Department of Commerce, after Thompson, 1965.)

resulting from measurements made on four halite specimens from Hockly, Texas. The experiments were carried out at a temperature of 306 K, and a confining pressure of 11·37 MPa (to convert to psi multiply by $1{\cdot}45138 \times 10^2$). Various differential load stresses were applied, as shown, varying from 9·65 MPa to 16·54 MPa. It is seen that under these circumstances, fairly uniform creep at different rates was exhibited by the specimens.

The tests just mentioned were carried out by Thompson (1965), who also investigated the effects of confining pressure only (Fig. 3.11). In this series of tests, differential stress was kept constant, and a range of confining pressures was applied. It can be seen that creep rate in halite is but little affected by variations in confining pressure alone.

On the other hand, the importance of temperature on the creep rate of halite was demonstrated by

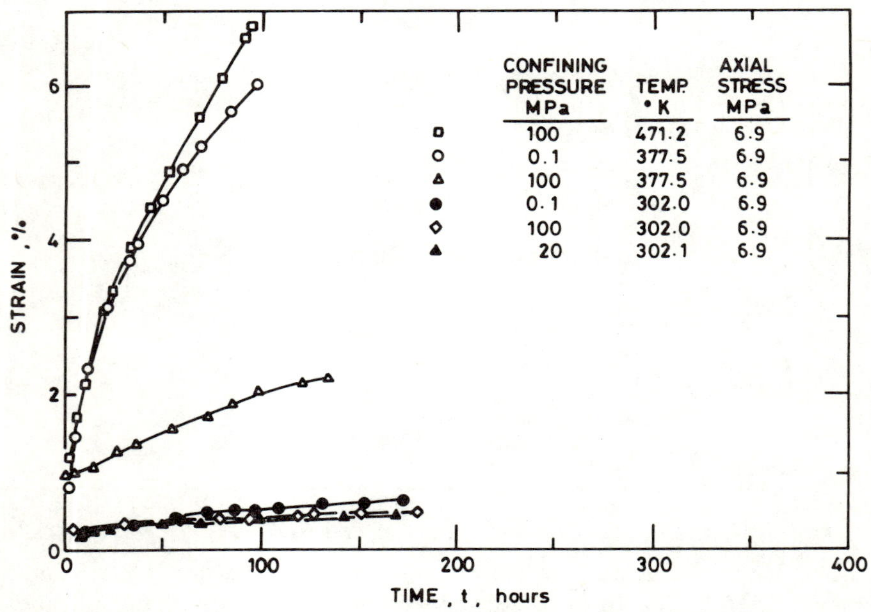

Fig. 3.12 Curves showing the importance of the effect of temperature on the creep rate of halite, in contrast to the relatively minor effect of confining pressure. (From NBS Monograph 167, 1981, courtesy of the National Bureau of Standards, US Department of Commerce, after LeComte, 1965.)

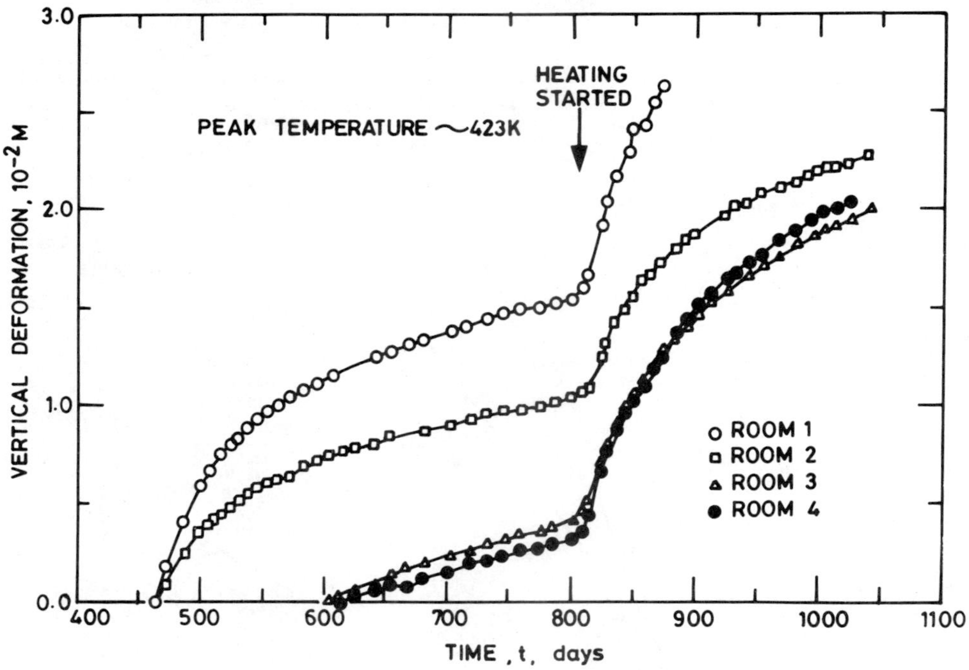

Fig. 3.13 Another set of curves demonstrating the importance of temperature effects on the creep rate of salt rock. A supporting salt rock pillar in each of four rooms in a salt mine was heated up to about 423 K at the time shown, with dramatic effects on the creep rate. (From NBS Monograph 167, 1981, courtesy of the National Bureau of Standards, US Department of Commerce, after Baar, 1977.)

LeComte (1965), in a series of tests whose results are shown in Fig. 3.12. These show the rather dramatic effect of temperature on halite creep rate.

The temperature effect is again seen in Fig. 3.13, which displays vividly the effects on the creep rate of halite of the heating up of a salt mine pillar in each of four rooms in the mine. Baar (1977) gives the field data from which the curves are drawn, and which show the rapid acceleration of creep rate following the application of the heat source, the temperature having reached a peak of about 423 K.

These results tend to underline the importance of temperature on the deformation behaviour of halite, and also to suggest that the influence of confining pressure on deformation parameters is rather small. On the other hand, field observations suggest strongly that differential pressure may also be an important factor in creep behaviour in salt rock. From a combination of laboratory experiments and field observations, it is possible to suggest that salt rock, due largely to free water effects (see later), temperature effects, and in a minor way to confining pressure, may be *potentially* mobile at only moderate depths. Actual mobility will depend upon the development of stress differentials at some stage. This seems to be supported by the measurable convergence rates of galleries in salt mines over short—one might say 'geologically instantaneous'—time periods, and at depths of only two or three hundreds of metres.

Creep in halite appears to take place in the intragranular dislocation/gliding mode (Odé, 1968); it would seem likely that diffusive mass transfer must also take place to varying degrees according to the temperature/pressure scenario which exists in particular salt bodies.

3.4.4 The Joffé Effect

It has been found that if microfractures which occur on the surface of halite crystals are removed by etching or chemical polishing (NBS Monograph 167, 1981), the crystal becomes more ductile. It is believed that the surface microfractures cause brittleness in the crystal. This phenomenon, known as the Joffé effect, may be the explanation for the fact that halite crystals, if held under water (particularly running water) can actually be deformed and bent by hand pressure alone. This odd effect has been noted by several workers. It is not known what relationship this may have to the processes of pressure solution related to halite in a polycrystalline mass, but would seem to be worth investigating in this context, and also in that of the section which follows.

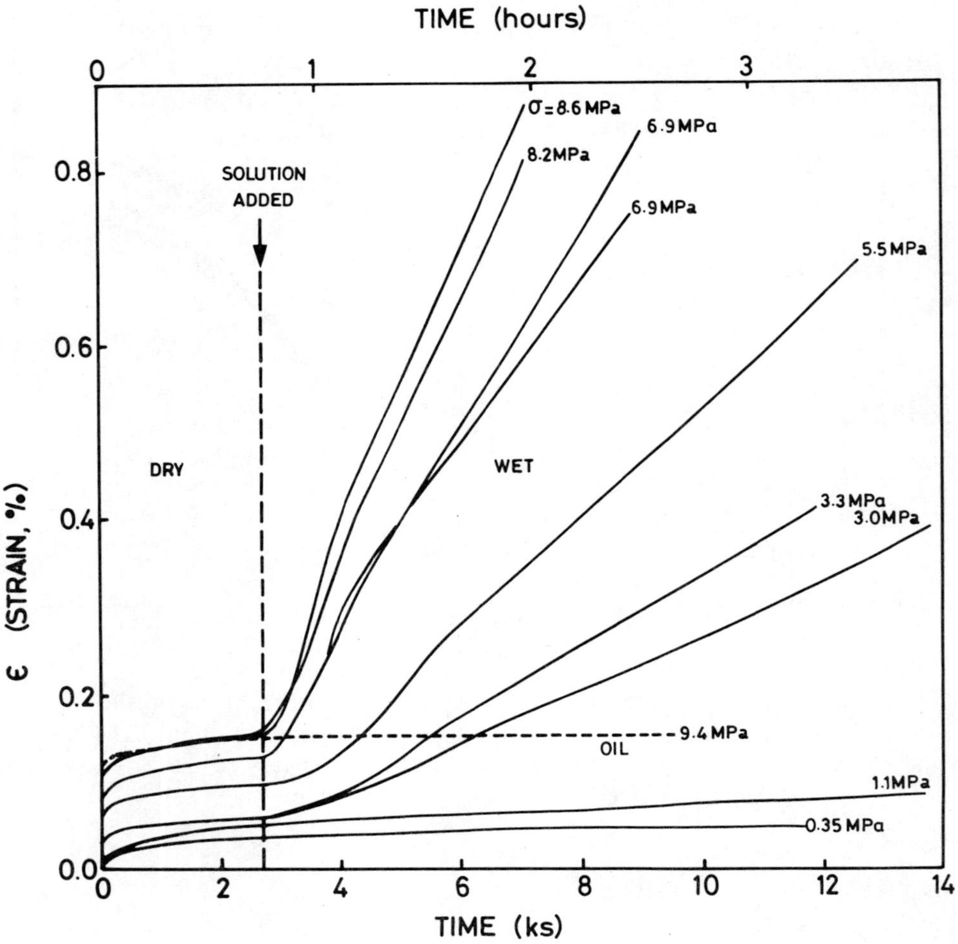

Fig. 3.14 Curves showing the effect of adding a saturated solution of NaCl to halite samples at atmospheric pressure and room temperature, with a range of differential stresses applied. Note the large increases in creep rate with increasing applied differential stress. (From Spiers *et al.*, 1982, courtesy of the authors, and of the Commission of the European Communities, Nuclear Science and Technology.)

3.4.5 The Presence of Free Water

The incidence and effects of liquid bubble inclusions have been studied in a range of minerals. Knipe and White (1978), investigating the effects of such bubbles in quartz, concluded that their presence produced important effects on the rheological behaviour of the mineral under stress.

Since it is known that primary halite commonly contains very large numbers of brine inclusions, often in the form of 'negative crystals' and frequently containing free gas bubbles, this aspect of study is of great potential importance in any consideration of the rheology of salt rock.

Recent work (Spiers *et al.*, 1982) indicates that the addition of a saturated solution of NaCl to a sample of fine-grained polycrystalline halite has a profound effect on its creep rate. Figure 3.14 shows a set of creep curves obtained in these experiments, in which a sample of the salt rock at atmospheric pressure and room temperature was dry loaded under a range of differential stresses as shown, from 0·35 MPa to 8·6 MPa. After 45 minutes, the solution was added, the dashed line indicating the time point at which this occurred, with startling effects on the creep rate, these effects being progressively more marked with increase in the differential load stress. As a control, experiments were conducted by adding evaporating oil instead of NaCl solution, and it can be seen that the oil had a negligible effect on the creep rate.

Spiers *et al.* conclude that 'wet' salt samples are much weaker than dry samples under the conditions obtaining in the laboratory. They believe that the rate of wet creep is governed by a grain boundary solution transfer process like pressure solution (see earlier in this chapter, and also Raj, 1982). They also suggest that the process may explain unusually high creep rates measured in the Iranian 'salt glaciers' (discussed earlier) during the wet season, as well as the low stress flow believed to have occurred in units such as the Boulby Halite of north-east England also mentioned earlier, which is known to have been wet when deforming (Talbot *et al.*, 1982).

The most recent work carried out by Spiers *et al.* (1986) demonstrates that the weakening effects mentioned can be caused by *minute* quantities of brine *inherent* in the salt (i.e. as inclusions), of the order of 0·05 vol%. They note that fluid inclusion arrays are common also on grain boundaries in salt, and have been interpreted by Urai (1983) to have formed by the breakdown of brine films.

It is interesting to compare Fig. 3.14 with Fig. 3.13, noting the accelerating effect on the creep rate of (a) the introduction of brine, and (b) the raising of temperature, in the system. Of particular importance is the observation that while the vertical scales of both sets of curves are of the same order of magnitude, there is a very large difference in the horizontal (time) scale. This suggests that the introduction of brine leads to an immediate and large acceleration in creep rate, while the temperature effect acts over a much longer period to produce a comparable increase in the creep rate. Thus, although the effects of the presence of small quantities of NaCl solution on halite deformation are still being studied, work done so far appears to suggest that the presence or absence of free water is an important—perhaps *the* most important—factor governing the deformation behaviour of salt rock. It seems likely that the free water and temperature factors operate by facilitating increased activity of pressure solution processes primarily, while differential stress operates principally through the intragranular dislocation/gliding mode of deformation, all these processes probably being inter-related to a large extent.

Summing-up the foregoing discussion, it is suggested that:

(i) Halite moves under differential pressure, once this passes some threshold value.
(ii) The threshold is lowered, and the creep rate increased, in the presence of free water—specifically, a solution of NaCl.
(iii) Increasing temperature also increases the creep rate.
(iv) Confining pressure, and some special mechanisms, are operative, but their effects are significant only in anhydrous environments. Elsewhere, these effects are obscured by the much greater responses to free water and temperature.

CHAPTER 4

Physical Model Studies, and the Development of Ideas Regarding Mobility and Diapirism in Salt Rock

4.1 INTRODUCTION

In order to understand how various current ideas on salt rock mobility and diapirism have evolved, it is useful and of some interest to look at the development of thought on these subjects, and at some of the experimental studies which have been carried out in attempts to throw light on the processes involved.

Serious speculation as to the nature and origin of diapiric salt structures began about the middle of the nineteenth century. In some areas such as Algeria, Portugal and Transylvania, structures due to the deformation of salt rock were observed and hypothesised upon (see, for example, Odé, 1968). Elsewhere, major evaporite basins still remained to be discovered—some, such as those beneath the Red Sea, Mediterranean and North Sea, not until the present century. In 1892, Jukes-Brown wrote of the Permian (or 'Dyassic') rocks of Great Britain without a single mention of salt rock, being completely unaware of the Upper Permian salt deposits, thousands of feet thick, which lay but a few miles offshore.

Where salt deposits which had become mobile were known, early ideas on salt movement often included the notion that actual melting of the salt had occurred under the influence of igneous intrusion or extrusion. Other theories ranged from the ingenious to the frankly bizarre. Little real progress was made until the early part of the twentieth century when more systematic information became available through mining, drilling and, latterly, subsurface geophysical methods.

When hypotheses began to be advanced which were in some measure provable, laboratory work using physical models was begun. Many significant contributions to theory have come from these studies, especially those which included a realistic approach to correct model scaling, and many of the suggestions prompted by them have later been confirmed by field observation, or by deep seismic data often combined with the results of drilling.

4.2 EARLY IDEAS

De Golyer (1926) in a symposium paper on salt structures lists over a dozen different theories of salt diapir origin for the Texas Gulf Coast features, with references dating from 1860 to 1924. These include the ideas that the features are of volcanic origin; islands in a Cretaceous sea; due to gas pressure; uplifted by forces of crystallisation or osmotic pressure; or are domes of tectonic origin.

Mention of the possibility of plastic flow in salt seems first to have been made in 1913 (although Reusch, as early as 1867, had found evidence of plastic deformation in halite crystals) and the idea gained general acceptance within a few years. At this time, the driving force producing the diapiric features was believed to be purely tectonic, i.e. due to compression and folding of the crustal rocks associated with the salt deposits. This seemed to be supported by alignments of salt diapirs parallel to orogenic trends in some areas. As will be discussed later in the book, there may be more substance to this concept than has been imagined in recent times.

The idea that the density difference between salt and sediments at depth may be of importance was first mooted by Arrhenius (1912) in studies of the German Zechstein salt area. At the time, this suggestion seems to have been largely ignored, or at most considered and discarded by many geologists, and it was to be several years before it was taken seriously.

While the concept of the importance of the density difference between salt and overburden sediments came to be generally accepted there was no consensus regarding the mechanisms involved in the rise of diapiric salt features, and a lively controversy raged, particularly in Germany in the second decade of this century, as recorded by Odé (1968). The main schools of thought held (a) that dissolution lows at the salt roof resulted in load stress differentials which caused the remnant salt highs to rise, (b) that tangential tectonic stress was the driving force for upward salt movement, or (c) that the pressure of the overburden thrust the salt upwards into zones of weakness, e.g. up faults. The controversy eventually resolved itself into opposed theories of the gravitationally controlled rise of salt versus the effects of tangential tectonic stresses. The latter idea fell from favour with the advent of the fluid mechanics hypothesis to be discussed later, although it has always retained some adherents.

The concept of salt diapirs being due to plastic flow and forceful intrusion initiated some of the earliest physical model experiments. These were not notably successful in reproducing phenomena observed in the field. They often employed rigid plugs which were forced up into model overburden materials, the force involved usually being greatly out of proportion to the size of the model, and the materials used therein. There was at this stage little or no appreciation of the importance of scale factors in the model in relation to real-earth materials and situations. As Nettleton (1955) mentions, the theory of model scaling had been studied by mechanical and civil engineers many years before. In the early years of the century, attention had been drawn to the importance of correct scaling generally in models, in work on the use of dimensional equations. The concepts involved were stressed for the specifically geological context by Hubbert in an important paper in 1937.

The model studies of Escher and Kuenen (1929) are of interest in that they are rather typical of early attempts to simulate the earth processes involved in salt diapirism. They demonstrated in a basic fashion the results of applying a heavy load vertically downwards upon a plastic substrate, with a manufactured hiatus in the model overburden allowing upward flow of the mobile material (Fig. 4.1).

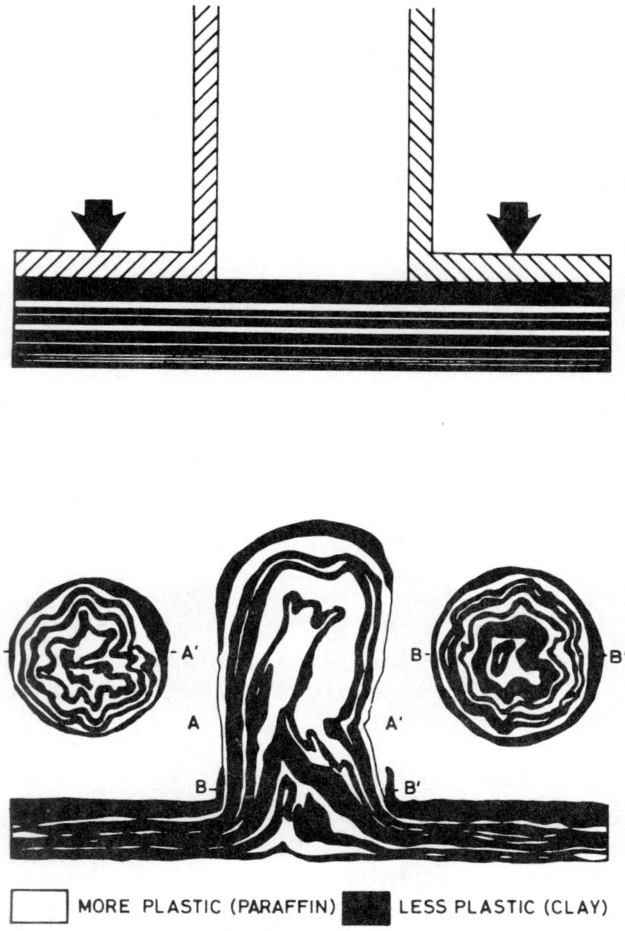

Fig. 4.1 Model study using thin layers of paraffin wax and clay as the mobile layer. The load stress of the overburden is represented by downward pressure on a flat plate which has a hole cut in the centre with a vertical cylinder attached to allow upward flow of the mobile material in response to the downward pressure, the arrangement being shown rather schematically here. Two sections across the model diapir are shown. Some similarity between the internal structures developed and those seen in certain real earth features was noted. (Redrawn, somewhat modified, after Escher and Kuenen, 1929.)

By using visually distinguishable layered plastic materials (paraffin wax and clay), an attempt was also made to investigate the internal structures which developed within the model diapir, and to try to relate these to structures seen in real-earth examples of salt diapirs.

It is easy at this later time to point out the shortcomings of such studies. They tend to beg the question of the initiatory stages of diapirism, and the fundamental processes of mass transfer and mass balance which were addressed by later workers such as Nettleton. They show little regard for model scaling either in the materials used or in the stresses applied in relation to the materials and dimensions of the model.

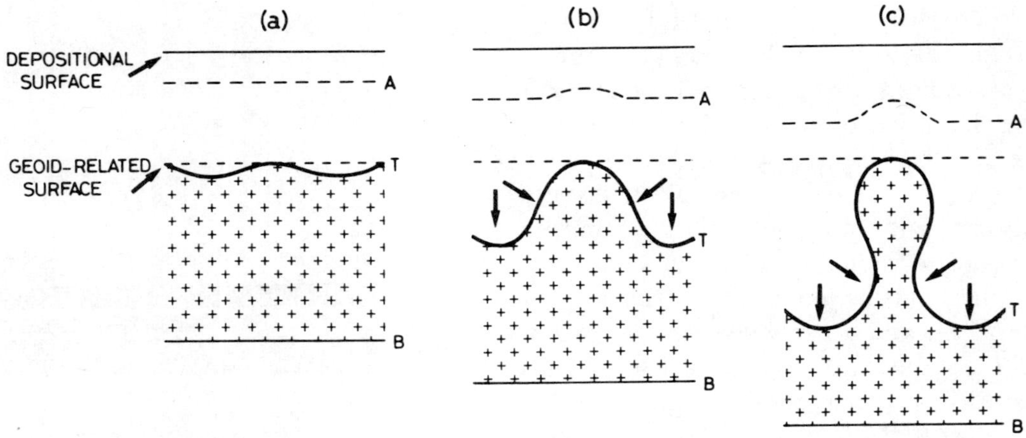

Fig. 4.2 Schematic diagrams summarising Barton's (1933) downbuilding hypothesis for the development of salt diapirs. Key: cross symbol, salt; overburden left blank; upper dashed line A is a reference horizon within the overburden; lower dashed line is the trace of a reference surface related to the geoid; solid lines B, T, are base and top salt levels. The sequence of events suggested is (a) a salt swell begins to develop for unspecified reasons; (b) basinal subsidence occurs beneath the load of continuing deposition, while crest of salt swell remains at same level as in (a), and components of overburden load stress perpendicular to salt flanks tend to compress the plastic salt mass laterally. (c) Salt diapir develops as an inverted 'teardrop' shape. Crest remains at the same level as the salt subsides, with flow into diapir continuing. Greater compaction effects above the salt 'lows' flanking the diapir lead to apparent uplift of horizon A above the diapir.

However, these were pioneer experiments which stimulated much thought and interest. Later workers studied the methods, adapted and improved them, and learned much from earlier errors. For this reason, the early experimenters deserve great credit. In this, as in other fields, much more is often learned from mistakes than from modest successes, particularly in the early stages.

4.3 'DOWNBUILDING' AND FLUID MECHANICS

The approach to the problem via the buoyancy of salt received attention from Barton (1933). As noted by Nettleton (1955) in his paper on the history of concepts of Gulf Coast salt dome formation, Barton, who had previously come to the relucant conclusion that the static thrust of the overburden—particularly in areas where the salt/sediment density difference is rather small—is inadequate to result in the forceful emplacement of diapirs, arrived at a compromise concept. He effectively turned the problem upside-down by suggesting an isostatic 'downbuilding' hypothesis, whereby the top of a developing diapir remains in the same position relative to the geoid, while the surrounding overburden sediments slowly subside under the weight of continuing deposition. Barton was satisfied that this process could overcome his previous reservations, since the amount of work involved would be much less than that required for active and forceful intrusion of salt through the overlying sediments. Figure 4.2 shows a simple sketch embodying the basic concept of downbuilding.

Certain examples of real-earth diapirs seem to lend support to Barton's hypothesis in that a remarkable lack of disturbance of the sediments surrounding the diapir is seen. However, many other examples exist where extremely severe disturbance of the overburden is evinced; many of the effects are compatible only with forceful intrusion of the salt. It seems feasible, however, that some diapirs may have occurred by downbuilding, and that the process may have operated over part of the growth history of others.

The geophysicist Nettleton (1934) was the first to attack the problem by way of fluid mechanics. Taking the view that both mobile substrate and overburden could be thought of as behaving like extremely viscous liquids, he conducted a number of experiments using two immiscible liquids. Typically, a model consisted of a layer of crude oil with density of about 1·0 as the model 'mobile substrate', and concentrated corn syrup, density of about 1·4, as the model 'overburden'. The original paper should be referred to for a detailed account of the experiments, but Fig. 4.3 shows a simplified sketch of the results of one series. It is of particular importance, since it suggests the formation of a *peripheral sink*, which is a depression in the upper surface of the mobile layer produced by the withdrawal of the mobile material into the diapiric feature. The

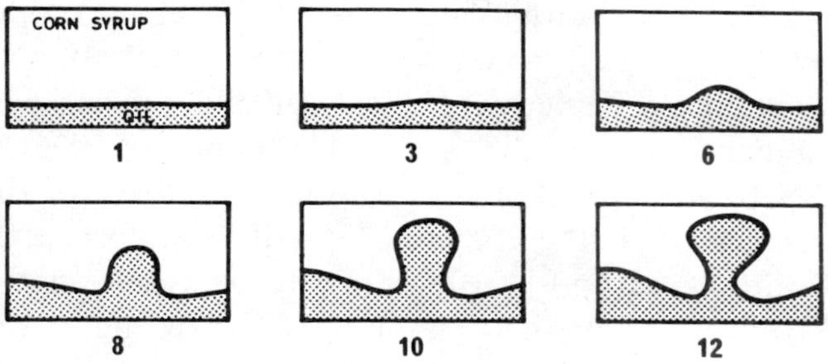

Fig. 4.3 Simplified sketch of one of Nettleton's experiments from the 1934 paper on model diapir formation using two immiscible liquids (heavy oil as the mobile source and concentrated corn syrup as the overburden). Elapsed time in minutes from the start of the experiment is shown. Note particularly the formation of the depression—the 'peripheral sink'—in the surface of the oil around the model diapir, due to flow of the oil into the feature. (After Nettleton, 1934.)

volume of material withdrawn from the sink is equal to the volume of material in the diapir above the level of the undisturbed upper surface of the mobile layer. Observations of this type of feature associated with salt diapirs in the field had already been made during drilling, and Nettleton's experiments served to explain the nature of these peripheral depressions.

A real-earth example of a peripheral sink development around a salt structure is seen in cross-section in Fig. 4.4, which is a migrated seismic section from the northern North Sea. It should be noted that the top surface of the salt in a peripheral sink is not a surface of deposition. A substantial thickness of overburden sediments will have been deposited above the salt before the latter begins to move, and so the contemporary depositional surface will occur in the section at some level above the top of the salt. This matter will be considered further in the chapters on diapirism.

Nettleton's approach through fluid mechanics was most important and seminal. It had the great attraction that it was amenable to treatment by an established body of mathematics. As a whimsical parenthetic comment, it is mentioned that geologists have learned, according to fluid mechanics theory, that they have their terminology wrong, since the diapir itself is to be regarded as the 'sink'. and not the peripheral depression (Ramberg, 1981). This information has been stoutly ignored to this day. As Nettleton himself recognised, however, this approach had some major defects. A liquid has an immeasurably small yield strength, and it is therefore impossible to create in the model overburden liquid the stress–strain systems with

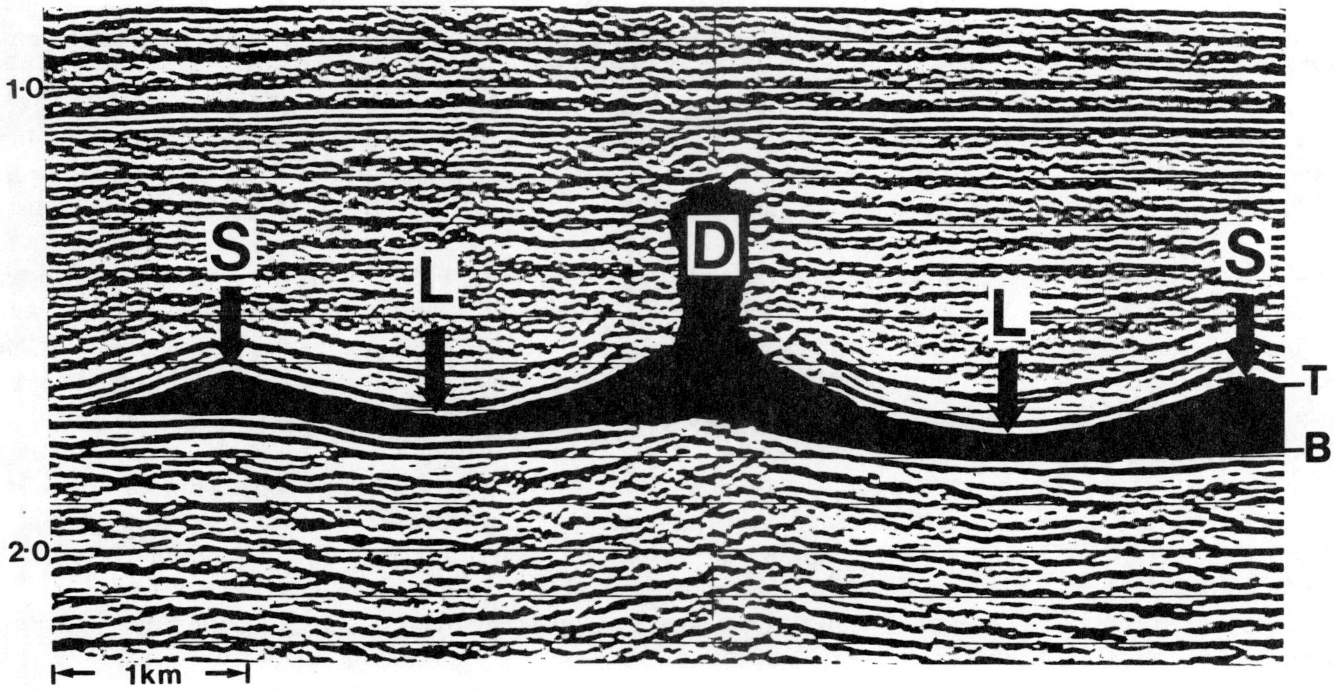

Fig. 4.4 Small primary diapir, D, with peripheral sink development, in cross-section. Trough of peripheral sink at L–L. A secondary pre-diapiric ridge or swell system has evolved at S–S. Note that L–L and S–S may be of annular form centred on diapir D. The swell system S–S has developed due to the pattern of salt flow (salt interval infilled in black) both outwards and inwards from the trough L of the peripheral sink. The salt supply was apparently exhausted before the secondary swells could develop into piercement stage. B and T mark base and top of the salt interval. (Courtesy of SS(E)L.)

resultant faulting and fracturing which are found in the real earth. It is questionable how close an analogy can be drawn between model overburden behaviour based on a liquid, and that in a real overburden with its increasing density and compaction with increasing depth. It seems at least possible that the competency of the overburden rocks may have an important influence upon whether a salt structure becomes a piercement feature (a diapir) or remains a non-piercement pillow structure.

Another aspect which may have led to misconceptions is that of the initiation of the model mobile features when two immiscible liquids are used. It was suggested by the work of Nettleton and others that, if the models were left in the density-inverted position, without any deliberate attempt to initiate a diapiric feature, after some elapsed time the latter would initiate spontaneously, either as spaced-out individual features or as waves, due to the unstable density configuration. This has led to the belief among some geologists that such spontaneous initiation of diapirs commonly occurs under natural conditions. In later model work it was found, however, that when overburden materials with small but measurable yield strengths were used in the models, the initiation of model diapirs had to be given a 'push'—usually either by removing some overburden material at preferred locations, or by building an artificial irregularity, a 'bump', on the top surface of the mobile layer. This is an interesting and possibly significant fact.

In the discussions which followed Nettleton's 1934 paper, several pertinent questions were raised, some of which still await answers. One, already mentioned, was the question of the competency of overburden rocks, and this was generally agreed to be of possible importance. Another question was why, in the light of the fluid mechanics hypothesis, there are areas with thick salt deposits where salt diapirism has **not** occurred. An answer put forward was that insufficient overburden of the requisite density is present in such areas to create the necessary gravitational instability. This is a matter which will be pursued further in the chapters on diapirism, but it can be mentioned here that areas of diapiric salt movement are known where the overburden present at the time of salt movement is believed to have varied between thicknesses of 150 m (500 ft) and over 1000 m (3250 ft). Conversely, areas with these thicknesses of overburden and sufficient density differences are known where no diapirism has occurred in an existing salt layer.

It was also pointed out that in regard to salt rock behaving as a fluid, intimate mixing might be expected to take place during movement whereas, in many salt bodies, deformed but definite stratification is commonly observed within the salt mass (even with laminar flow, some diffusive boundary mixing occurs).

Nettleton drew attention to the importance of the time factor in the rheology of salt rock; processes operating far too slowly to be recognisable in the laboratory, even in a human life span, may well produce major effects over geological periods of time. Possible processes of this nature must remain as unknown quantities in the study of salt movement.

Dobrin (1941) also studied models based on fluid mechanics.

4.4 MODELLING WITH MATERIALS OF FINITE STRENGTH

An important series of model studies were carried out by Parker and McDowell (1955) in which materials with finite strengths were used, so that stress–strain effects in the model overburden in response to movement of a mobile substrate could be studied. Asphalt was used as the mobile substrate, and various weak barium sulphate muds or granular material of slightly higher density than the asphalt were used for the overburden. Many of their results are still of great interest and importance, and the original paper should be studied; only a brief summary of some of the more significant findings are made here.

An additional point of difference between studies of this type and those using two immiscible liquids should not be overlooked. Using model materials of finite strength, it is possible to differentiate between the early, pre-piercement stages of mobile layer uplift, and the later, diapiric piercement stage; with immiscible liquids, such a distinction is not possible. This has a considerable bearing on attempts to relate model behaviour to the processes which take place in nature.

After modelling a large number of diapirs, Parker and McDowell came to the conclusion that there was a relationship between the size of a model diapir, as expressed by, for example, its diameter, and the thickness of the mobile source layer from which it developed. A relationship of this kind had been recognised by Nettleton, and was also the subject of later study (Ramberg, 1981).

Another observation of importance was that the development of a model diapir could be slowed, halted and restarted by varying the thickness of overburden material. This relates directly to field observations of the history of many salt diapirs which indicate frequent

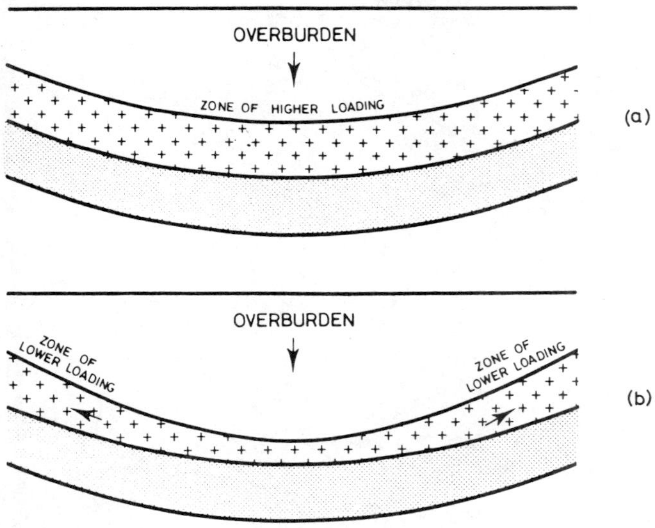

Fig. 4.5 Sketch model of a simple saucer-shaped basin, in cross-section with a potentially mobile layer (a) before and (b) after critical overburden loading has developed. Cross symbol—mobile layer.

variations in the rate of growth. Such variations can usually be correlated with changes in the rate of deposition, either regional or local, and hence to rates of basin subsidence.

As noted previously, with model materials of finite strength, some intervention was required in order to initiate a diapiric feature. The methods used for this were (i) the manufacture of a positive irregularity on the surface of the mobile layer, (ii) varying the thickness of the overburden material, (iii) varying the density of the overburden material, and (iv) manufacturing a 'fault' of normal type in the overburden and mobile material, with a throw which does not completely sever the mobile layer. It was noted that method (i) is unlikely, under normal circumstances, to have a real-earth analogue. Methods (ii) and (iii) produced flow in the asphalt layer from beneath locations of higher load stress towards regions of lower load stress. If the overburden thickening is of simple basinal form, as in the sketch of Fig. 4.5, when the thickest overburden at the basin centre has exceeded the critical value of bulk density to produce a density inversion at the potentially mobile layer (cross symbol), and the latter has actually become mobile, movement will take place. There will be a tendency for the mobile material to flow away from the centre towards the basin margin, where the overburden loading is lower.

The eventual outcome of such movement in a basin with mobile salt will be that all salt over an appreciable area in the basin centre may withdraw towards the margin, leaving the overburden resting on the erstwhile salt floor, as in Fig. 4.6.

It should be remembered that we are here dealing not with diapirism (overburden piercement) but with lateral stress-relief flow of a mobile layer.

If, as opposed to a simple basinal configuration, the overburden loading above a mobile substrate varies in some other way, a different pattern of stress-relief movement of the mobile substrate can be expected. In the model studies of Parker and McDowell, if a thicker or denser overburden layer with a linear margin

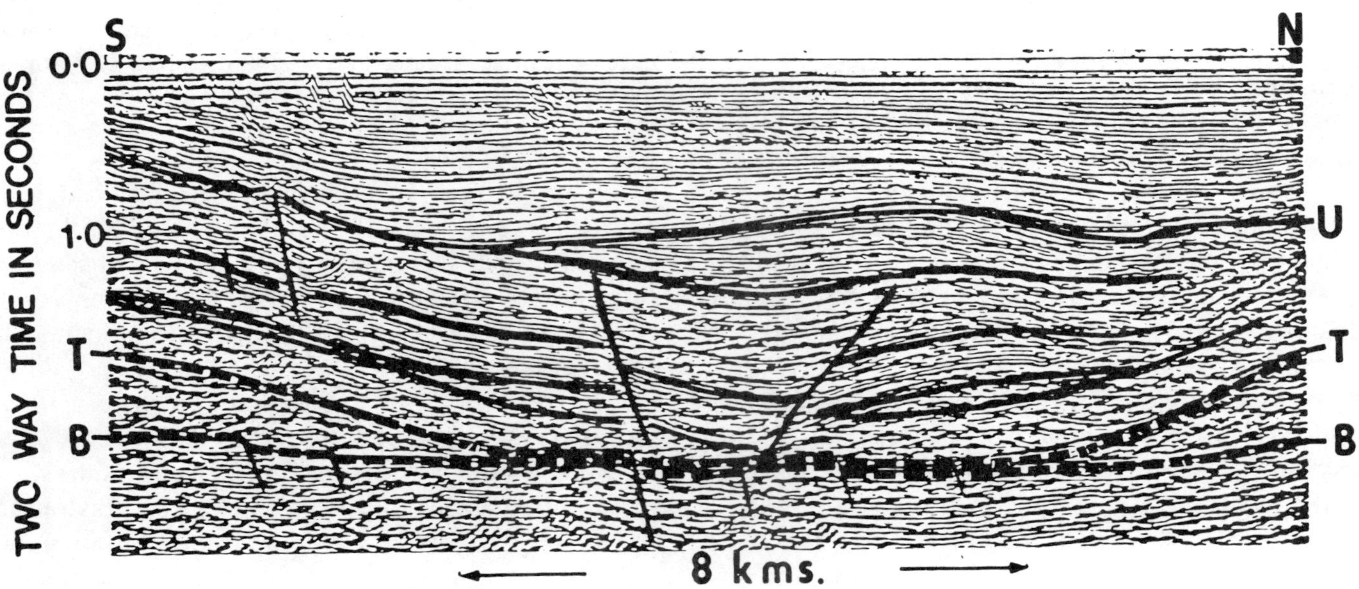

Fig. 4.6 Migrated section from centre of Southern Zechstein salt basin of British North Sea. Salt interval is B–T. Section runs across short axis of an elongate salt withdrawal basin. Note complete absence of salt from central 8 km of section, and graben feature in centre marking zone of earliest complete withdrawal of salt. (Courtesy of SS(E)L.)

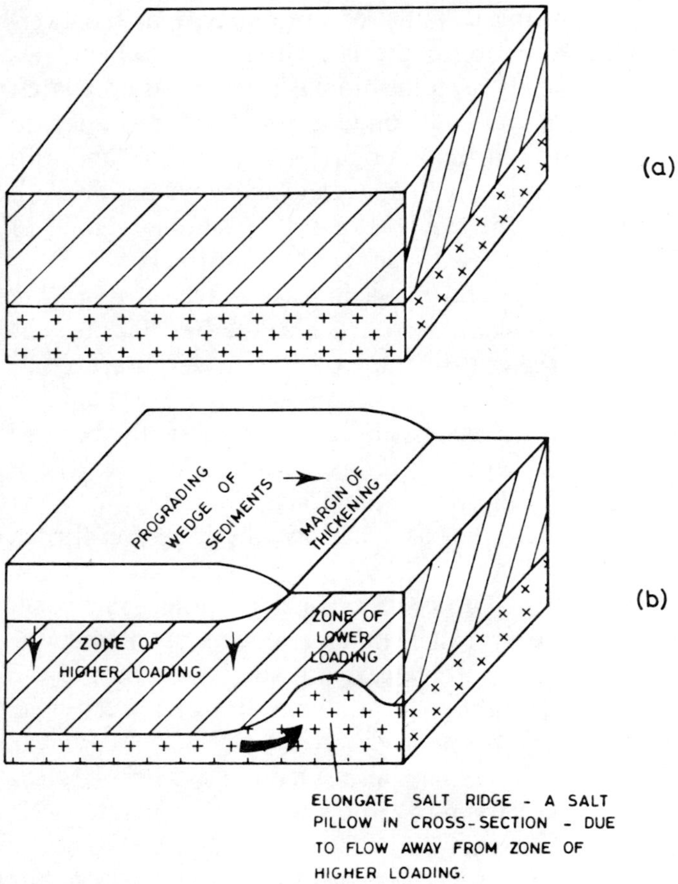

Fig. 4.7 Block diagrams sketching the effect on (a) a uniform salt layer (cross symbol) overlain by a uniform overburden (cross-hatched) of (b) a prograding wedge of later sediment.

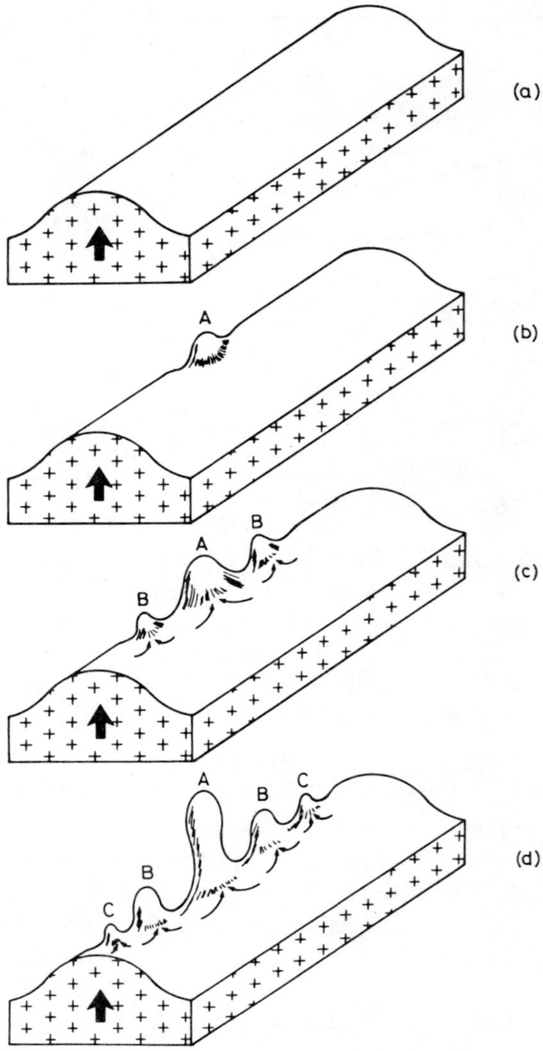

Fig. 4.8 Cartoon showing progressive development of diapiric processes from the crest of a salt ridge of the type shown in Fig. 4.7. A primary feature, A, leads to later development of secondary (B) and third-stage (C), features, due to a flow pattern related to later-discussed features (see Figs. 4.10–4.12 and related text).

covered only part of the model, the asphalt tended to flow into a ridge or ridges parallel to the margin of thickened or denser overburden. The block diagram in Fig. 4.7 shows a model which is analogous to this situation, with a mobile layer such as salt (cross symbol) of originally uniform thickness lying beneath an overburden (cross-hatched) of uniform thickness, the upper surface of which is supposed to be a depositional surface (upper sketch, a). If a wedge of sediment, such as a delta or other environmental type, progrades across the overburden layer, the increased loading produces a stress differential which will cause the underlying salt layer, if it is potentially mobile, to flow into a ridge parallel to the margin of thickened overburden (lower sketch, b). This type of mechanism was also noted by Ramberg (1981) in model experiments.

Where such a ridge of mobile material rises in this way, instability may often develop due to slight irregularities in the rate of rise of the ridge crest resulting from variations in competence of the overburden materials caused by, for example, faulting or lithology changes. It is likely that a pattern of salt flow which causes a number of these spine-like diapiric features to occur along the ridge crest will ensue, as suggested by the cartoons in Fig. 4.8 (reference will again be made in Chapter 7 to the two simple models discussed above).

It is of interest to note that Trusheim (1960) suggested a process of this kind in an illustration in his paper on the North German salt area, a simplified sketch of which is seen in Fig. 4.9. The illustration shows the development of a salt wall at the right, supposed to be near the basin centre. Individual diapirs decreasing in size and, finally, salt pillows/swells (left) develop towards the basin margin. The diapirs are shown as rising from ridges of salt.

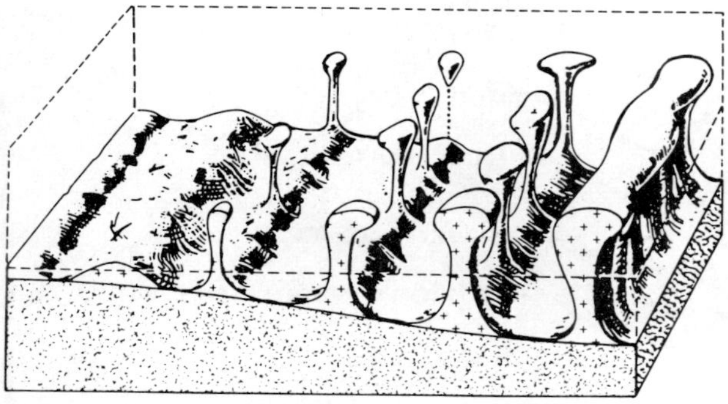

Fig. 4.9 Development of a salt wall (right), diapirs (centre three rows of features), and pillows/swells (left) from thick salt near basin centre (right) to thin salt at basin margin (left) with the diapirs developing as spine-like structures from salt ridges. Salt shown by cross symbol. (Redrawn after Trusheim, 1960.)

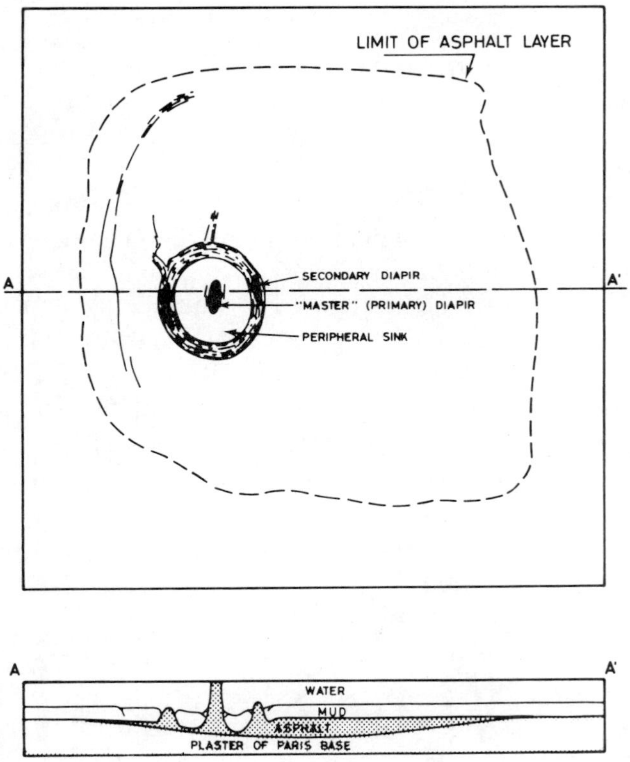

Fig. 4.10 Simplified sketch of the model experiment (plan view at top, cross-section at bottom) in which Parker and McDowell (1955) produced a concentric diapir group round a central primary diapir. Over an extended time period, further concentric sinks and diapirs with increasing radii developed (see Fig. 4.11). The ring of broken lines within which the secondary diapirs develop (plan view) is an annular graben which forms before the emergence of the asphalt secondary diapirs above the mud surface. The graben is produced by stresses in the mud overburden associated with the subsidence effects related to the developing sink. (After Parker and McDowell, 1955, courtesy of the AAPG.)

An example of a similar process, but where the mobile material consists of prodelta clays rather than salt, is afforded by the work of Morgan *et al.* (1968) on the 'mudlump' islets in the mouth of the Mississippi River. These are interpreted as diapiric clay spines rising to the surface from flow ridges of prodelta clays produced by differential loading resulting from an advancing load of fluvial sediments at the present surface.

Referring back to the four methods mentioned for the initiation of model diapirs, real-earth examples of fault-associated flow (method iv) have been identified, and examples from the North Sea (Jenyon, 1985b) will be discussed in Chapter 5.

Perhaps one of the most interesting and significant series of experiments by Parker and McDowell involved studies of a model consisting of an asphalt substrate in a shallow bowl-shaped depression formed in a plaster of Paris base (see Fig. 4.10). The model was covered by a water layer about 5 cm deep, to which was added a thin slurry of powdered barium sulphate mixed with water. Settlement of the solid particles produced a thin layer of weak mud over the model, which acted as a slightly denser overburden. After settlement was complete, some water was removed, but a thin water layer was left so as to avoid disturbance of the mud surface and also to prevent drying-out and shrinkage. Model diapir formation was initiated at preferred locations by removal of some of the mud overburden to create differential load stress.

Typically, at such a location, a slightly elongate model diapir formed, with the asphalt flowing into the diapir and causing a depression—a peripheral sink—to form around it. Early movement of the asphalt was indicated by fracturing and graben formation in the mud overburden. The pressure pattern in the mobile layer, as calculated previously by Nettleton, developed in such a way that, after a time, asphalt was flowing not only inwards towards the 'diapir', but also outwards away from the trough of the peripheral sink towards its margin. This outward flow includes a vertical component of movement due to the difference in depth of the asphalt surface at the deepest part of the trough, and at the periphery of the sink. After some time, a diapiric ridge of somewhat uneven and discontinuous form rose around the sink margin, resulting in elongate and arcuate model diapirs in annular configuration around the central, master diapir.

This secondary diapiric ridge, due to the inward flow of the asphalt outside the circular ridge, resulted

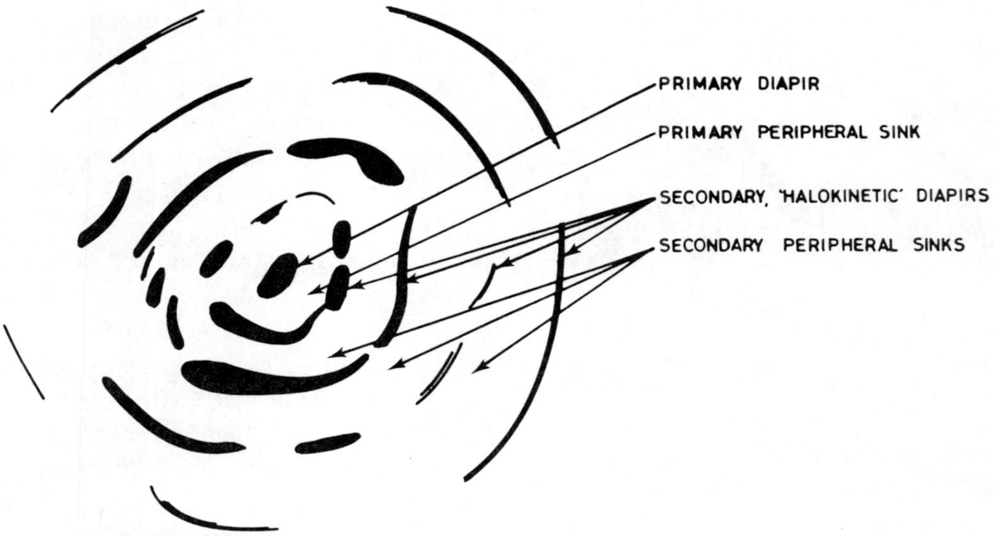

Fig. 4.11 Freehand sketch of a photograph of later-stage developments in one of the concentric diapir groups in Parker and McDowell's model studies, as shown in Fig. 4.10 (After Parker and McDowell, 1955.)

in a second annular peripheral sink, on the outer margin of which a second diapiric ridge eventually developed. This process continued, producing a series of alternating sinks and ridges of annular form around the central master diapir, as indicated in the sketch of Fig. 4.11.

Parker and McDowell made a distinction between the mechanism of formation of the central 'master dome' and the secondary ridges which developed at the margins of the progressively formed annular sinks. They noted that the master feature may come into existence due to various causes, but that the secondary ridges develop due to the 'disturbance of the condition of equilibrium between (mobile layer) and (overburden) accompanying the growth of the master dome and the development of its peripheral sink'. In other words, the 'secondary domes' or ridges would, in the real earth, be due to a purely gravitationally controlled, i.e. halokinetic, mechanism.

At the time the studies were made (1955), the suggestion that such a mechanism could operate under natural conditions was advanced as being hypothetical only. No unequivocal examples of such diapiric developments were known, although the Spindletop Dome in Texas was mentioned as a possible example of the 'master dome' of a group. Mathematical treatments of salt movement dynamics and patterns of features were produced by Daneš (1964), Biot and Odé (1965), Selig (1965) and Selig and Wermund (1966). The latter paper is of particular interest in that the authors suggest that, apart from the concentric group arrangement suggested by the work of Parker and McDowell, a similar mechanism with an elongate salt ridge as the 'master' feature could lead to the development of parallel alignments of secondary ridges and sinks 'younging' in both directions away from the primary feature.

This type of mechanism is strongly suggested in a well-known group of diapirs and salt pillows northwest of Hamburg, Germany (see sketch-map in Fig. 4.12). This group, which has figured in a number of papers, e.g. Sanneman (1968), strongly suggests the operation of the progressive mechanism described, with its geometry being the result of the formation of an elongate linear primary feature—a salt wall—in the centre of the group, rather than a circular or ovate

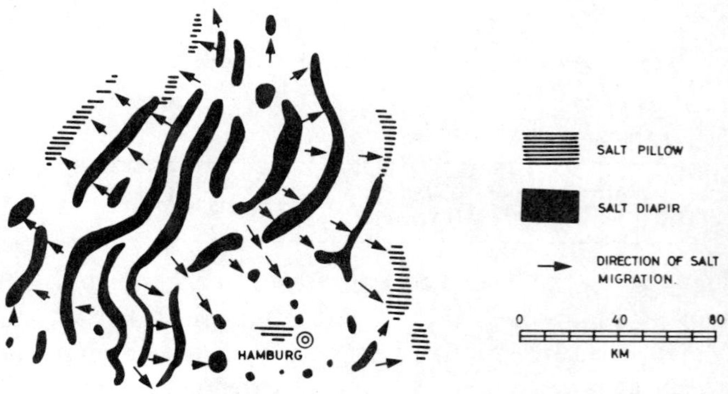

Fig. 4.12 Sketch of a diapir and salt pillow group north of Hamburg, with salt flow directions as determined by Sanneman (1968). Although of less regular form than the model of Fig. 4.11, the flow patterns and features bear a strong resemblance to the model groups. In fact, similar flow patterns can develop from elongate primary diapirs, producing linear secondary sinks and diapirs rather than concentric groups, as will be discussed in a later chapter on diapirism. (After Sanneman, 1968, and others.)

feature. Subsequently, other probable examples of such groups have been identified, and this subject will be discussed again in Chapter 8, in the section on 'patterned' diapirism.

The studies of Parker and McDowell into effects of deformation on model overburden materials produced structures which have, in general, been found to have counterparts in the real earth. These effects include both radial and peripheral faults in the overburden above sub-circular model diapirs, and parallel, linear faults with graben formation above elongate model diapirs.

Currie (1956) carried out model studies which are of interest in that an attempt was made to model sedimentation contemporaneously with model diapir growth and the overburden faulting attendant upon piercement. However, the model mobile source used was, in effect, unlimited quantitatively so that peripheral sink and associated subsidence effects were not developed, since the main objective was the study of the graben faulting which developed above the crest of the model diapir.

4.5 LATER PHYSICAL MODEL STUDIES

Important work has been carried out by Ramberg (1981) in this field. Using two immiscible liquids—usually with heavy oil as the mobile layer, and syrups or salt solutions of different concentrations as the overburden layer—in a similar way to Nettleton, he noted the effects of heterogeneities such as air bubbles on the surface of the mobile layer liquid as being equivalent on a small scale to thickness variations in the mobile source layer. With liquids, these are sufficient to initiate model diapir development due to the differential pressure head present at the location of such a heterogeneity. Conducting similar experiments with completely air-bubble-free source layers of oil, he observed: 'On the originally plane but gravitationally unstable source layer surface, waves with horizontal axes start to form . . . soon, rather evenly-spaced culminations develop on these anticlinal ridges, the culminations changing into circular diapirs and rising rapidly to the surface of the overburden.' The similarity of these spine-like upgrowths from the ridges, and the explanation suggested by Morgan *et al.* (1968) for the Mississippi mudlumps, was mentioned earlier.

Apart from these experiments, Ramberg conceived the ingenious idea of exercising much greater control over the model ratio of time, and over gravitational effects, by centrifuging some models. By varying the rate of spin, and the elapsed spin time, it was possible to produce a wide variety of flow and diapiric features.

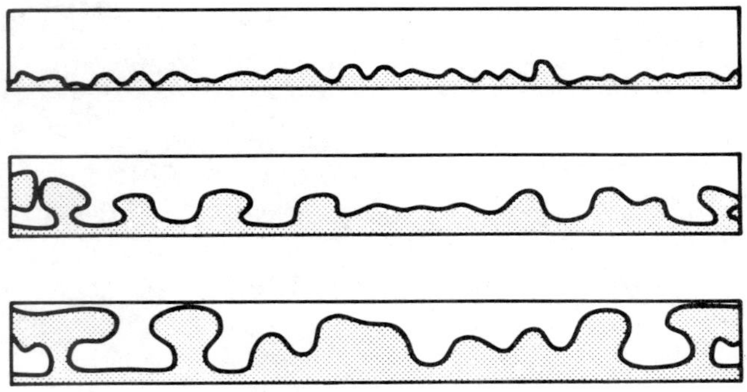

Fig. 4.13 Three profiles through silicone putty models; densities: lower ('mobile') layer, 1·13 g/cm^3; upper ('overburden') layer, 1·36 g/cm^3), the profiles being 9 cm long. Ramberg centrifuged these models to show that the dominant wavelength of the diapiric features resulting from the density inversion (and hence also the diameter of the diapirs) is related to the original thickness of the mobile source layer. (After Ramberg, 1981, courtesy Academic Press.)

In these centrifuged models, Ramberg used various silicone putties and modelling clays of different densities as mobile source and overburden layers. Related to earlier suggestions by Nettleton and Parker and McDowell, Fig. 4.13 shows a simplified sketch of a model demonstration of the relationship between the thickness of the source layer and the size/spacing of diapiric features, involving the statement from dynamic theory that the dominant wavelength of features in a buoyant layer is proportional to the buoyant layer thickness in models consisting of the same kinds of materials. The somewhat irregular appearance of the models is due to straight sections being taken through a model where the diapiric features are spaced irregularly in the horizontal plane, the vertical section thus passing through the features at various distances from their centres. These observations are clearly relevant to previous work which attempted to relate diapir diameter and sink diameter to mobile layer thickness. An idea of the relevance of this to real earth structures is given in Fig. 4.14, which shows the difference in wavelength and size of salt features where the salt source layer differs in original thickness. It is also noted in passing that, as a corollary to this, as might be expected, in an area of uniform salt layer thickness there is a tendency towards roughly even spacing of any salt structures which form. Figure 4.14 is indicative of this. Ramberg also confirmed the earlier observation, by Parker and McDowell and others, that the diameter of the peripheral sink round a model diapir was six to eight times the diameter of

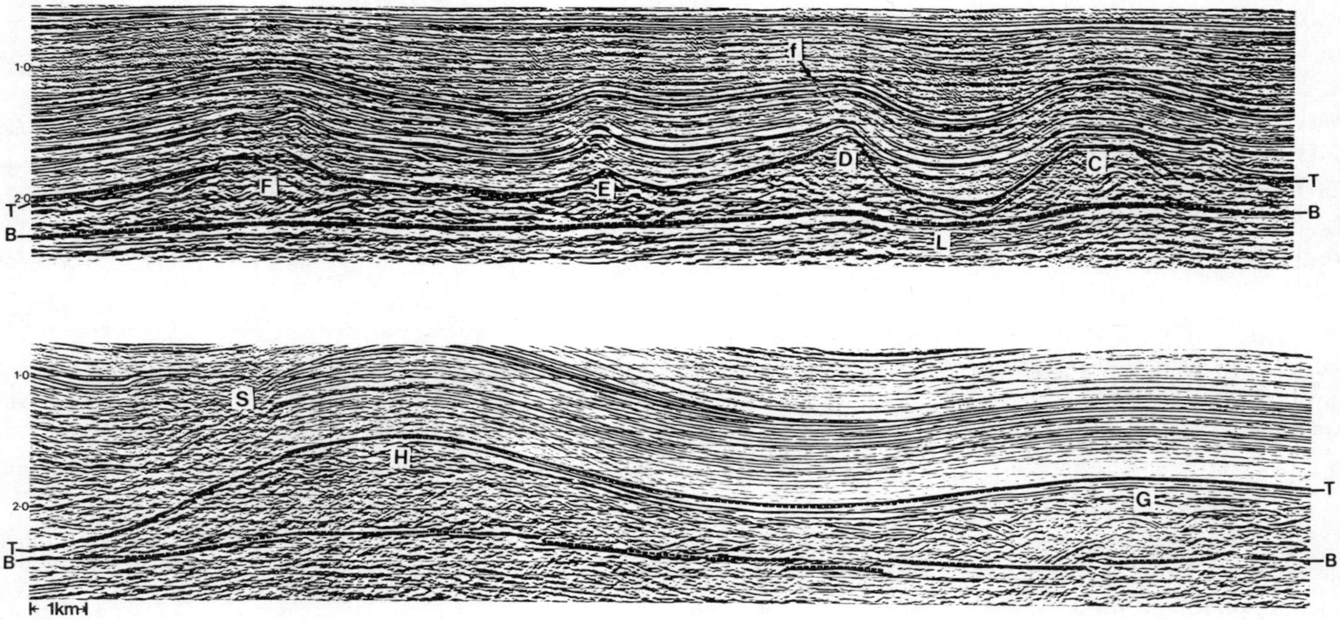

Fig. 4.14 Relationship of wavelength and size of salt features to original thickness of salt source layer. The salt interval is between B and T. In the lower illustration, two salt swells G and H have developed in a salt layer estimated to have been originally 600 ms (TWT) thick. In the upper illustration, the original time thickness of the salt over most of the section is estimated at about half this, i.e. some 300 ms. Four salt swells/pillows have developed which are much smaller than those in the lower illustration. Features C, D and E, from centre to right of section, are equally spaced. The wavelength increases between E in the centre and F on the left, as the salt here was thickening towards the basin centre, which is off section to the left. In the upper illustration, note the fault 'f', showing a fault-plane reflection, which has resulted from the marked subsidence of the overlying sequence into the salt withdrawal basin between features C and D; also the 'pull-down' velocity anomaly at base salt level at L due to the considerable thinning of the salt in this withdrawal basin. An interesting point about the lower illustration is the presence of events which indicate slumping of the overburden at S; the steep flank of this asymmetric salt swell has here exceeded the angle of rest of undercompacted Mesozoic sediments. (Courtesy of SS(E)L.)

the diapir, regardless of the thickness of the mobile source layer.

Ramberg's work on the initiation of model diapirs is of great interest. As discussed previously, it is clear that, where immiscible liquids or other materials of negligibly small yield strengths are used as overburden materials in the model, effectively spontaneous initiation of model diapiric uplifts occurs. However, when overburden materials of small but measurable yield strengths are used, an initiatory bulge (among other methods) is required to initiate diapiric growth; this may be small for materials of low strength, but increasingly large for stronger materials. Ramberg introduced the term 'plunging strength' in order to compare the behaviour of different materials. This is defined as 'the minimum *force* needed to push a cylindrical solid plunger of given cross-section into the plastic overburden body' sufficient to initiate diapiric rise when the model is centrifuged. Experiments with low strength materials such as silicone putty as mobile material, and strain-hardened painter's putty as overburden, the minimum height of an initiating bulge of 1 cm^2 cross-section was determined as only 0·2 cm. However, with increasingly strong overburden materials like modelling clay, plaster of Paris and concrete, much higher bulges were required up to more than 2 cm high, i.e. ten times the height required in the putty model.

It would seem from this work that, as far as piercement structures are concerned, the real earth quantity equivalent to Ramberg's 'plunging strength' is of primary importance to diapir initiation. The measurement of such a quantity is fraught with obvious difficulties, and (to the writer's knowledge) has not been attempted. Estimates may be made, but it is clearly difficult to arrive at correct scaling factors between models using the materials quoted and, say, a salt layer roofed by a thick sequence of highly compacted anhydrites or carbonates beneath several thousands of feet of clastic overburden of various rock types.

The matter is of some importance; the model work

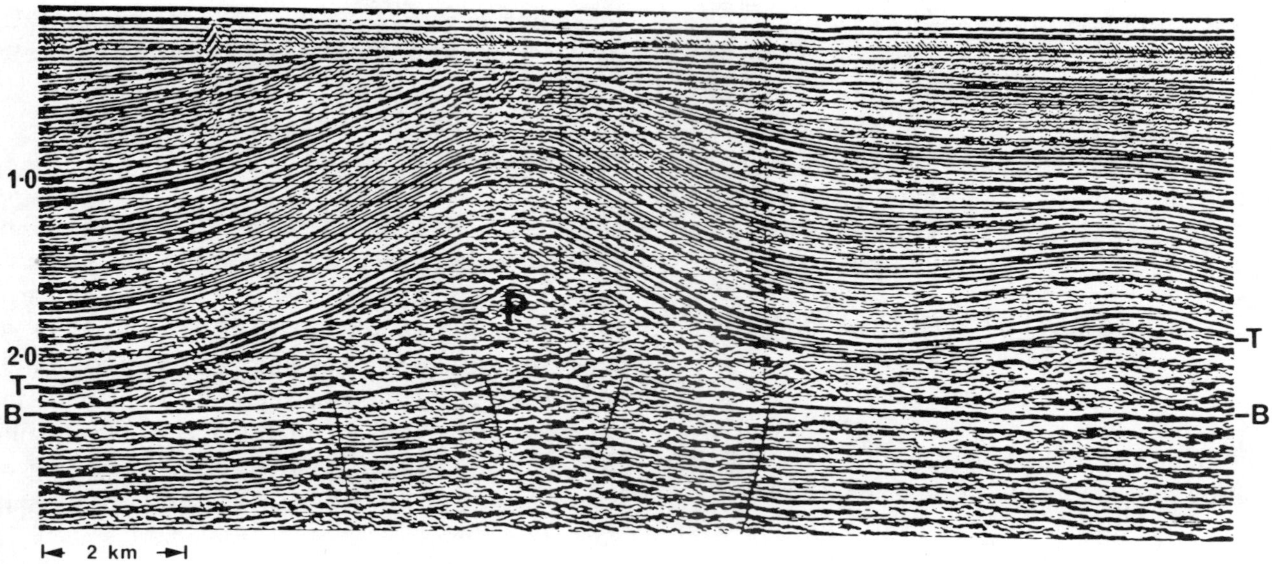

Fig. 4.15 An example of a large salt pillow, P, in the North Sea. The crest of the pillow is probably more than 1000 m above the top of the adjacent salt layer, and yet no piercement has taken place. Base and top of the salt interval are marked by B and T. Original salt interval thickness is estimated at about 1000 m. Present overburden thickness is 3000–4000 m. Such features testify to the strong influence which overburden competency must have on salt piercement. (Courtesy of SS(E)L.)

suggests that the competence of the salt roof is one of the controlling factors for diapir initiation.

We must agree with Parker and McDowell that any natural equivalent of the bulge or positive irregularity in the model, regarded as a very localised feature, is highly unlikely. It should be noted that an initiating irregularity must, in accordance with the work of Ramberg, and our understanding of diapirism as a piercement process, be a *piercement* irregularity, rather than simply a swell. This is borne out by observation of deep seismic data, where frequently large salt pillows, the crests of which are hundreds of metres above the level of the source layer, show no evidence of piercing the overburden. An example is shown in Fig. 4.15. Thus developed, such pillows should have passed through the piercement stage if all that is required is a sufficiently large differential head of pressure between the diapir top and the top of the source layer. However, in many examples the overburden at the crest of the structure has remained quite intact, at least on the macroscopic scale.

Thus, the suggestion is that the plunging strength equivalent in the real earth may be very large in some cases. Often, in fact, some stretching must occur in the overburden. Recently, careful measurement was made within the plane of a migrated seismic section in the dip direction of the length of the top and base horizons of a salt layer which has produced a series of swells and pillows in the overburden over a distance representing about 15 km. The roof sequence confining the salt layer, in this case a sequence of anhydrites and dolomites, showed a linear 'stretch' factor of 4% over the base salt horizon, and showed no visible evidence of fracturing or piercement. The salt features present were relatively gentle, and no doubt other more extreme examples could be found.

The reality of an initiatory irregularity in the real earth may lie in fault features affecting a salt layer. A fault affecting the top of a salt interval could be the equivalent of a piercement perturbation, while a fault affecting the base of the salt could, due to the differential loading on the basal part of the salt interval across the fault, lead to upward movement (although not necessarily to piercement).

The implication generally appears to be that one possible inhibitory factor as regards diapirism is the competence of the overburden. Another such factor discovered by various workers is that if the overburden loading in a model exceeds a certain value, it is impossible to initiate model diapirs. If these two factors hold good under natural conditions, then circumstances will occur where, without the intervention of some other factor, stress relief in a mobile salt layer must be confined to lateral movement of the salt away from regions of higher stress towards those of lower stress. Alternatively, where no effective stress differential exists, the salt would remain static, although potentially mobile. In such circumstances, should the overburden subsequently be faulted, the immediate stress-relief focus of the fault zone would cause the salt to inject diapirically, producing subsidence effects in the overburden typical of the post-piercement stage of diapirism, but with an absence of the pre-piercement stages of salt pillow formation and the accompanying effects in the overburden. Such situations are known. Sanneman (1968) discusses diapirs of this type in the area of north-west Germany.

4.6 DEFORMATION MECHANISM IN PHYSICAL MODELS

As discussed in the chapter on deformation mechanisms in halite, these involve basically intracrystalline dislocation and gliding within the space lattice, combined with pressure solution effects both within and outside the individual crystals. Although the movement of salt rock is often referred to as 'flow', this is a non-rigorous use of the term. Flow implies interparticulate movement, which cannot possibly occur in an interlocking mass of halite crystals. On the other hand, interparticulate movement is the norm in the materials commonly used in physical model studies. This difference in the mode of deformation may not be significant in the context; however, the writer is not aware of any studies directed towards determining the possible differences in behaviour under stress of materials which deform in these different ways.

4.7 SUMMING UP

Through laboratory work, field studies and model experiments, a certain basic consensus has been arrived at among most geologists as regards movement and diapirism in salt rock. There is general agreement that (i) salt uplifts and diapirs derive from source salt layers which were laid down as evaporite deposits; (ii) salt movement stems from the property of halite (and other salts) of deforming plastically under applied stress; (iii) the gravitational effects produced by a heavy overburden overlying a more buoyant salt layer are of importance in determining the form of movement of the salt in many areas; (iv) in some areas, tectonic stress has had at least some influence on the formation of salt structures, particularly in the initiatory stage. Parker and McDowell (1955) mention, effectively, the same areas in which there is general agreement, and the position has not really changed since that time.

On the other hand, there are certain points on which there is still no general agreement. Seni and Jackson (1983) note that: 'Controversy surrounds the emplacement history of diapirs, and hinges on whether the dominant processes were intrusion . . . extrusion . . . alternation between intrusion and extrusion . . . 'downbuilding' . . .', or related very closely to the depositional history of surrounding sediments, which is the view taken by Seni and Jackson for the East Texas diapir province. The question of the initiation of piercements is another matter on which there is no consensus in certain areas. There is also disagreement as regards the physical parameters of halite deformation in the real earth. The evidence in some areas brings into question even the basic importance of density inversion at the salt layer. Some of these points will be taken up in more detail in the chapters on diapirism.

CHAPTER 5

The Salt/Fault Association

5.1 INTRODUCTION

There is frequently, perhaps usually, a close relationship between faults and salt features, whether swells, pillows or diapirs are concerned. A great many salt structures are elongate rectilinear or curvilinear bodies in plan. Even where they are subcircular in plan and spine-like in form, they are often found to have risen from elongate salt ridges or massifs at greater depths.

A fault is typically a linear, more or less elongate feature. Although not all elongate salt structures are necessarily related to faulting (they can be secondary resulting from a pattern of salt flow associated with a primary feature, as will be discussed in the chapter on Diapirism, Part 2, or formed under the influence of compressive tectonic stress), there is a clear suggestion that faulting may be an important causative factor in the upward movement of salt rock in many cases (Fig. 5.1).

Much can be learned about salt movement mechanisms from the study of the relationships between faulting and mobile salt behaviour. The simplest aspect of this is where a fault provides an upward stress-relief route for salt which is under great overburden loading, tectonic stress, or the stress applied by lateral movement within the salt interval. This is certainly an important factor, but it is by no means the only one.

The present chapter will examine various processes which occur when a salt interval or structure is affected by faulting, and some suggestions will be made as to how certain of these processes may be of wider significance in the context of salt mobility and diapirism.

5.2 BASIC EFFECTS OF FAULTING ON A MOBILE LAYER

This discussion will begin by considering two cases, neither of which involves either salt or faulting, but which serve to illustrate certain matters basic to the understanding of the behaviour of a mobile substrate in the presence of faulting.

The first case concerns the initiation of the 'mudlumps' in the mouth of the Mississippi River as described by Morgan *et al.* (1968). The mudlumps are small clay islets which rise above water level at various locations in the mouth of the river. For present purposes, we are concerned only with the early initiatory processes which eventually produce these features, and Fig. 5.2 shows in diagrammatic form the interpretation given by Morgan *et al.* to the effects seen.

A load of fluvial sediments has prograded across an underlying prodelta clay sequence, compacting and depressing this as it advances. The load differential has caused underlying clays to become mobile and flow laterally from beneath the zone of higher load stress towards the zone of lower loading beyond the advancing edge of fluvial sediments. Just beyond this edge, the prodelta clays formed a thickened ridge, implying upward as well as lateral movement of the clays. Three factors may have influenced this upward movement: (i) stress relief in an upward direction afforded by the smaller overburden loading beyond the edge of advancing sediments; (ii) the mobile, laterally flowing clay becoming increasingly viscous as the overburden loading decreases, producing a 'self-damming' or buttress effect, with concomitant upward thickening; and (iii) the introduction of a vertical component of movement into the lateral flow by the upslope at the edge of the compaction depression.

The sketches of Fig. 5.2 represent a longitudinal section down-river from left to right. The mudlumps are actually diapiric clay spines which develop later by upward growth at various locations from the top of the clay ridge seen in cross-section. These spine-like upgrowths eventually break surface and rise above the water to form the small islets.

The behaviour of the mobile clay as it moves out

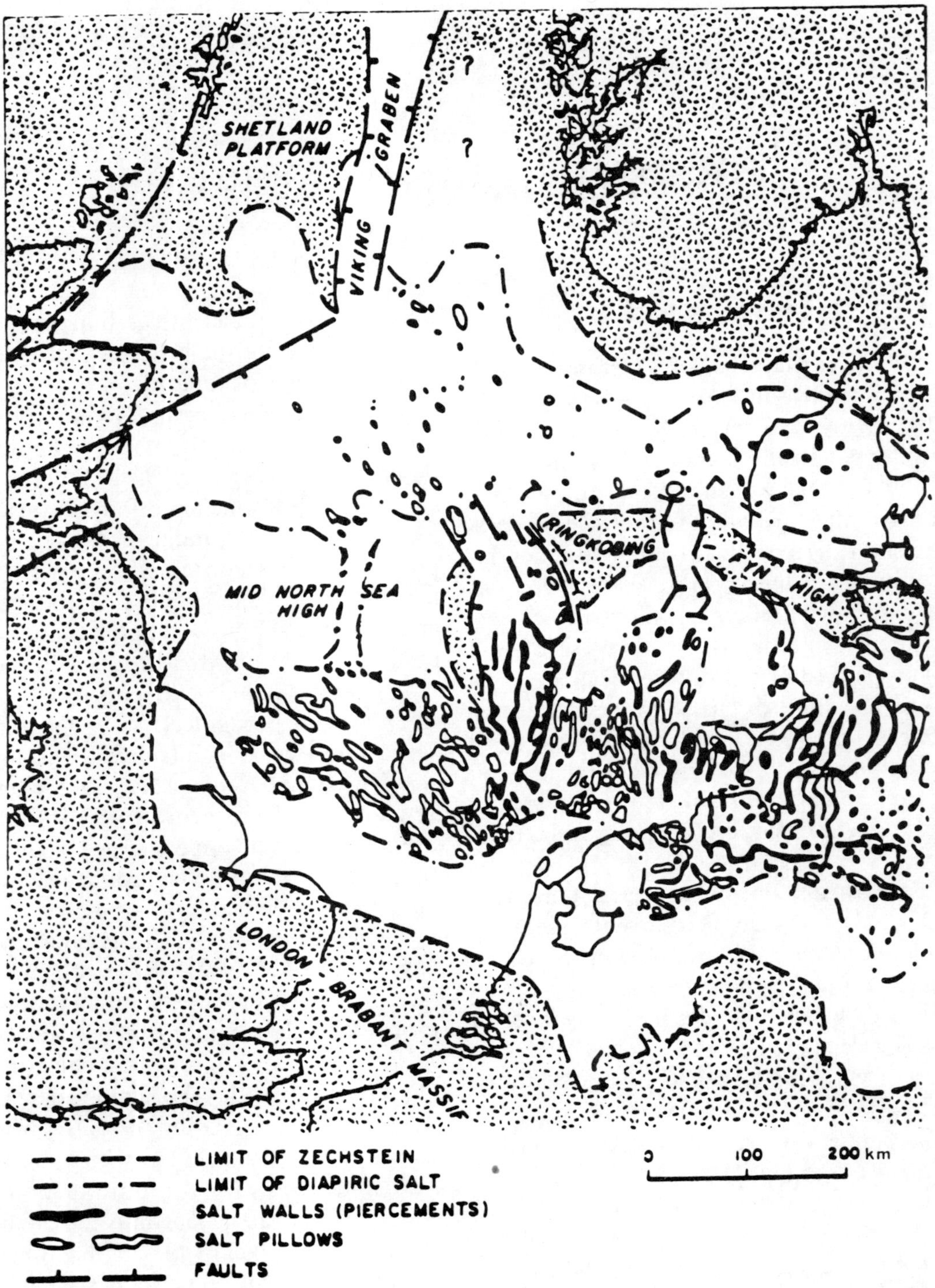

Fig. 5.1 The different alignments of salt features in the areas south and east of the Mid North Sea High indicate the strong relationship between salt movement and tectonic elements—expressed by major faults—in these areas. (Slightly modified from Taylor, 1984, by courtesy of the author and Blackwell Scientific Publications.)

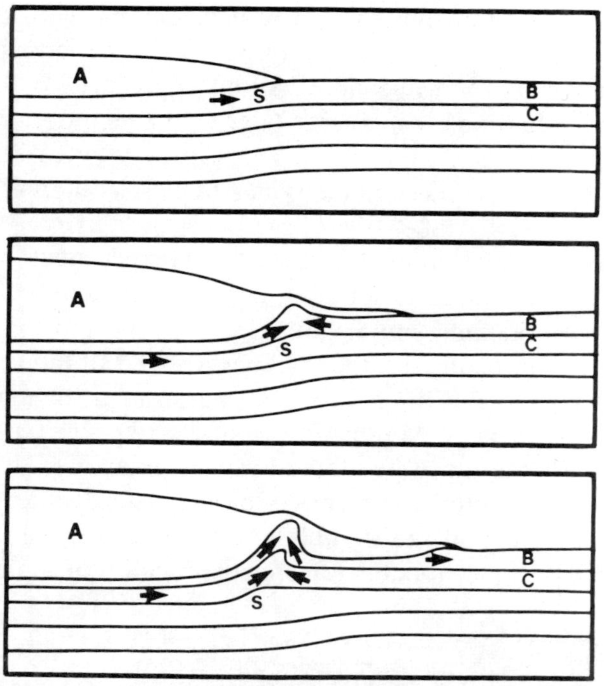

Fig. 5.2 Diagrammatic sketches showing the early stages of formation of diapiric clay ridge from which the mudlump islets subsequently rise as diapiric spines. Key: A, advancing load of river sediments; B and C, prodelta clays; S, upslope at the margin of the compaction depression produced by the advancing alluvial sediments. (Simplified after Morgan *et al.*, 1968, courtesy of the AAPG.)

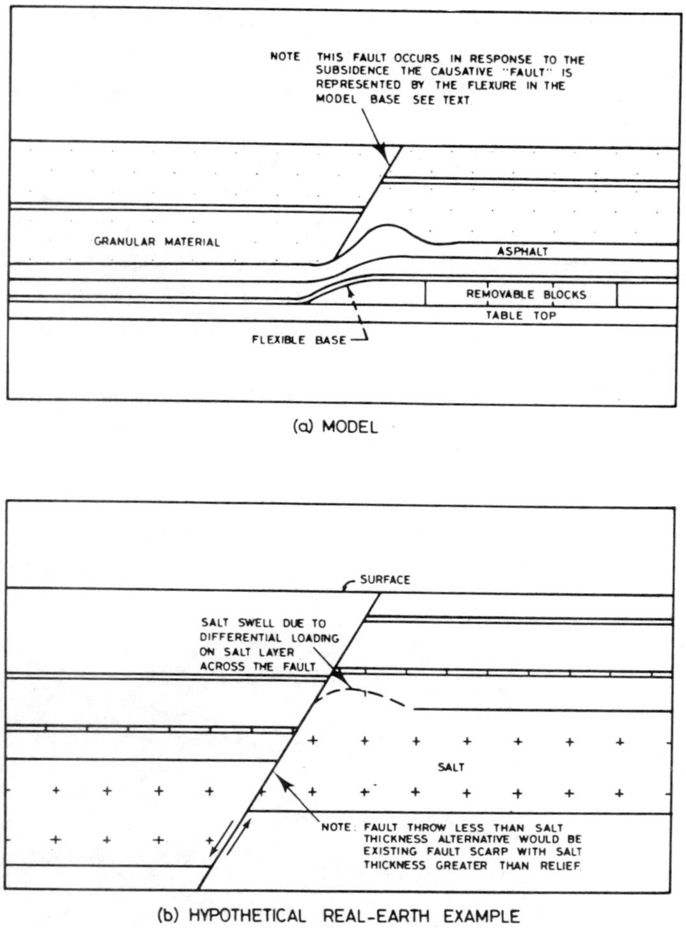

Fig. 5.3 Diagrammatic sketches illustrating (a) the effects in a model of a simulated 'fault' on the behaviour of a mobile substrate (asphalt) with an overburden of granular material, when the mobile layer is not completely severed by the fault, and (b) the hypothetical equivalent real-earth situation where the mobile layer is a salt rock interval. (After Parker and McDowell, 1955, courtesy of the AAPG.)

from beneath the zone of higher overburden loading, and the possible analogy between the upslope at the edge of the compacted zone and a fault in the base of a mobile layer, are points which should be noted particularly.

The second case concerns an experiment carried out by Parker and McDowell (1955) as part of a larger programme of model studies into pillow and diapir growth which is referred to several times in this book. The particular experiment is illustrated diagrammatically in Fig. 5.3 (upper sketch). Essentially, the model consists of a surface of removable blocks covered by a flexible sheet on which is spread a layer of asphalt representing a mobile layer, and an overburden of granular material.

In order to model the behaviour of a 'faulted' mobile layer, underlying blocks on the intended 'downthrow' side were removed so that the asphalt above this subsided, although still continuous across the 'fault zone'. The latter, because of the nature of the asphalt, took the form of a ductile shear zone rather than an actual fault.

It should be noted that, at this stage, the thickness of the granular overburden material on both sides of the model remained the same. In order to simulate either later depositional, or erosional, levelling of the surface, additional granular material was added over the downthrow side until the surface was level again. After this, a linear ridge-like thickening—seen here in cross-section—was observed to form in the asphalt on the upthrow side of the model. This was clearly due to movement of the mobile asphalt from beneath the higher overburden loading of the downthrow side towards the lower loading of the upthrow side.

The point was made by the authors that the fault throw would have to be less than the mobile layer thickness for this to occur. If the fault completely severed the mobile layer, two quite independent systems would exist on each side of the fault zone, with no possibility of flow between them (discounting the possibility of mobile material injecting the fault plane and re-establishing contact).

Since the properties of the asphalt made it impossible to fracture along a plane surface so as to simulate a real fault situation, the authors did not claim that they had modelled exactly a phenomenon of salt movement in the earth. However, they did suggest what the results might be under real earth conditions with salt as the mobile layer, illustrating this as in the lower sketch of Fig. 5.3 and categorising it as a 'hypothetical case of dome initiation'.

One particular point of interest is that, in the model, the growth of the asphalt ridge occurred by thickening on the *upthrow* side only of the model fault; presumably this is dictated by the constriction of the asphalt layer by the 'fault zone', and may be explicable by some effect related to Bernoulli's Theorem.

This model study shows some interesting points of similarity as regards the behaviour of the mobile layer and its relationship to overburden loading, with the 'mudlump' case discussed previously. The possible significance of these observations in regard to real earth situations will emerge later in the chapter.

5.2.1 Faults Affecting the Whole of a Salt Interval

The relevance in real terms of the experiment just described will, in a salt area, depend on a number of factors, the most important being the type and throw of any faults present, the thickness of the salt interval, and later depositional/erosional patterns in the overlying sequences. Major dip-slip and strike-slip faults, where large throws or lateral movements have taken place, will often completely separate the salt interval across the fault zone, unless the salt is very thick. Also, it is questionable how far the rheidity of asphalt

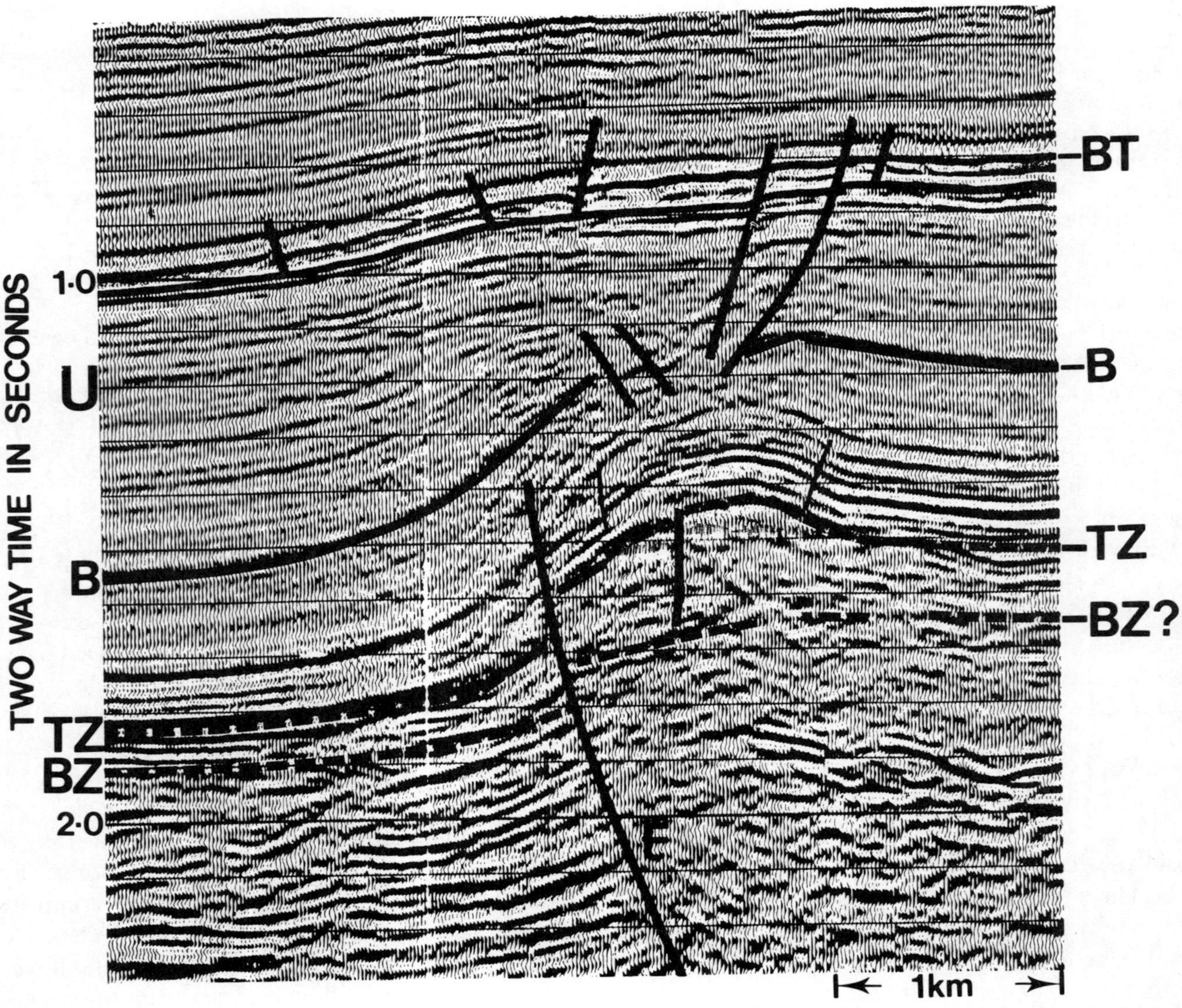

Fig. 5.4 Migrated seismic section believed to be a real-earth example of the mechanism suggested by Parker and McDowell (1955) as illustrated in Fig. 5.3. (Courtesy of SS(E)L.)

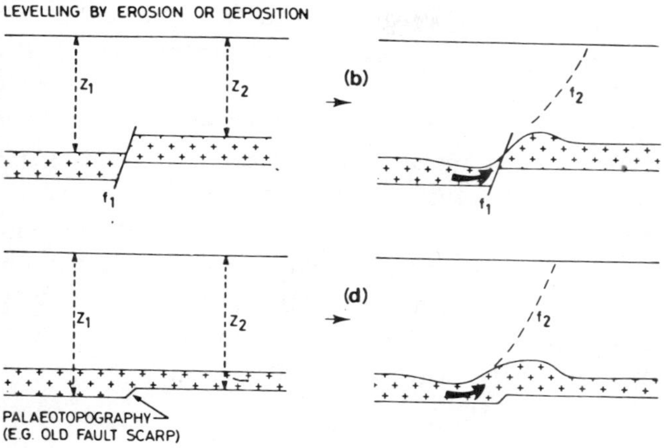

Fig. 5.5 Sketch models of the fault/salt mechanism illustrated in Figs. 5.3 and 5.4 (sketches (a) and (b)), together with those of similar effects related to a slightly different structural situation where a salt layer covers an already existing fault scarp or other buried topographic feature (sketches (c) and (d)).

is in accurate model ratio to that of other materials used, and to that of salt. As noted earlier, the deformation mode and the conditions for deformation of materials such as asphalt are different from those of salt rock.

In spite of these factors, features have been identified in seismic data from the North Sea which appear to confirm the hypothesis put forward, and there seems no reason why the physical processes involved should be confined to this area. In Fig. 5.4, a migrated seismic section is seen in which a large fault, which has probably been reactivated by strike-slip movement, is seen to cut across a salt interval without completely separating it. The Zechstein interval (BZ–TZ) is beginning to thicken into the salt basin to the right, although probably only about half of the Zechstein on the 'downthrow' side of the fault is salt rock, the rest consisting of marginal lithofacies.

From the observed effects here, it is believed that flow of salt rock from the downthrow to the upthrow side across the fault zone has produced an antiform pod of thickened salt, as seen in cross-section on the upthrow side of the fault.

There are reasons for believing that this salt feature was originally somewhat thicker; the shape of the later horizon 'B' suggests that some fracturing and dissolution has occurred over the crest of the feature, with subsequent collapse effects in the overburden. The Base Tertiary event BT shows fault features

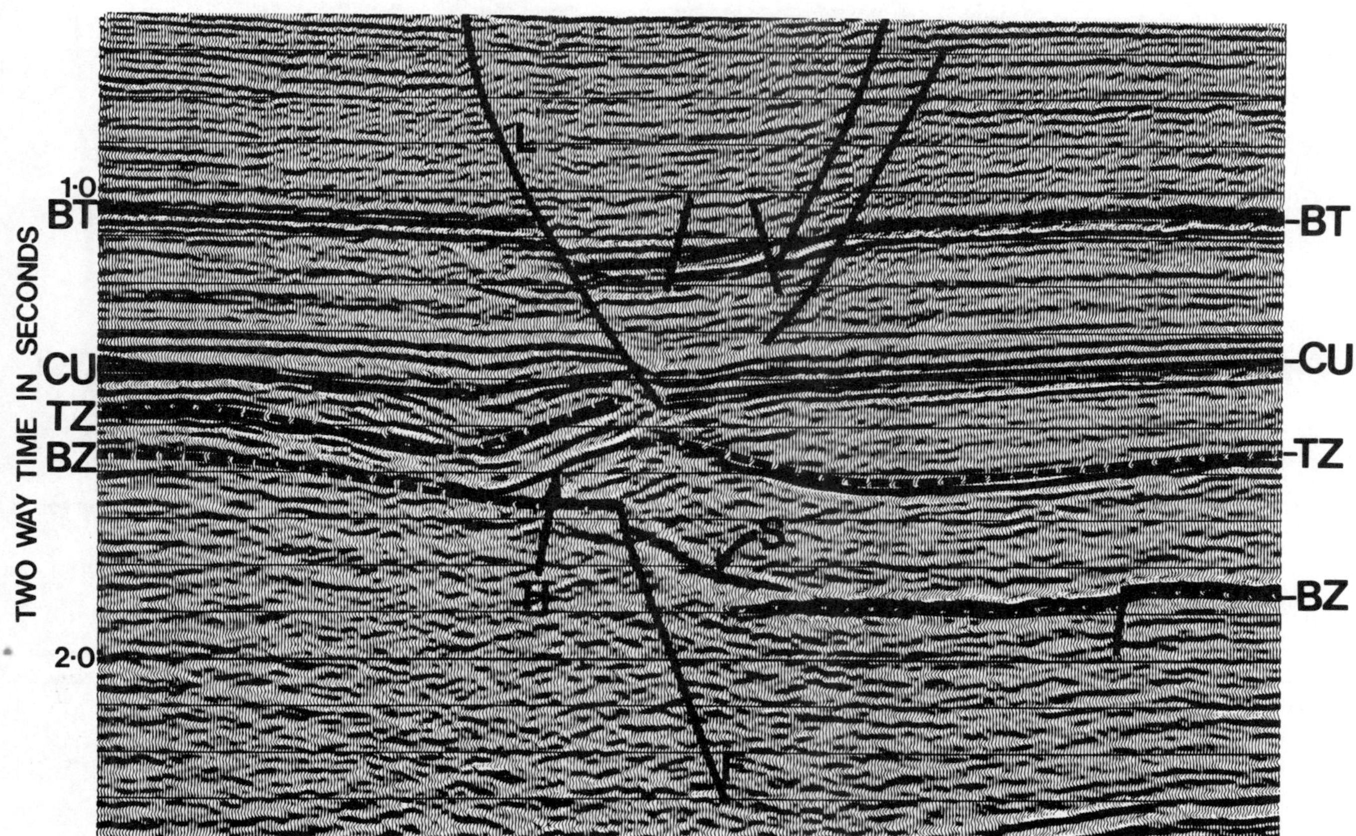

Fig. 5.6 Migrated seismic section showing a real-earth example of the mechanism suggested in Fig. 5.5, sketches (c) and (d). BT, Base Tertiary; remainder of key in text. (Courtesy of SS(E)L.)

implying that the latter processes were still operating in early Tertiary time. It is also possible, as suggested by the marked flattening of reflection events in the B–BT interval above the main fault F that the fault mechanisms here may have been influenced to some extent by subsidence effects due to withdrawal of salt from the downthrow side of the fault; the flattening of the events is unlikely to be related to the fairly localised dissolution collapse at and above the crest of the salt swell.

The situation illustrated by this example is quite similar to the hypothetical case presented in the second part of Section 5.2, previously, and the general resemblance of Fig. 5.4 to both upper and lower sketches in Fig. 5.3, even down to the younger fault configuration, is quite striking (see Jenyon, 1985b).

The importance of the presence of an unconformity above the faulting, or thicker/more dense overburden above the downthrow side of the fault, in this type of situation should be stressed, since later levelling of the surface by some process is necessary to develop the differential loading which drives the salt movement mechanism. In the case of Fig. 5.4, an unconformity exists at U within the B–BT interval, and another at the contact of the Base Tertiary (BT) with the underlying Chalk. It is possible that each of these induced a separate phase of movement of the salt, and there is some evidence for this in the rather faint reflection configurations at the unconformity U above, and to the left of the position of fault F.

5.2.2 Faults Affecting only the Base of a Salt Interval

In Fig. 5.5, the upper sketches (a) and (b) represent diagrammatically the situation which has just been described, with a salt interval (cross symbol) partially separated by a fault, f_1. Later levelling of the surface

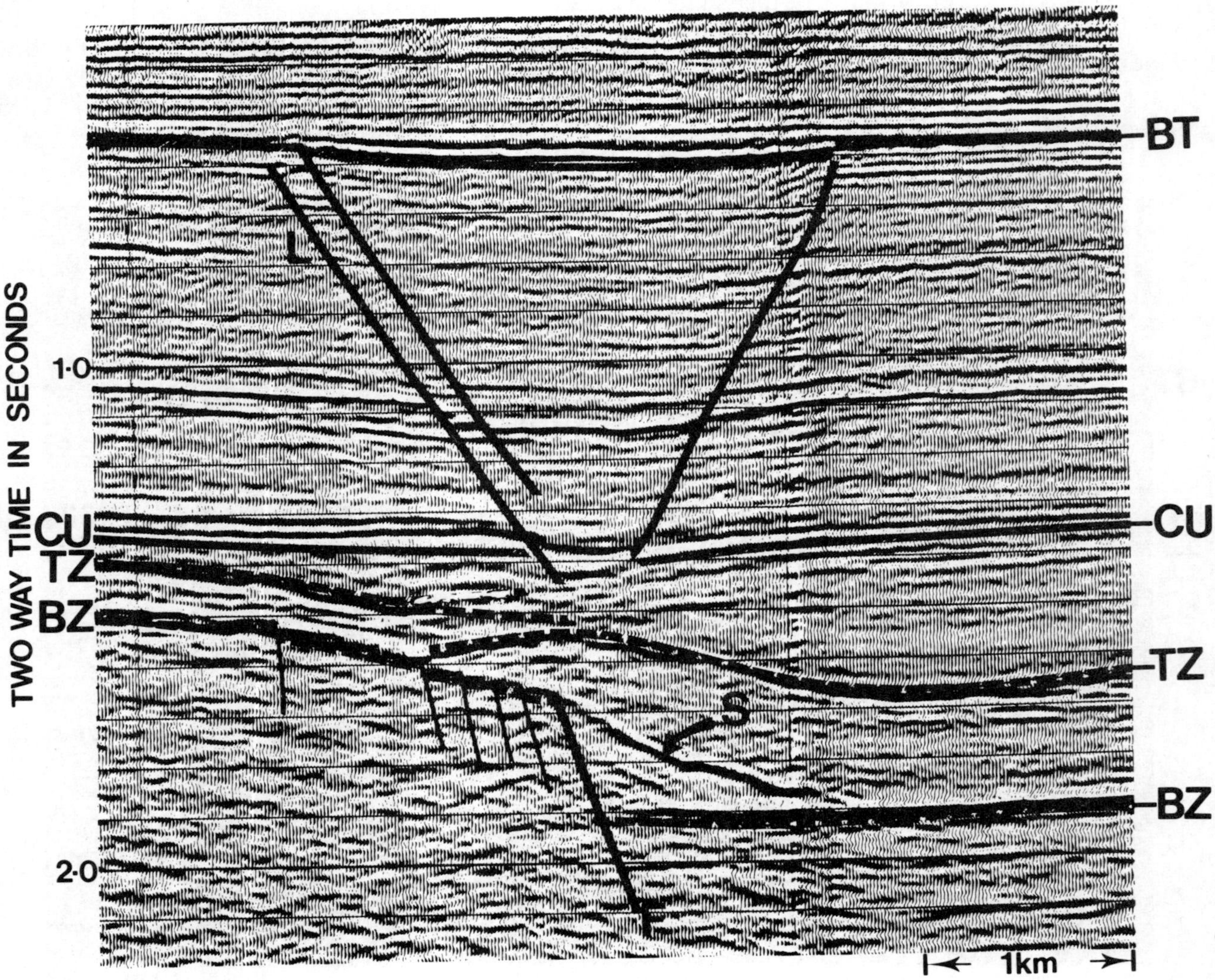

Fig. 5.7 Another example section across the same fault as that shown in Fig. 5.6, but in a different location. Key as in Fig. 5.6. (Courtesy of SS(E)L.)

by erosion or deposition has resulted in a load stress differential (due to thickness difference Z_1–Z_2). The resulting flow of salt into a swell on the upthrow side of the fault is indicated by the arrow, and a listric fault f_2 is the result of the overburden subsidence accompanying salt withdrawal from the downthrow side.

A similar effect may be achieved in a somewhat different structural situation, as sketched in Fig. 5.5(c) and (d). In this case, a salt interval has been laid down across a pre-existing fault, or buried topographic feature. In this case, there is an inbuilt differential loading situation on the *lower* part of the salt interval (Z_1–Z_2) which may cause the lower salt to become mobile and to flow to the 'upthrow' side of the buried feature to form a salt swell as shown, again with a listric fault, f_2, developing in response to withdrawal subsidence.

It was noted in the previous case that the constriction formed by the fault zone appeared to have the effect—seen in the model, and in the real earth example of Fig. 5.4—of limiting the formation of the salt swell to the upthrow side of the fault. In the case of an old fault, or buried topography, this constriction does not exist, and therefore presumably, as suggested in Fig. 5.5(d), the salt swell would be at liberty to develop more or less symmetrically above the buried feature.

Seen in Fig. 5.6 is a migrated seismic section from the Mid North Sea High area; shown is an old fault (F) of early- or pre-Zechstein age which was discussed by Jenyon *et al.* (1984). It is of unusual interest since it is believed to be a peripheral shear produced by the intrusion of a granite batholith deep in the Mid North Sea High off to the left of the section. For present purposes, however, it is of interest since basinal Zechstein salt (seen in the BZ–TZ interval in the right-hand half of the section) passed as a much thinner development directly on to the already existing 'upthrown' platform to the left, together with the shelf lithological units which now make up the BZ–TZ interval. The salt originally extended further on to the platform than it does at present but, owing to edge-dissolution (see Chapter 6), has retreated basinwards, and now terminates in the dissolution slope, H, with

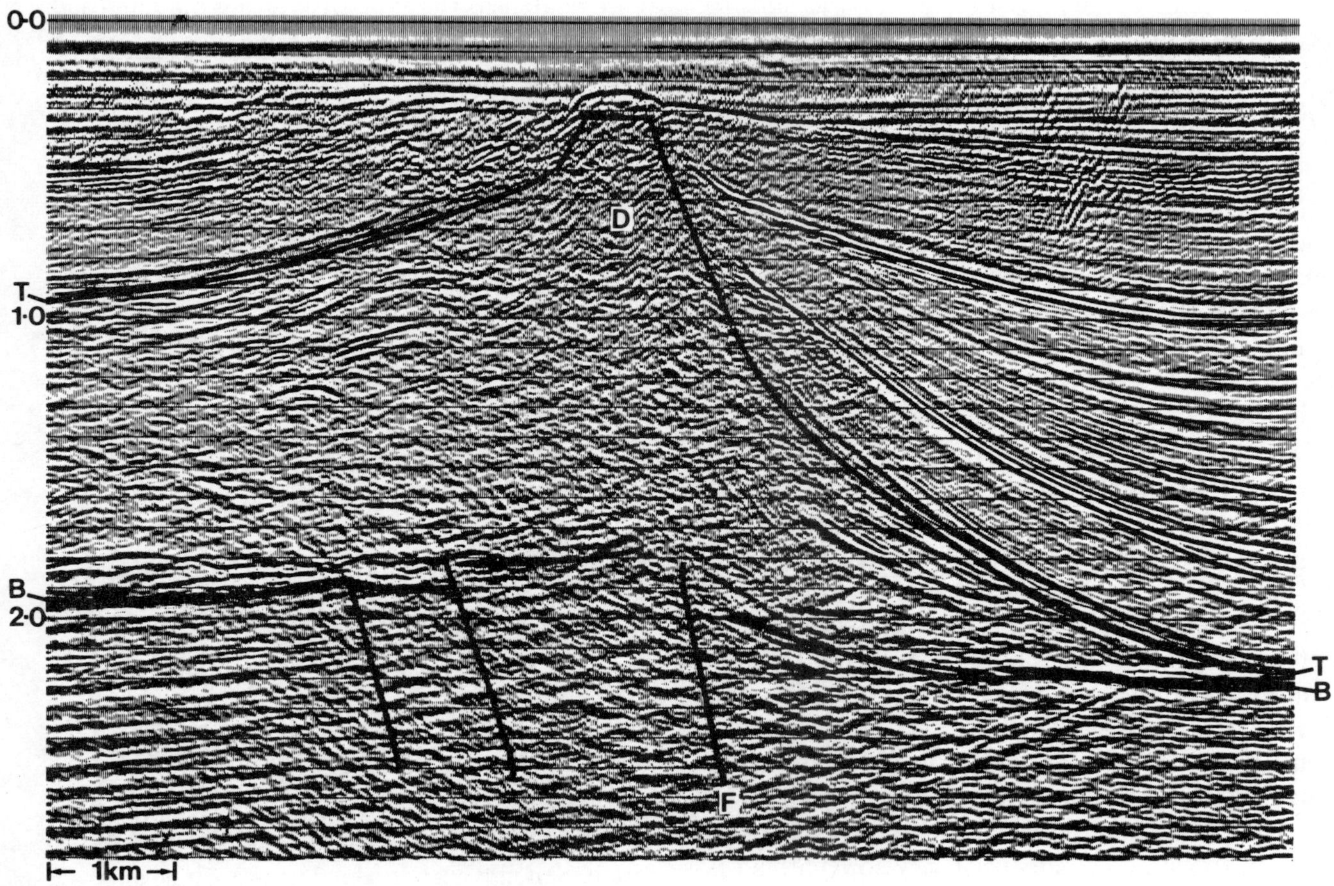

Fig. 5.8 Another feature (D) believed to be due to the differential loading mechanism across a fault, F. This structure developed as an asymmetric pillow (note thinning and onlap configurations of Mesozoic section to the right, indicating syndepositional movement), only becoming a piercement at a late stage. (Courtesy of SS(E)L.)

the related collapse structures being seen in the TZ–CU interval.

The interpretation given to the present situation is that the lower part of the basinal salt on the downthrow side of the fault F became mobile due to its greater depth and overburden loading than the salt on the upthrow side. Flow of the salt towards the upthrow side began early in Mesozoic time, as indicated by the somewhat fragmentary evidence of stratal reflections onlapping against the salt swell in the TZ–CU (CU, Late Cimmerian unconformity) on the downthrow side of F. Note also the slight downlapping configuration of these onlapping events below the unconformity, which indicates considerable salt withdrawal and subsidence before the Late Cimmerian erosional phase. Salt must have invaded the upper section and was probably extruded at the contemporary surface before being removed by the erosion. The well-marked uplift of the top salt seen by following the TZ horizon from the right-hand end of the section towards the left is clearly indicative of the flow process which occurred.

Salt movement continued after Late Cimmerian times and, due to the development of the sink, or withdrawal basin on the downthrow side of F, and the resulting subsidence of the overburden related to the geometry of the fault scarp, a listric growth fault, L, evolved which subsequently affected all levels up to a Miocene unconformity (just off the top of this section). The listric curvature of the main fault L induced some antithetic faulting which resulted in the graben feature seen in the shallower section. The fault L is clearly analogous to the shallow 'fault' in the granular material of the model in Fig. 5.3(a). Rollover structural dip in the CU–BT interval and above, seen in Fig. 5.6 to the right of L, is a classic feature often related to growth faulting, and which can be of great importance in the trapping of hydrocarbons. The mechanism of salt withdrawal, overburden subsidence and growth fault development, seen here on the small scale, can be of great significance on the much larger scale of continental margin geology, and will be mentioned again in this context later in the chapter.

Another example across the same fault but in a different location is seen in Fig. 5.7. In this migrated section example, the top of the salt has been less disrupted than in the previous example, and shows a smooth curve of thickening. The distinct upward swell of the salt top shows symmetrical development across and above the scarp zone of fault F, there being no fault constriction in the salt interval to cause swell development to be confined to the upthrow side, as in the example of Fig. 5.4. Similar subsidence effects, together with a younger fault L and an antithetic fault related to it, are seen in the shallower section. The fault L appears to show no listric curvature here, but the fault trace as marked may be too straight, or the curvature too subtle to resolve, since the antithetic fault occurrence implies some curvature of the master fault. The event marked S in this and the previous illustration is of some interest. It is interpreted as a slump surface indicating mass movement of material down the fault scarp. The flattened segment of the event at its lower right-hand extremity suggests that it is resting on a thin layer (about 90 m at this scale) of early Zechstein salt, while the stratal reflections above show that the remaining bulk of the Zechstein salt laps on to the event. This situation is clear evidence that the fault F cannot have occurred later than very early Zechstein times, and could have occurred prior to the onset of Zechstein deposition, with the mass movement episode having taken place due to renewed movement of the fault in the early Zechstein.

The examples of Figs. 5.4, 5.6 and 5.7 indicate how, in the general situation described, the shape of the salt uplift related to the type of faulting may assist in the

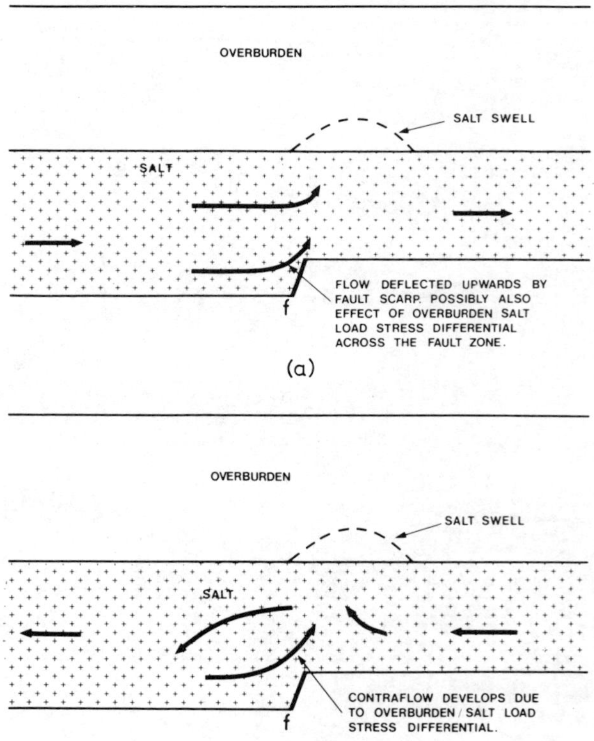

Fig. 5.9 (a) Sketch showing suggested effects of a fault scarp on salt moving horizontally towards the upthrow side of the scarp. (b) As in (a), but with horizontal flow component in salt towards the downthrow side of the fault scarp. This suggestion is considered to be rather speculative.

dating of the fault movement. See also Fig. 5.8 for another example of the process.

So far, the situation relating to a base-salt fault has been considered as though the salt was static, although potentially mobile. Where lateral movement is taking place within a salt interval which has a basal fault, the same differential load process across the fault zone could occur, but also there may be an element of vertical movement induced by simple physical deflection upwards of the salt by the fault scarp. Where there is a component of lateral movement in the salt towards the upthrow side of a scarp, the hypothetical scenario of Fig. 5.9(a) may exist, with upward deflection and differential loading both operating to produce a salt swell on the upthrow side. Where there is a component of movement laterally towards the downthrow side of the fault, the more speculative suggestion embodied in Fig. 5.9(b) is put forward to indicate how a salt swell could still result on the upthrow side of the fault. In this case, operation of the lateral flow may combine with the differential load stress present to produce a kind of 'eddy current' effect which may lead to the formation of a salt swell. Whether this would be located as shown here, or more symmetrically with respect to the salt scarp, is again a speculative matter. It is believed that there must be a strong case for the combined process suggested by the sketch in Fig. 5.9(a) occurring in nature; that suggested by Fig. 5.9(b) is much less secure and, may not occur. The processes suggested as operating by differential loading rely upon the reality of mobile salt seeking a route to lower load zones.

5.2.3 Faults Affecting only the Top of a Salt Interval

The effects discussed under this heading are in some circumstances the most difficult to study or illustrate since, by their nature, the evidence of faulting is destroyed, or at least severely distorted, by the upward passage of mobile salt in the region of the fault zone, and often the subsequent development of a diapir in the location of the erstwhile fault. This results because a fault which affects the top of a salt interval that is overlain by a superincumbent load of denser overburden—and hence under very high pressure—will act

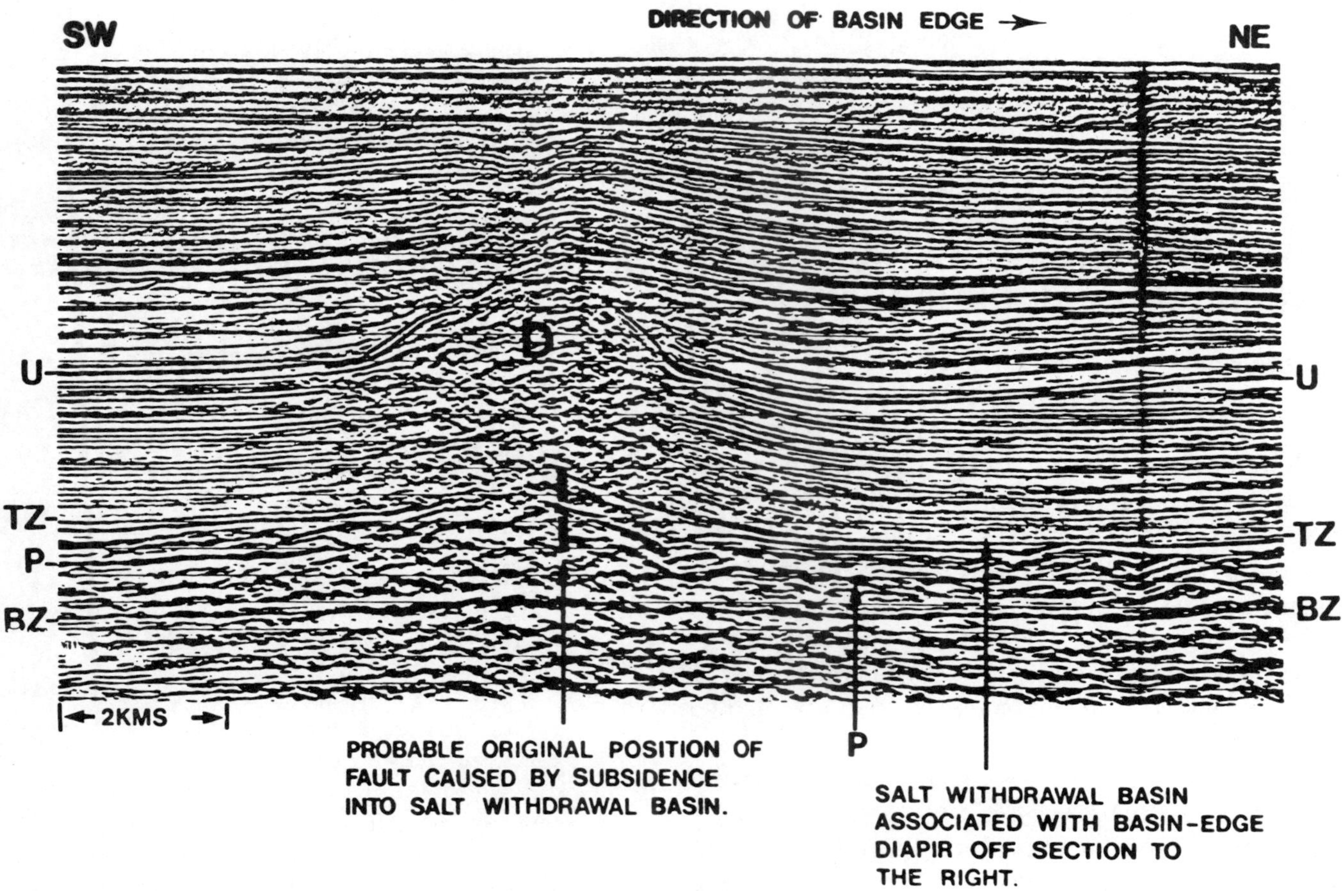

Fig. 5.10 Small diapiric feature, D, believed to be the result of the breaching of the top of the salt interval by a fault caused by overburden subsidence into the withdrawal basin to the NE of the diapir. (Courtesy of SS(E)L.)

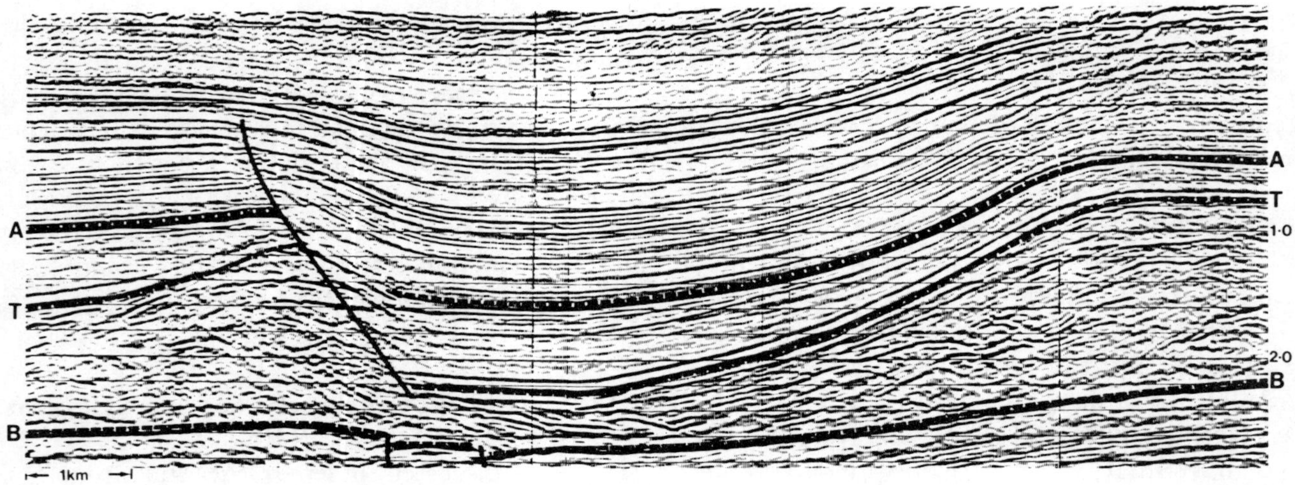

Fig. 5.11 Large fault in the top of a salt interval produced by subsidence into a withdrawal basin. (Courtesy of SS(E)L.)

as a zone of weakness, and an immediate focus of stress relief for the salt if the latter is in a mobile state.

Three situations in which faulting of the top of a salt interval is likely to occur are:

(i) Breaching of the top salt at the margin of a salt withdrawal basin caused by overburden subsidence in the basin.
(ii) Shear faulting developing at the crest of a salt structure due to overburden uplift by the rising salt.
(iii) Overburden faulting due to a tectonic pulse.

In Fig. 5.10, a migrated seismic section is seen in which a small diapir has developed (D), probably due to the breaching of the salt interval top related to subsidence into a salt withdrawal basin (seen to the right of D). The estimated original position of the fault caused by the subsidence is marked as a dashed line beneath D. The salt interval is within BZ–TZ, and it is clear that there is no evidence of the faulting having affected the base of the interval. Evidence of the initial lateral flow of salt to the SW out of the withdrawal basin can be seen in the toplap truncation of reflections beneath the unconformity U on the NE side of the diapir and, on this same side, the effects of subsidence due to salt withdrawal can be seen from U up to the shallower unconformity near the top of the section. This example is of considerable interest for other reasons, and will be seen again in the chapter on Diapirism, Part 2.

Another example, which is seen in a different context in Chapter 6, is that shown in Fig. 5.11, with the salt interval B–T, and an unconformity at A. Although here some signs of minor, probably strike-slip, faulting can be seen at the base of the salt interval, the major feature present is a fault of very large throw (over 500 ms two-way time) affecting the top of the salt, T. Note the thickening of the salt interval towards the fault on the upthrow side. It seems likely that strike-slip reactivation of the relatively minor faulting at base salt level, together with increasing overburden loading, may have initiated the salt movement from beneath the withdrawal basin; some salt flowed to the

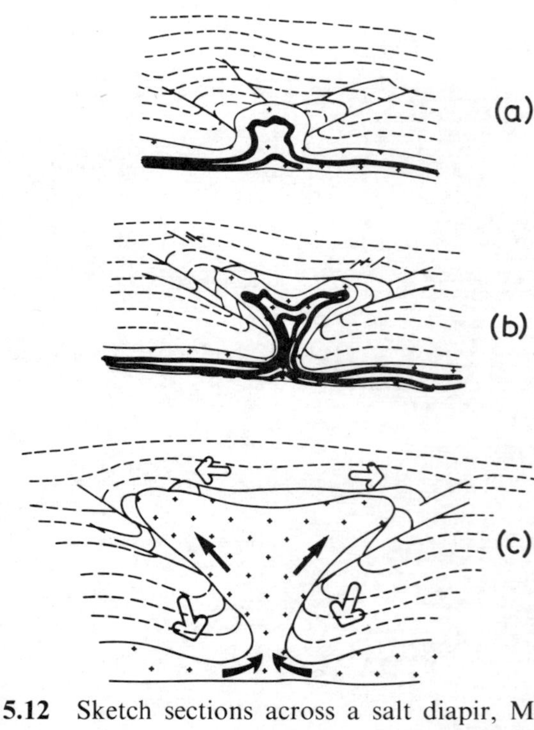

Fig. 5.12 Sketch sections across a salt diapir, Murguia, Spain, showing three stages of injection of salt into two sets of shear zones caused by the stresses set up in the overburden by the rising salt. In sketch (c), solid arrows show directions of movement in the salt, and open arrows directions of movement in the overburden. Cross symbol is salt which includes competent bands (black). Conjugate shear sets in the overburden may be common at some stages of diapirism. (Redrawn after Baar, 1977, and Lotze, 1957.)

upthrow side of the developing major fault, but the bulk of the salt moved to the right, into the salt swell, progressively as the overburden loading increased. The short, straight reflection events seen dipping steeply from the right-hand side of the major fault trace at and below the level of A are reflected refractions. The major fault, which is a slightly listric normal type, and the salt flow to the right into the swell, represent a situation on the small scale which is analogous to certain continental margin structures, as will be seen later in this chapter.

Shear fractures can develop in the overburden rocks at the crest of a rising salt structure. These may form a conjugate set in response to the vertically-upwards stresses developed by the upthrust of the salt. If only one side of the conjugate set develops, a 'trapdoor' diapir may be the result. Figure 5.12 shows in diagrammatic form development in this way. Sketches (a) and (b) show two stages in the process, in which the rising salt has induced conjugate shears disposed antithetically across the crest of the feature, each set dipping at about 30°. The salt has injected the fracture zones, roughly following the track of the two shear directions. Sketch (c) shows a later stage, with directions of movement in the salt (solid arrows) and the overburden (open arrows) indicated. In Fig. 5.13, an example from the southern North Sea shows an incipient development of this kind. A salt pillow is seen in cross-section; the rising salt has caused shear stresses to develop at its crest, and the top of the salt interval has been breached with the salt injecting zones along the two shear directions to form the double-antiform shape in the overburden above the salt crest by uplift deformation of the overlying sequences. Another example from nearby, with the same characteristics, is seen in Fig. 5.14.

Sometimes, the shears in one direction are better developed than those in the other direction in the conjugate system, in which case the salt injects the overburden obliquely, lifting it in a manner reminiscent of an opening trapdoor. Such 'trapdoor' diapirs are mentioned and illustrated by Trusheim (1960).

Kupfer (1968) advanced the hypothesis that shear surfaces developed actually within the salt itself in a

Fig. 5.13 Seismic section across a salt swell which is breached at its crest. The two antiform shapes in the overburden above the salt crest indicate the presence of a set of conjugate uplift shears as seen in the sketches of Fig. 5.12. (Courtesy of SS(E)L.)

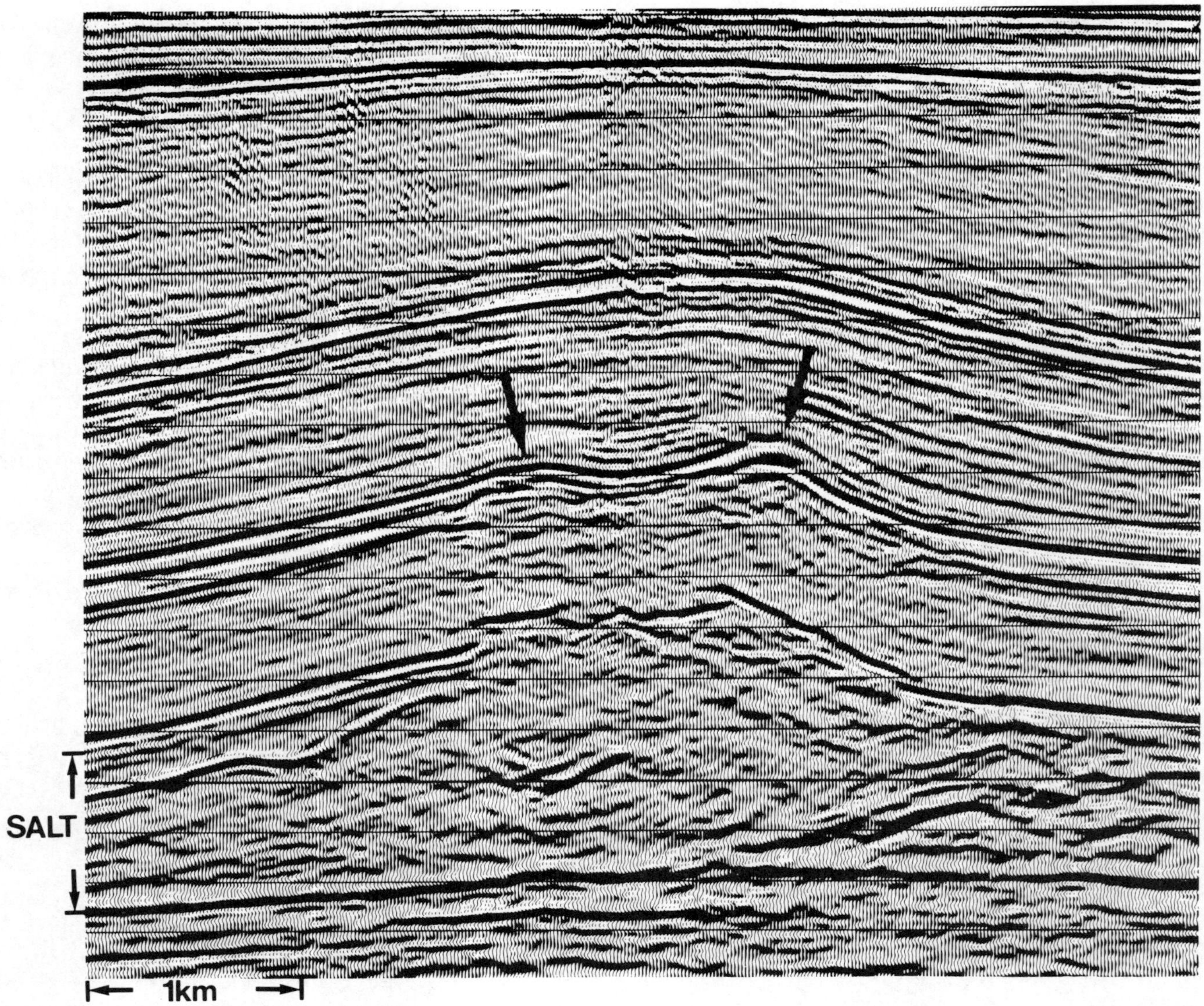

Fig. 5.14 Another example similar to those in Figs. 5.12 and 5.13, with two antiform shapes developed in the overburden above the crest of a breached salt swell. (Courtesy of SS(E)L.)

diapiric feature, and that individual parts of the salt mass moved upwards ahead of the body of the salt—presumably in the elastic solid state—causing disruption of the overburden and any already existing graben feature (see Fig. 5.15). These leading masses of salt were called 'spines' by Kupfer. Evidence of fractures in salt masses is available from mining operations in salt—Kupfer mentions several examples from the north east and north central USA—where the fractures have been re-sealed by more coarsely crystalline halite.

Whether in this way, or as plastic injections, such salt spines rising above the main salt body are frequently observed by their effects on the overlying rocks (Fig. 5.16), although it should be noted that, in the case of an elongate salt feature, the leading mass may be an elongate vertical slab, rather than the 'spine' it will resemble in cross-section. In individual cases, it will be difficult or impossible to decide whether injection of overburden shears by mobile salt, or uplift of spines by intra-salt shearing, is the correct model. Where the diapiric spine has risen obliquely from the main salt body, the former model would seem the more likely, although the converse is not necessarily true; later salt movement may mask the evidence of initial obliquity of the rising salt.

The top of a salt interval may be affected by overburden faulting due to tectonic stress. If this occurs when the salt is in a potentially mobile state but has not commenced any upward movement, no initial salt swell or primary rim syncline will have developed before the stress-relief upward diapiric flow of the salt in the fault zone takes place. The normal multi-stage halokinetic process of diapir formation,

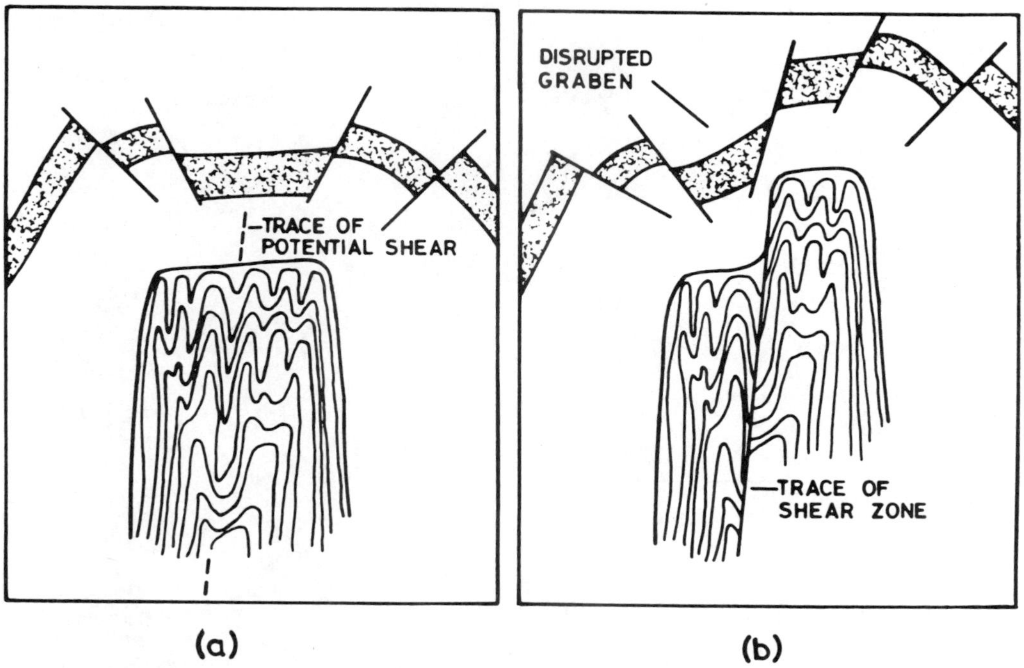

Fig. 5.15 Sketches suggesting the possible mode of formation of diapiric spines leading the main salt mass in a rising diapiric feature, as advanced by Kupfer. The spines are one reason for the frequent disruption of crestal graben associated with salt structures. (After Kupfer, 1968, courtesy of the AAPG.)

Fig. 5.16 An example of the effects seen in the overburden due to the development of spines above a rising salt mass. Salt interval is between B and T. (Courtesy of SS(E)L.)

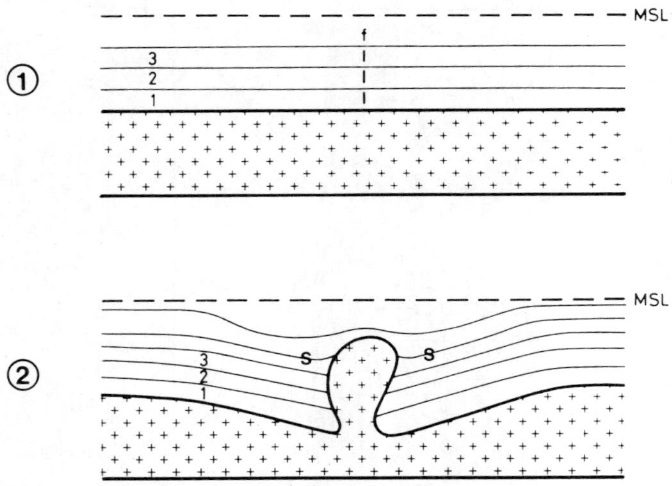

Fig. 5.17 Sketch model showing (1) the occurrence of an overburden fault, f, above a salt interval which is potentially mobile, and (2) the development of a diapir with an associated secondary rim syncline, S, but with no primary rim syncline development. This situation may arise because of the lack of an initial salt swell before the salt injected the fault zone. Cross symbol is salt rock.

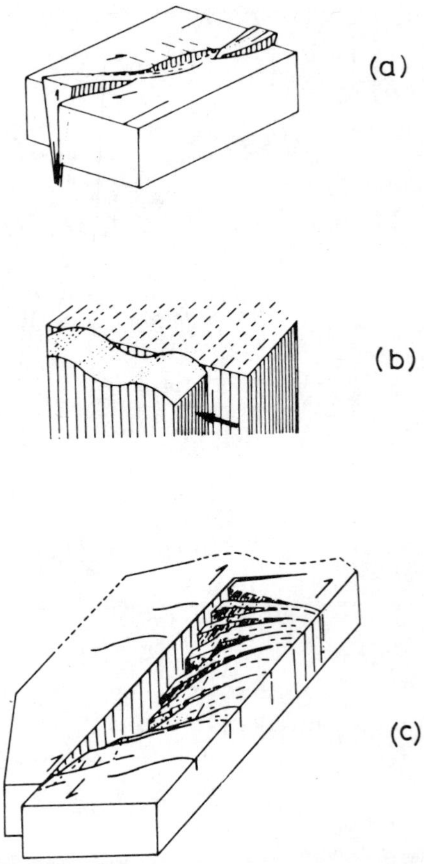

Fig. 5.18 Some effects associated with strike-slip faulting. See text for explanation. (From Hills, 1972; sketches (a) and (c) after Kingma (1958); sketch (b) after Cotton (1956); courtesy of Chapman & Hall, London.)

discussed in Chapter 7, will be 'short-circuited' as it were. Rapid subsidence of the overburden in the immediate vicinity of the diapir will lead to secondary rim syncline deposition. As suggested in the sketch diagram of Fig. 5.17, a diapir will thus evolve with an associated secondary rim syncline, but with no primary rim syncline sediments present. Sanneman (1968) suggests that diapirs of this type may be present in the Zechstein of the NW German area, resulting from overburden faulting caused by a tectonic pulse. The final form of the diapiric feature in Fig. 5.10 is believed to be of this kind, although the causative salt movement in this case was lateral flow within the salt interval, and from one side only.

5.3 STRIKE-SLIP FAULTING AND SALT

Strike-slip faulting produces some unusual and complex effects in the geological section which can be difficult to interpret in seismic data. Some reasons for this are suggested in diagrammatic form in Fig. 5.18, which shows three possible situations related to this type of faulting. In Fig. 5.18(a), shattered rock slivers within the fault zone are seen which have ridden up and down during the movement, which is one reason why the sense of the throw frequently changes along a strike-slip zone. Clearly, the effects on overlying, or later-deposited sediments would be very involved. Figure 5.18(b) shows what may happen where one flank of the fault zone is subjected to greater compressive stress than the other. Folding with axes normal to the fault zone has developed, which again would give the appearance of changing throw direction on any seismic lines normal to the fault zone. The sketch of Fig. 5.18(c) is of interest since it suggests the configuration at the termination of a strike-slip fault zone. The lateral movement must begin and end somewhere, and one way for this to occur is by the formation of a series of *en echelon* shears, stepping down as shown. Such a situation has been suggested on the portion of time-structure map seen in Fig. 5.19, to the right of the fault marked S, which is believed to be a right-lateral strike-slip fault located in an area of extensive strike-slip movements in the Southern Basin of the North Sea.

It can be imagined that, where a salt interval overlies rocks which have undergone this type of faulting, the results may be complex. One of the most striking effects often seen is that resulting from the ability of salt rock in a potentially mobile state to absorb the effects of underlying strike-slip movements; the salt

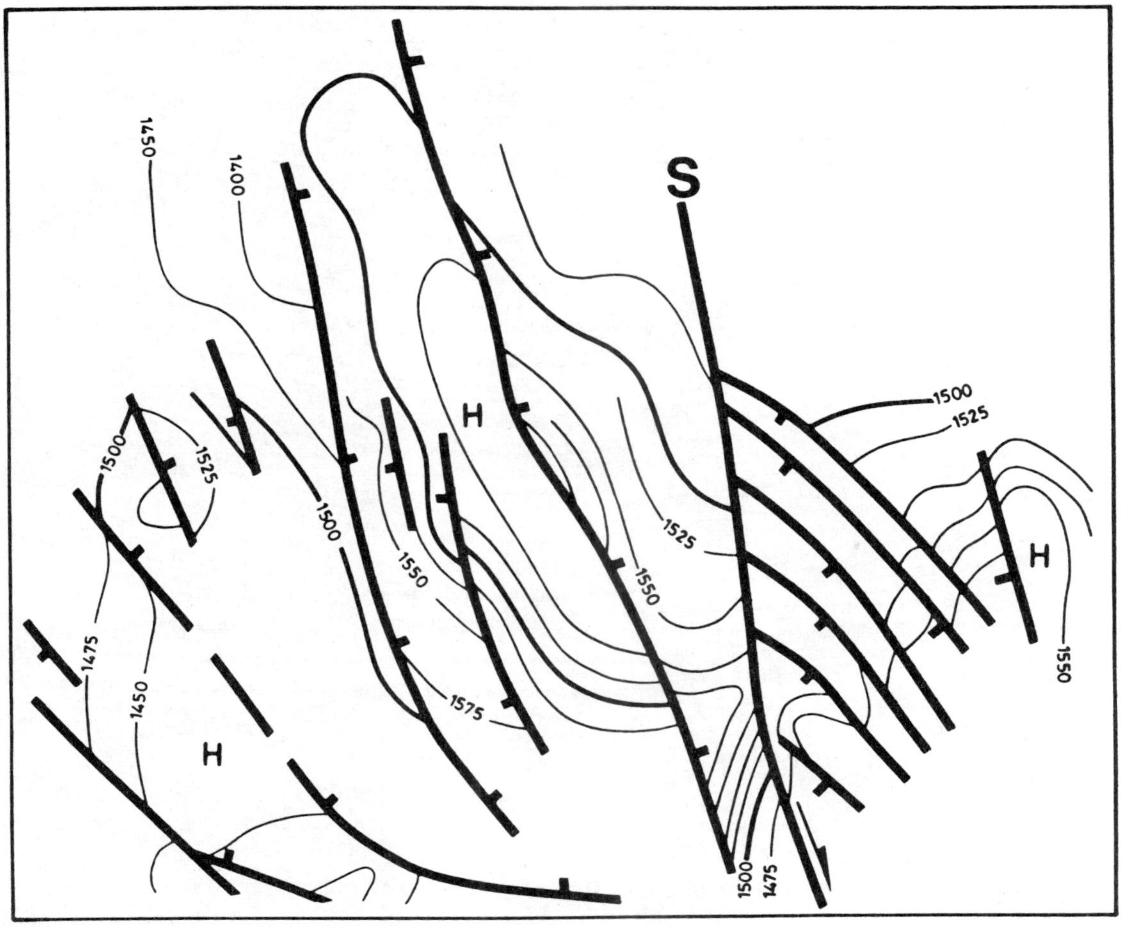

Fig. 5.19 Part of a time contour map constructed from seismic data in the Southern North Sea, showing a fault situation (to the right of fault S) believed to be similar to that in the sketch of Fig. 5.18(c). (Courtesy of SS(E)L.)

acts as a decoupling layer, preventing the effects of the faulting from being transmitted to the overburden.

In Fig. 5.20, an example of this is shown, with the base of the salt layer at Z and the top at T. This interval is the Zechstein (Upper Permian), whilst R marks the base of the Rotliegendes (Lower Permian) in this area. The strike-slip fault zone marked on the section has affected the Lower Permian interval and the base of the Zechstein, Z, but the top of the Zechstein, T, shows no evidence of disruption at all. The salt in the Z–T interval has accommodated the horizontal movement of the fault completely. The subsidence effects in the overburden above T indicate that salt has moved away from this zone in both directions, leaving only a relatively thin salt interval.

A similar example is seen in Fig. 5.21, the lettering for which is the same as in the previous illustration. Again there is no disruption of the top of the Zechstein above the fault zone.

Figure 5.22 gives another example which is of a much more exaggerated effect. Here, a strike-slip fault F–F, which is believed from isopach studies to have undergone left-lateral movement of some 10 km, has brought a thick section of Permian on the right (BR, Base Rotliegendes; BZ, Base Zechstein; TZ, Top Zechstein) into juxtaposition with a thin section on the left. In this instance, the salt of the Zechstein Z2 cycle (interval BZ–P on the right) has accommodated the horizontal fault movement, with little or no effects being seen in the overlying Z3 and Z4 cycles (interval P–TZ). It is probable that the Zechstein interval BZ–TZ to the left of the fault consists mainly of marginal facies rocks, there being little or no salt present. The thickness changes of both the Zechstein and Rotliegendes intervals across the fault, together with the lack of disruption of the top of the Zechstein, are diagnostic of strike-slip faulting in these circumstances.

A point which must be considered is the reason for the lack of salt movement across the fault zone due to differential loading on the lower part of the salt interval—the process discussed earlier in the chapter. In the last example seen (Fig. 5.22), this is probably due to the lack of salt on one side of the fault zone,

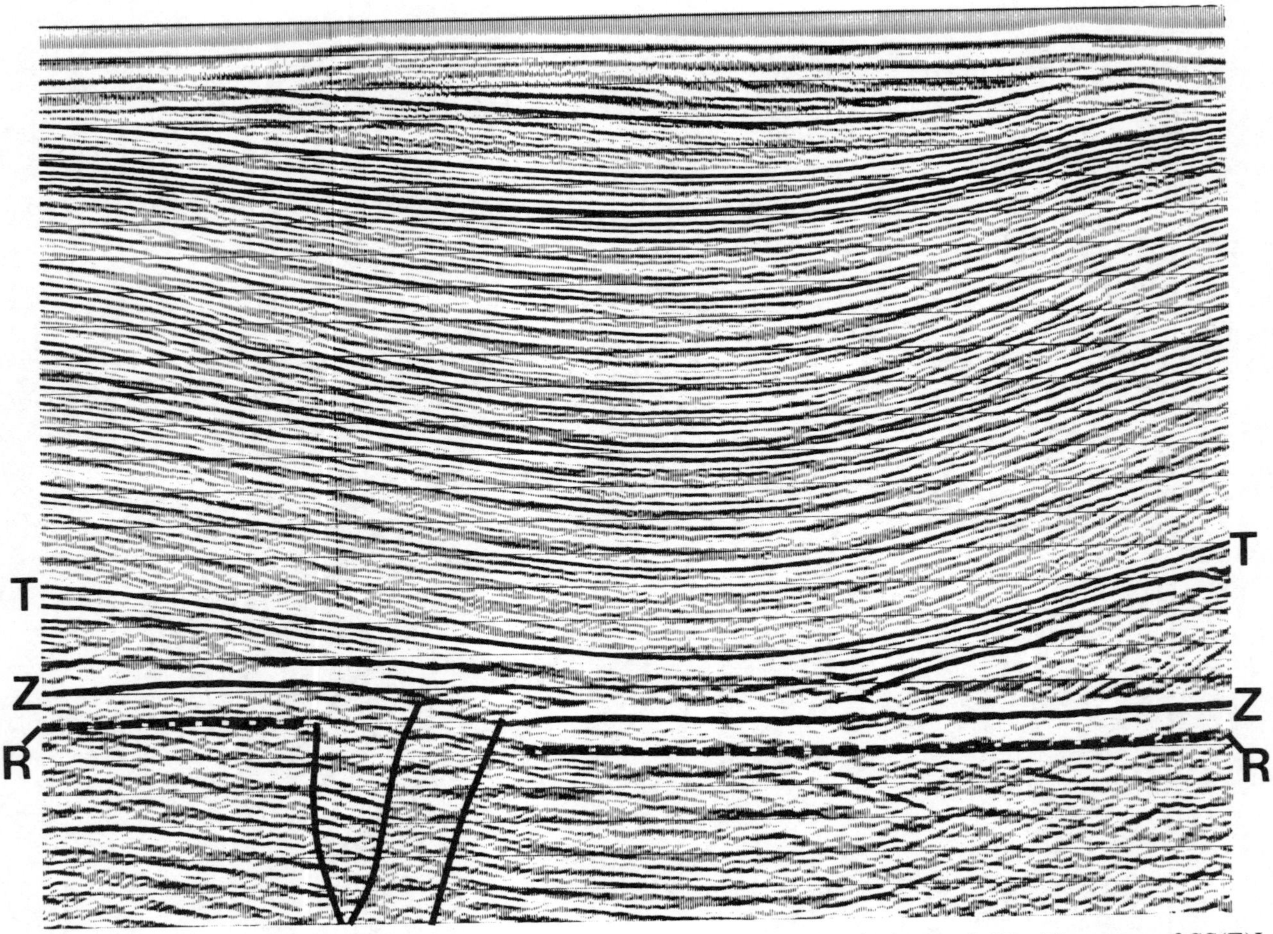

Fig. 5.20 Strike-slip fault zone. Note lack of effect at the top of the Zechstein interval (T). (Courtesy of SS(E)L.)

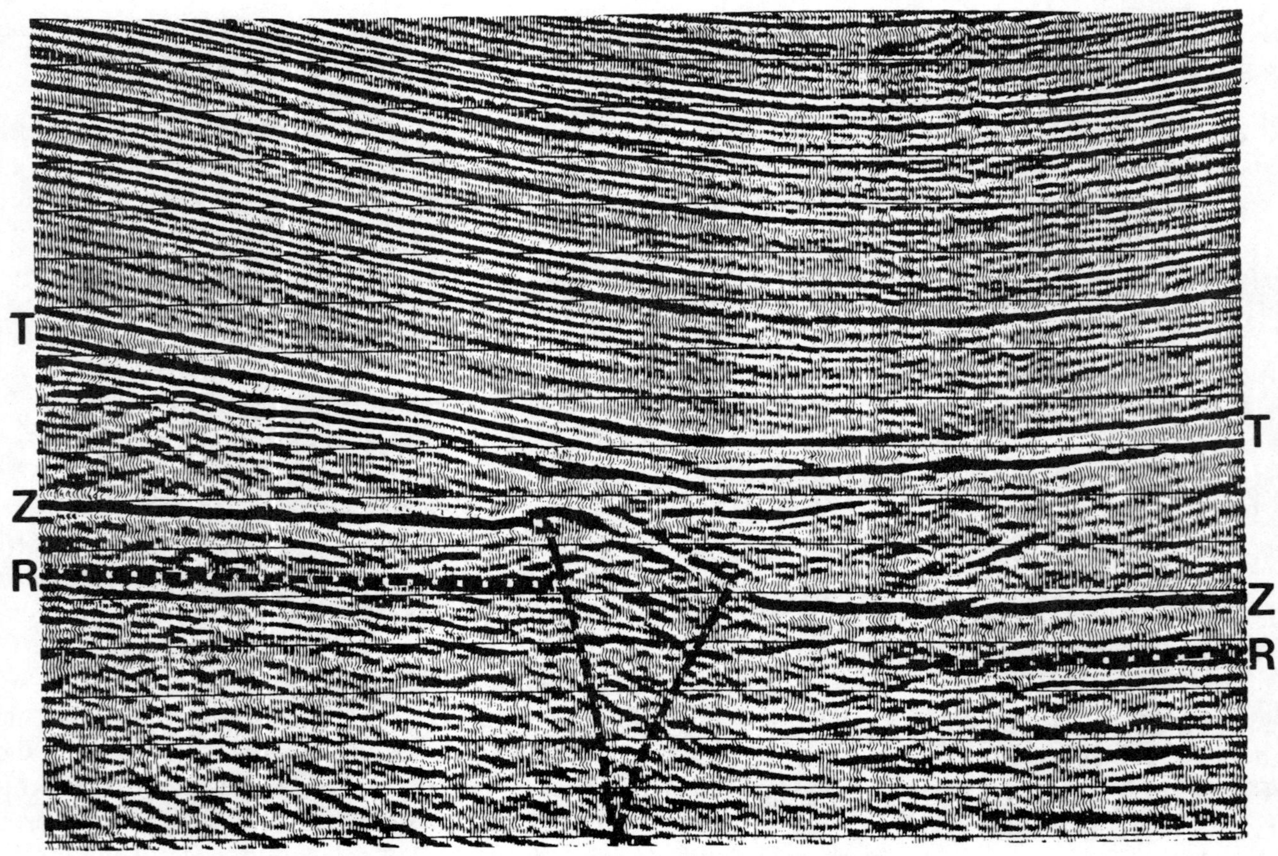

Fig. 5.21 Strike-slip fault zone. (Courtesy of SS(E)L.)

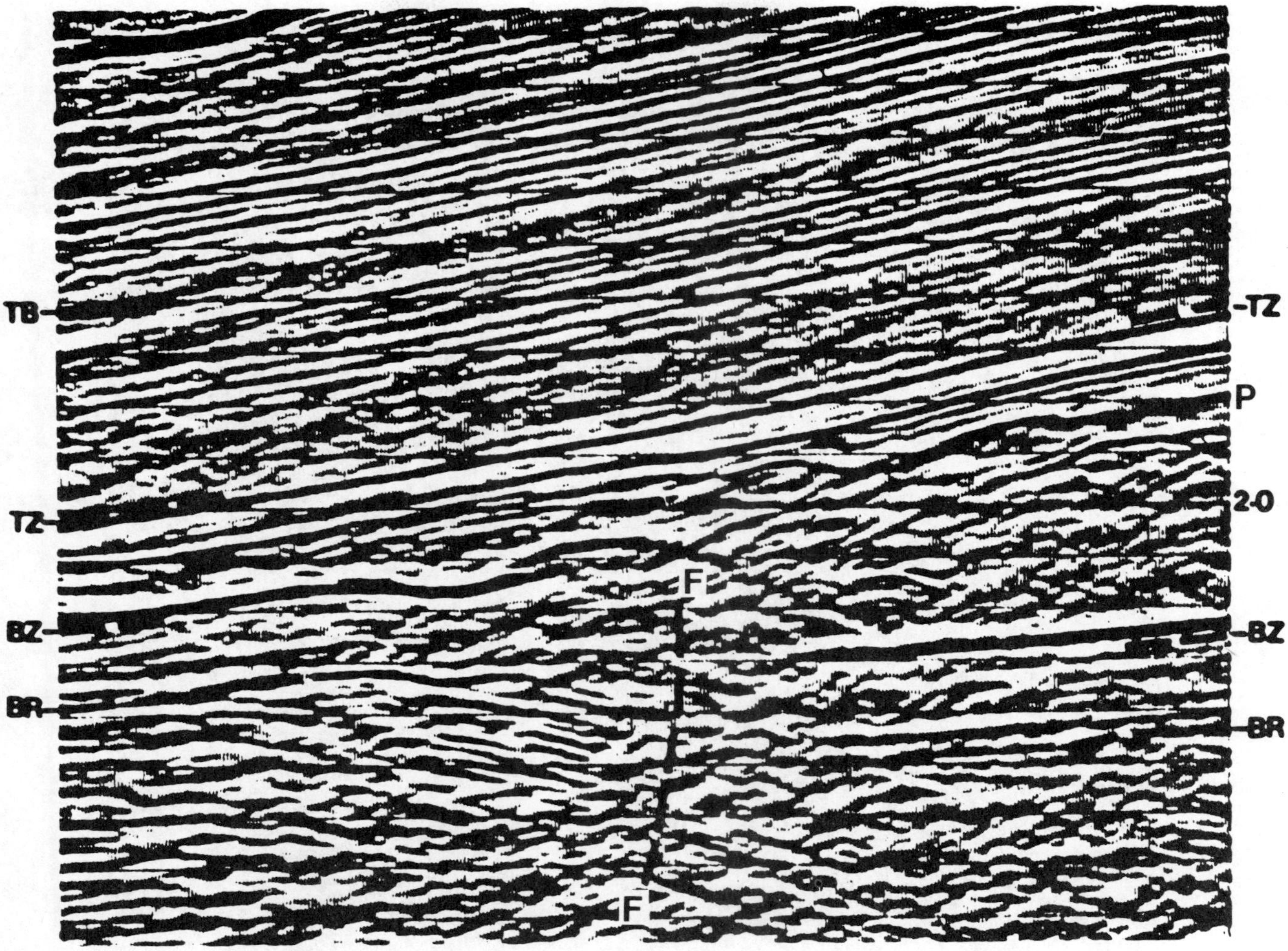

Fig. 5.22 Marked effects of a large strike-slip fault affecting the lower, but not the upper, part of the Zechstein interval. (Courtesy of SS(E)L.)

since the flow process requires a salt interval to be still connected across the fault. In the previous examples, relatively thin salt is present; bidirectional flow away from the fault zone accompanied by overburden subsidence acting like a vice upon the salt interval, combined with very competent salt roof rocks of the confining sequence, are factors militating against the operation of the differential loading mechanism.

5.4 OVERBURDEN FAULTING RELATED TO SALT MOVEMENT

In the foregoing sections, various cases of fault-related salt movement have been considered where frequently the faulting has a causative role. Situations will now be examined in which the faulting is often less directly associated with a salt layer or structure, but is rather an effect in the overburden of the movement of underlying salt.

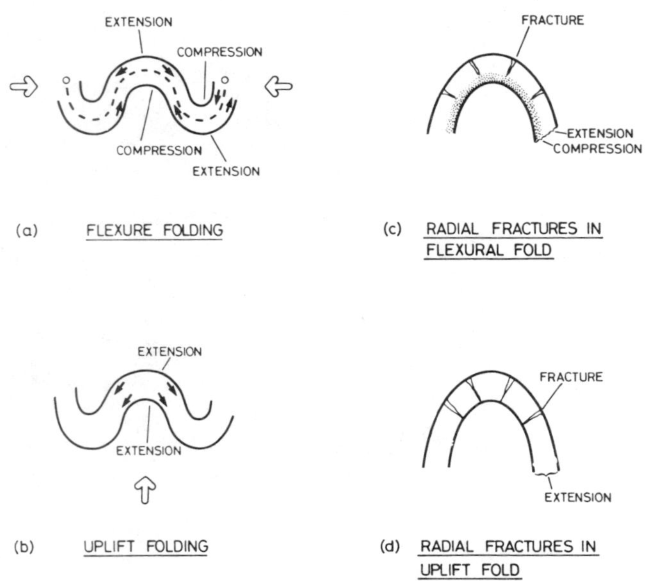

Fig. 5.23 Sketches indicating the stress pattern and fracture developments in flexural folding (sketches (a) and (c)), and uplift folding (sketches (b) and (d)).

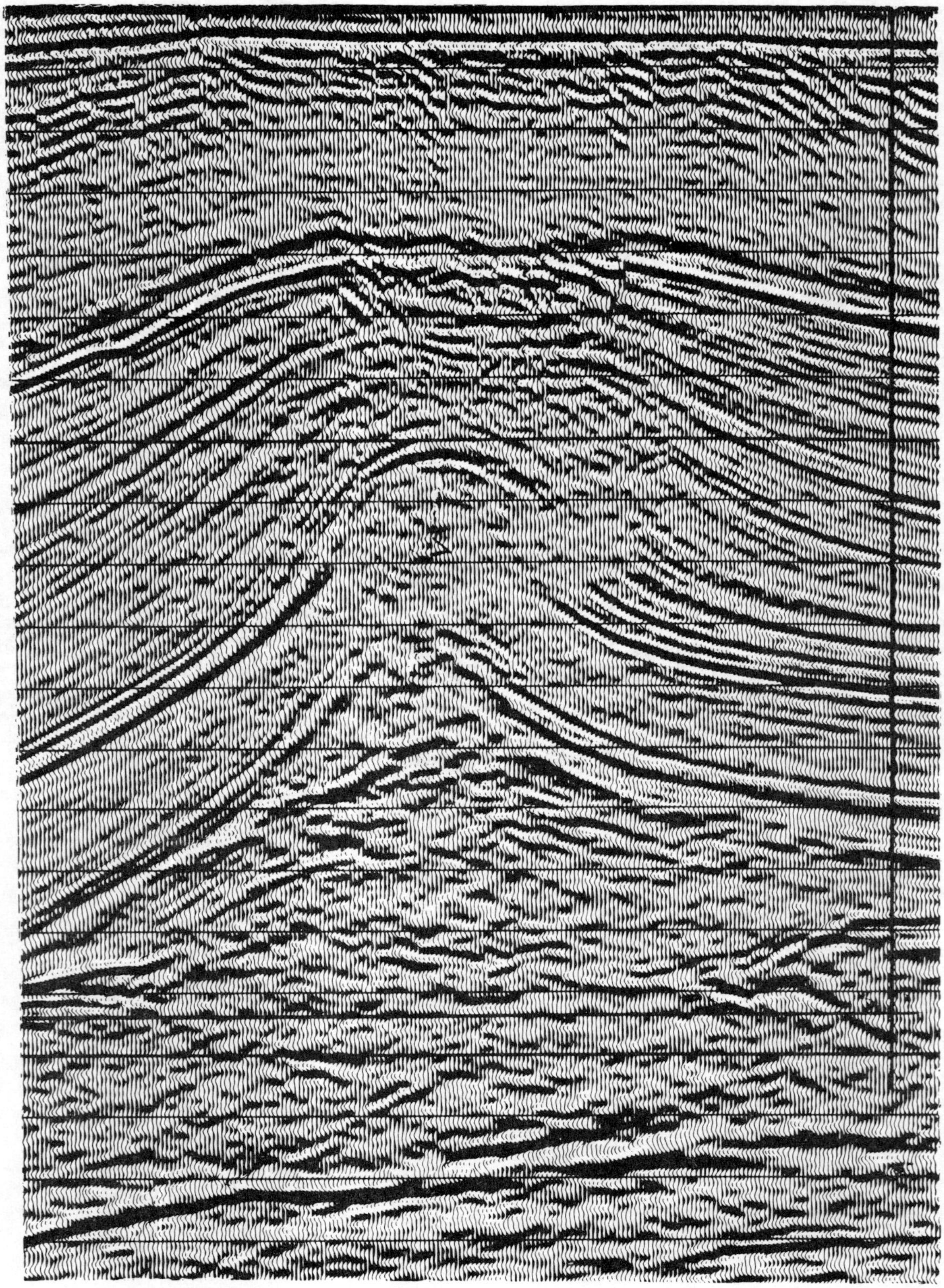

Fig. 5.24 Example showing the early stages of graben formation in the crest of an overburden uplift fold above a salt swell. (Courtesy of SS(E)L.)

5.4.1 The Development of Stress Zone Fractures

In Fig. 5.23, the contrast is recalled between the stress pattern in a flexural fold produced by compressive stress, and that in an uplift fold produced, for example, by a rising salt mass. In both the sketches of the figure, the applied stress is represented by open arrows and the resultant stress by solid arrows.

In sketch (a), the anticlinal fold in the centre develops with extensional stress acting in the crest of the fold, and compressive stress in the core. Somewhere between these two stress zones there is a surface of null stress (0–0), above which extensional stress increases upwards, and below which compressive stress increases downwards. The position of the null stress surface will depend on the geometry of the fold, the type and direction of application of applied stress, and to some extent on the rock types involved.

In the adjacent synclines, with respect to upward and downward directions, the converse will be the case—although if regarded as an inverted anticline, the syncline has a similarly developed stress pattern.

With an uplift fold, however, as sketched in Fig. 5.23(b), the situation may be quite different. If pure uplift of the overburden sequence by rising salt is concerned, with no element of tangential compressive

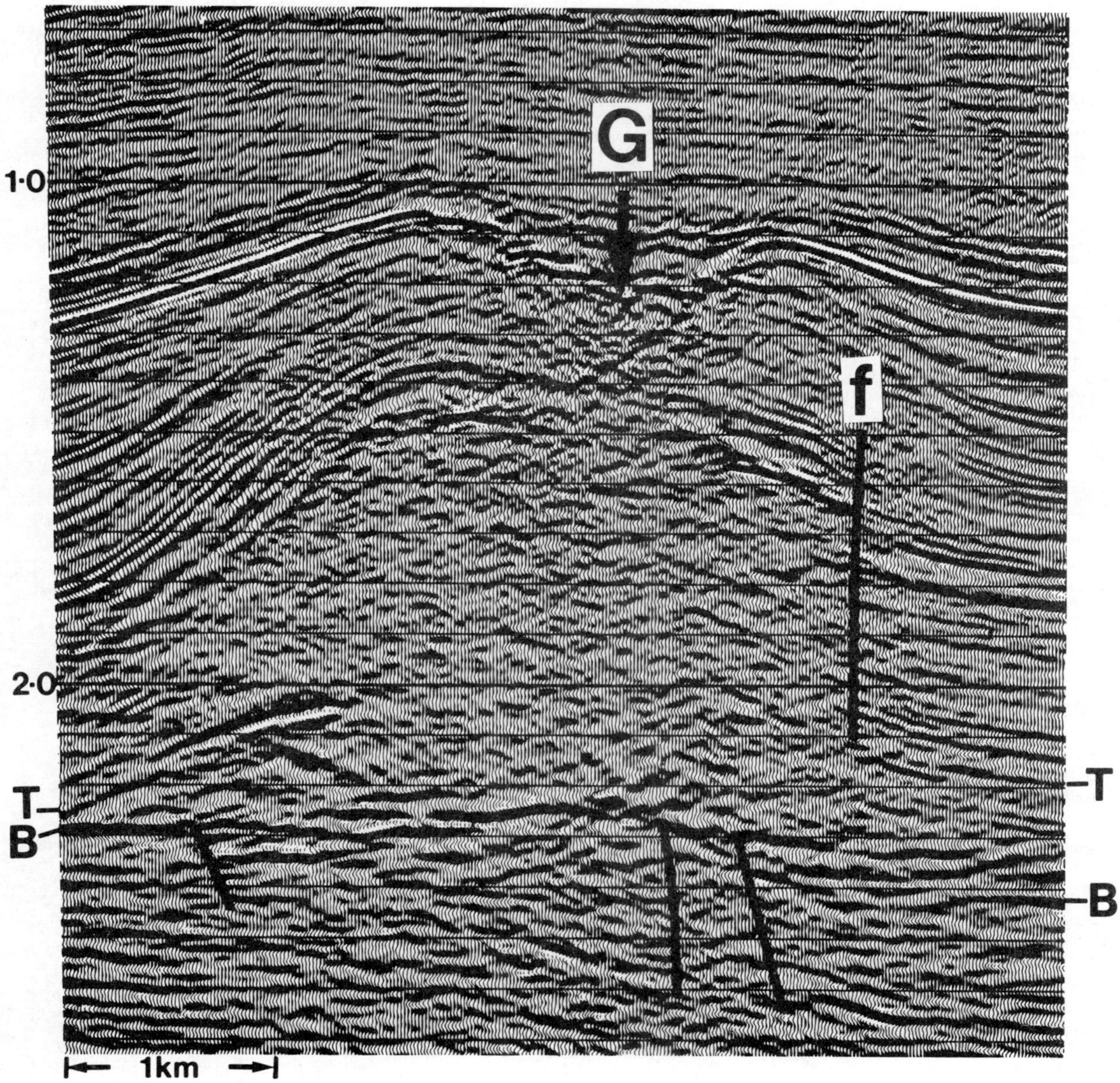

Fig. 5.25 A later stage of graben formation. Extensional fractures have extended down to the salt in the top of a rising salt body, salt has injected the fractures, and circulating undersaturated water has dissolved this out, and also salt in the top of the feature, resulting in overburden collapse and the formation of a large collapse graben. (Courtesy of SS(E)L.)

stress, all parts of the fold may be under extensional stress due to the continuing upward pressure of the salt. Either no compressive stress zone, or at most only a very tenuous one, may develop towards the core of the fold.

This difference may be of importance in a salt area. If (in cross-section) radial faults/fractures occur over the crest of a flexural fold (Fig. 5.23c), they will be inhibited from extending far down to the fold core by the zone of compression (which may develop its own system of brittle failure, but will be highly compacted). With an uplift fold, however, the compressive zone may be completely absent, with the extensional stress tending to open up fractures in the overburden folding, and any faults or fractures which in cross-section radiate from a point in the fold core can extend downwards to the core, as in Fig. 5.23(d). If this takes place, they may be injected by salt under high pressure. This process often leads to dissolution and removal of the salt both in the fractures and in the top of the salt pillow or diapir from which they radiate.

5.4.2 Graben Formation

This process may be followed in the three examples which are shown here. Figure 5.24 illustrates an early stage of the application of extensional stress in the overburden due to the rise of an elongate salt swell, seen here in cross-section. Fractures have developed in the overburden leading to the formation of a shallow graben. Note that if the traces of the bounding faults were produced downwards, they would intersect approximately on a line joining a crestal point on the fold with the centre of curvature of the latter. At this stage, the shallow graben is purely a structural feature.

When extensional stress across the fold crest increases due to the continued rise of the salt, the fractures will eventually extend down to the salt top, when injection of salt into the fractures will take place. If a circulatory system of undersaturated water is established, the salt injected into the fracture/fault system will be dissolved and removed, as will also eventually the salt in the upper part of the salt feature. As

Fig. 5.26 A large collapse graben above the crest of a salt pillow which has become diapiric at a late stage. (Courtesy of SS(E)L.)

the support provided by this salt disappears, the overburden will collapse; this will be aided by the already existing disruption in the structural graben above the salt crest. An example of this is seen in Fig. 5.25, with a dissolution collapse graben, G, clearly in evidence above the crest of a salt pillow, the top of which has a 'deflated' appearance due to dissolution removal of the salt.

In this example, the configuration of the reflection events on each side of the graben, contrasted with the flattened top of the pillow, show that the latter was formerly higher than it is at present. The fault 'f' on the right flank of the salt pillow is probably an uplift shear caused by the original rise of the salt, rather than a normal gravity fault. In fact, this 'pillow' is actually a diapir, having pierced the lower section, but uplifted the higher section which is concordant with the salt top.

A more extreme example of collapse graben formation is seen in Fig. 5.26. A large elongate salt pillow, seen in cross-section, has caused fracturing in the overburden above the salt crest, and has invaded the fractured overburden diapirically. The early swell formation occurred prior to the erosional phase represented by the unconformity, U, as seen by the truncation of reflection events beneath U on the left-hand side of the feature. This truncation is not in evidence on the right of the feature, which indicates that the swell development was asymmetrical; this is a frequent phenomenon in the Southern North Sea Zechstein Basin, and will be discussed further in Chapters 7 and 8. Pillow development resumed after the erosional phase of U, with events showing truncation again at the higher unconformity, C. In the post-C period, the diapiric invasion, dissolution and collapse took place to produce the major collapse graben. Whilst graben subsidence continued, general uplift of the pillow was still taking place, as seen by effects at the shallower unconformity about 300 ms above the high point of horizon C on the left of the graben. Concurrent processes of uplift and dissolution of this type are often seen.

The example of Fig. 5.27 is of interest; a similar process has taken place to that in Fig. 5.26, but with a less strongly developed collapse. The initial diapiric invasion of the overburden by salt from the crest of

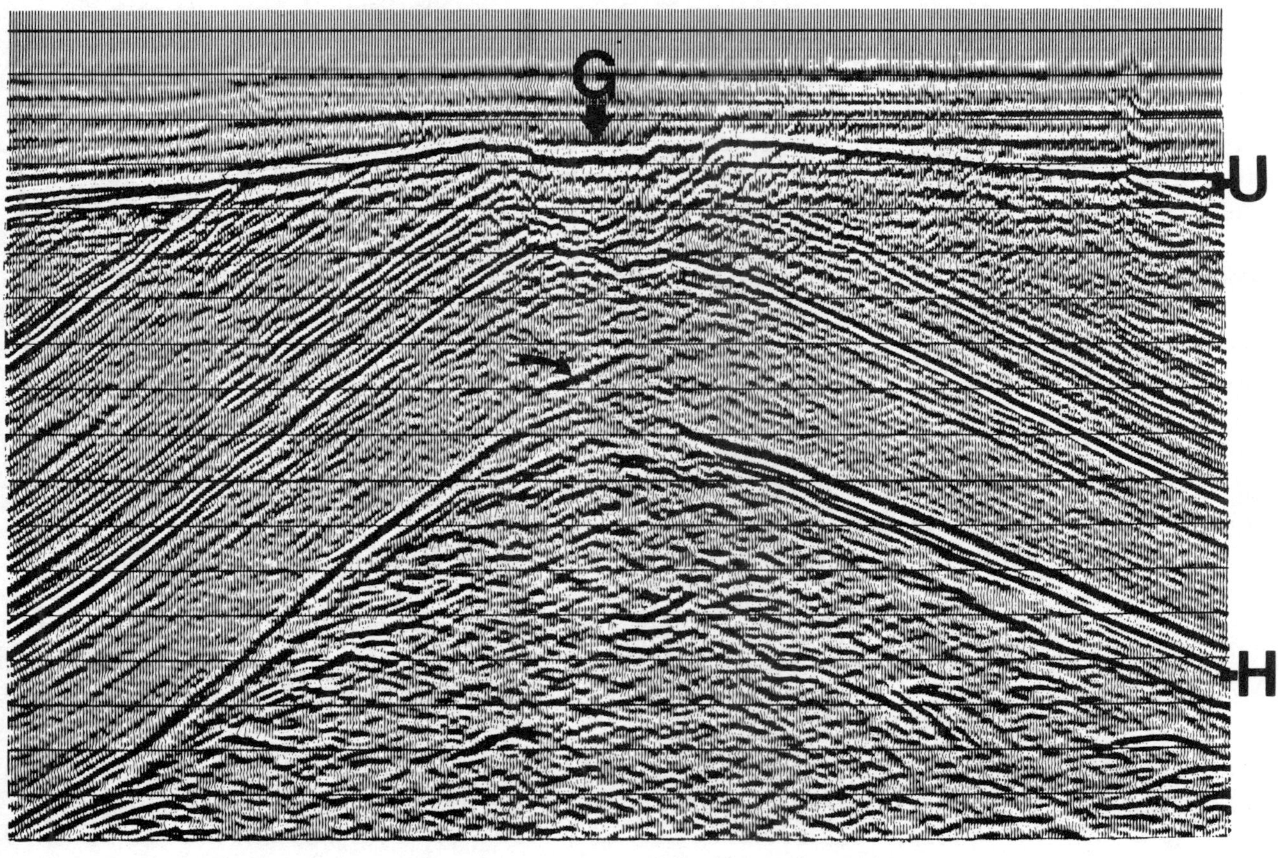

Fig. 5.27 Small collapse graben above the crest of a large salt pillow which was breached to inject a rather minor salt piercement up into the overburden at a late stage. The arrowed event is thought to be a shard of competent rocks from the salt roof which has been rafted upwards during the salt piercement. (Courtesy of SS(E)L.)

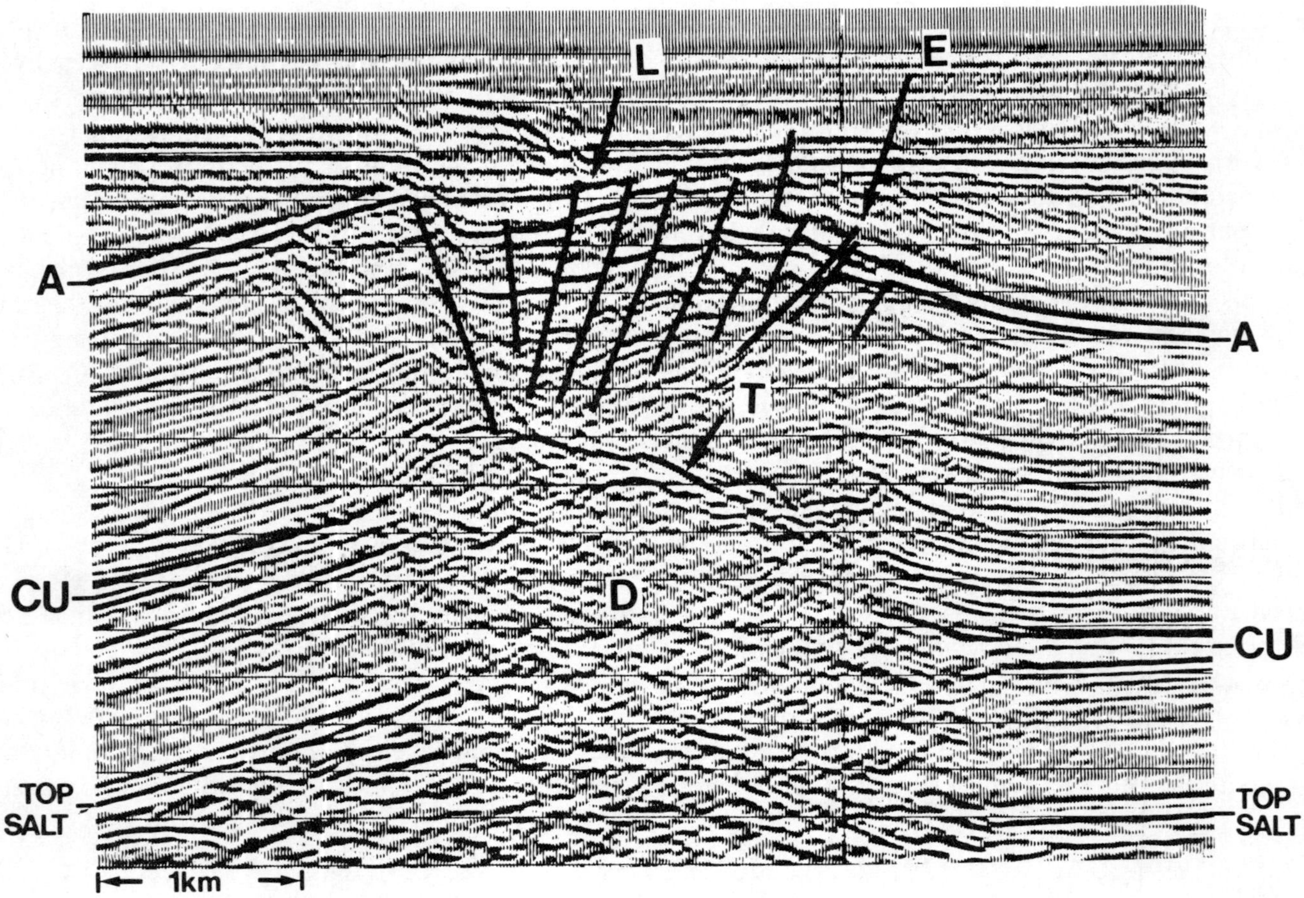

Fig. 5.28 Major step-faulted graben structure. (Courtesy of SS(E)L.)

the pillow has 'rafted' a shard of the competent anhydrite/dolomite beds which form the top of the Zechstein interval (at and above H, the salt top), carrying this up above the crest (arrowed event). The main collapse occurred prior to the erosional phase of U, but subsidence into the graben has continued to a lesser extent in post-U times.

In Fig. 5.28, another salt pillow which became a piercement at a late stage by a similar process is seen. In this example, pre-CU (unconformity) events on the left show truncation beneath this unconformity, although the same effect is not seen on the right, indicating again an initially asymmetrical salt swell. Steepening dips of reflections towards the salt feature on both sides in the CU–A interval show that the salt swell grew into a sizeable pillow during this period. Early overburden fracturing (E) began prior to the shallower erosional phase (U), and the fracturing continued through to post-U times, as evidenced by the step faulting and subsidence effects which 'young' progressively to the left of fault E. The strong event marked as T is interpreted as the response to the top of a cap-rock development at the crest of the salt. As the latter was dissolved and removed, a zone of insolubles—probably mainly of anhydrite—was left behind, subsiding and thickening as more salt was dissolved. The other strong fragmentary event below T may mark the base of this cap-rock sequence and the present top of the salt rock proper. Shallow unconformity U shows by its shape that the salt uplift ceased before this erosional phase, and the later shape of the structure was controlled entirely by dissolution collapse. L is latest step-fault.

5.4.3 Sub-circular Salt Features and Faulting

The examples shown in Figs. 5.24 to 5.28 are all cross-sections in the vertical plane of the short axis of elongate salt structures, with the graben in each example forming a long, trough-like feature running perpendicular to the plane of the section. When the salt feature concerned is sub-circular in plan view rather than elongate, the fracture/fault pattern developed is different. Although a short, trough-like collapse graben may be present more or less centrally above the salt crest, there is usually a more marked development of a system of fractures radiating (in plan view) outwards from a point above the centre of

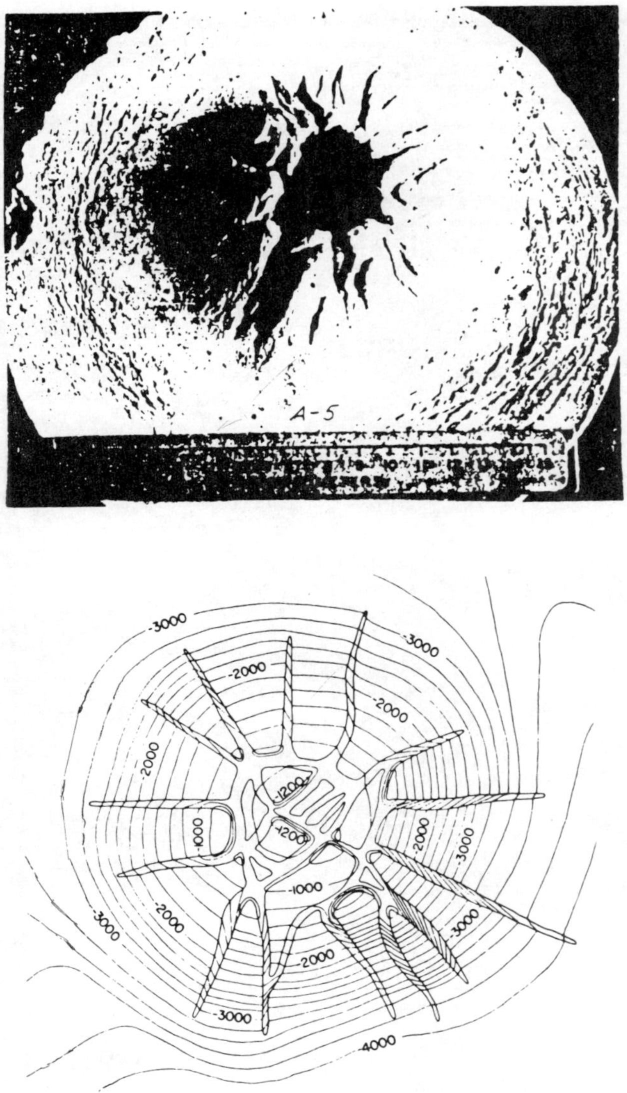

Fig. 5.29 Radial fracture pattern above a rising salt feature, Clay Creek salt dome, Texas, USA (lower sketch), compared with similar features developed in a model (upper sketch). (Upper illustration from Parker and McDowell, 1955; lower illustration from Gussow, 1968; both by courtesy of the AAPG.)

the salt body. There is a tendency for triangular flaps of overburden to be thrust upwards and outwards by the rising salt. In Fig. 5.29, there is a striking similarity between the fracture pattern developed in a model diapir experiment (upper sketch) carried out by Parker and McDowell (1955), using an asphalt mobile layer and a mud overburden, and a real-earth fracture system occurring above the crest of the Clay Creek salt dome in Texas (lower illustration) as sketched by Gussow (1968) after Parker and McDowell (1951). In the latter sketch, a short, trough-like graben is seen centrally situated, surrounded by radial fractures very similar to the effects seen in the model. This process is of interest, as it may in some cases be the answer to the question of what happens to the overburden which is displaced by a rising salt body. Where the overburden is not very thick, and consists of relatively incompetent rocks, the uplift of such triangular flaps, followed by erosion, may be the explanation of this problem.

5.4.4 Peripheral Faulting Related to Subsidence

Overburden subsidence, whether due to dissolution collapse or to salt withdrawal, can lead to gravitational normal faulting in some circumstances.

In Fig. 5.30, overburden subsidence into a large dissolution collapse graben, G, has led to the appearance of a normal gravity fault, F, at the periphery of the zone of subsidence. This fault is marked by an excellent fault-plane reflection over part of its trajectory, and can be traced beyond this almost up to the overlying unconformity. Another, less well marked, fault parallel to F is present to the right of the latter.

Another peripheral fault of similar type, marked by a rather fragmentary fault-plane reflection, is present in Fig. 5.31, but this time is due to overburden subsidence into a salt withdrawal basin. The fault shows listric curvature which is rather typical, and may be compared to the 'younger' fault, L, seen in Fig. 5.6, which is of essentially the same origin.

In the usual circumstances where a salt withdrawal basin is located adjacent to a salt structure, a combination of stress-zone faulting and the peripheral subsidence faults just discussed can be present, as seen in Fig. 5.32. Of the faults numbered 1, 2 and 3 at the level of horizon A above a small salt swell (T, top salt; B, base salt), fault 1 shows a faint fault-plane trace directed towards the crest of the salt swell, and is a normal fault produced by the extensional stress zone associated with the salt uplift. Faults 2 and 3, however, have traces which are directed not towards the salt crest, but down the peripheral zone of the subsidence into the withdrawal basin to the left of the salt swell.

5.5 Growth Faults and Salt

Attention has been drawn on several occasions in this chapter to listric growth fault forms which are often related to overburden subsidence resulting from salt withdrawal movements or dissolution collapse effects.

A rather classic example is seen in Fig. 5.33, where

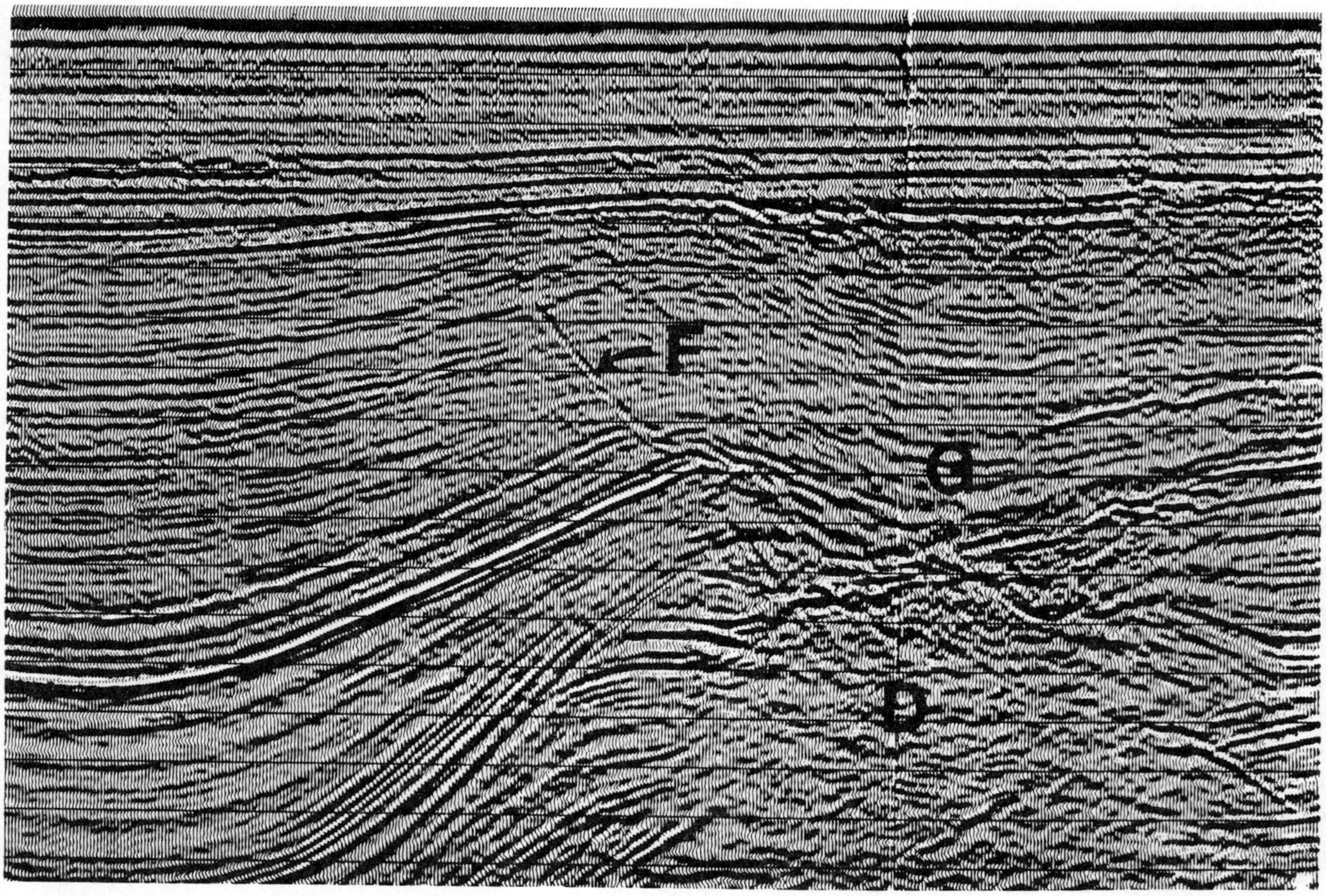

Fig. 5.30 Peripheral subsidence fault, with fault-plane reflection event F, related to formation of collapse graben G at the summit of a salt diapir, D. (Courtesy of SS(E)L.)

a small diapiric uplift at the edge of a shallow withdrawal basin has resulted in a growth fault, A. The movement of this fault seems to have been somewhat irregular, but it shows the general decrease in throw upwards, syndepositional effects of rollover and local thickening in towards the fault on the downthrow side, and antithetic faulting, B, which are all characteristic features of this type of faulting.

The development of growth faulting associated with subsidence has a much wider significance than that suggested by the relatively minor structures illustrated in this chapter so far. Such faulting is often an important element in continental margin tectonics, where salt basins are found along the strike of the continental shelf or slope. The importance of the processes discussed in relation to the small features illustrated so far is that they are often recognisable in much larger scale structures. One possible situation is shown diagrammatically in Fig. 5.34, where a basin filled with evaporites is seen on a continental slope, the evaporites consisting mainly of salt rock (cross symbol). The loading of a later sedimentary sequence prograding seawards across the shelf as seen in sketch (b) causes the basinal salt to move away from the advancing zone of higher loading, and to become diapiric at the seaward margin of the basin. Further deposition, and subsidence effects due to the salt withdrawal, lead to the occurrence of a gravitational normal fault of growth type above the landward margin of the basin, as in Fig. 5.34(c).

Of the offshore salt basins known to exist along the eastern margin of the North American continent, only two contain salt which has undergone appreciable movement including diapirism, namely the Scotian Basin adjacent to the Maritime Provinces of Canada, and the Carolina Trough. The latter area has been studied by Dillon *et al.* (1982) and Cashman and Popenoe (1985). The Carolina Trough is a long, narrow marginal basin paralleling the coast, which was the location of evaporite deposition of Jurassic age. Later sediment loading in the Jurassic and Cretaceous caused migration of the underlying salt ahead of the advancing wedge of deposition, and where the basement shallowed at the seaward margin of the

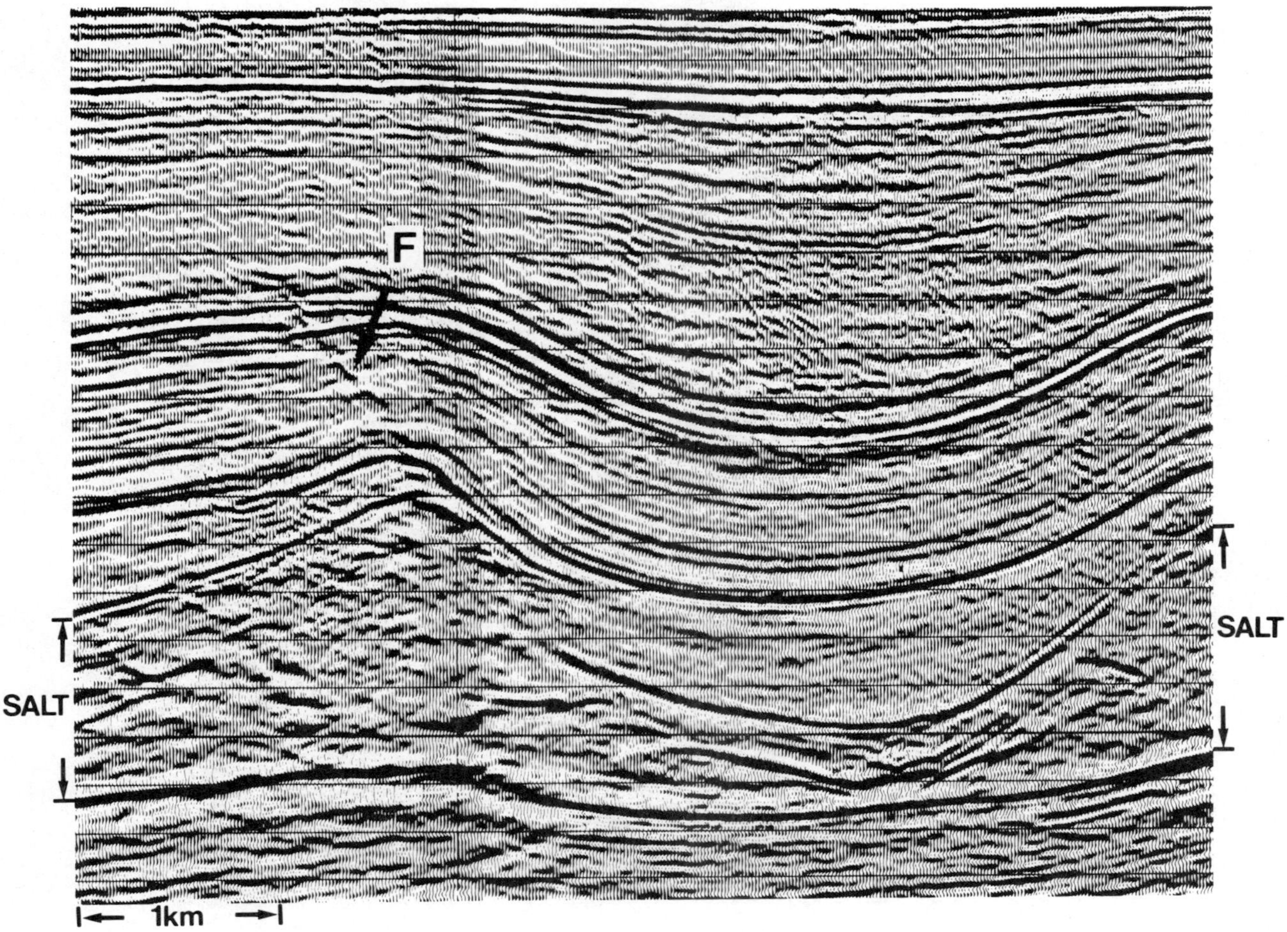

Fig. 5.31 Peripheral subsidence fault marked by fragmentary fault-plane reflection F, associated with subsidence into a salt withdrawal basin. (Courtesy of SS(E)L.)

Trough, a line of diapirs formed. As salt withdrew from the basin into this zone of diapirism, removal of support from the overlying sediments led to subsidence, and the formation along the landward margin of the basin of a growth fault in the overburden, on the model of that shown in Fig. 5.34(c). Figure 5.35 is a sketch map showing the general situation. Clearly, there is nothing startling or unusual about the mechanisms involved in this situation in the light of processes discussed earlier in this chapter.

Another situation, which is in a sense the converse of the Carolina Trough example, is modelled in the sketch of Fig. 5.36, where a salt interval (sketch (a), cross symbol) overlies a basement slope, and above the salt the overburden thickens downslope. In sketch (b), salt is shown moving upslope towards the zone of lower overburden load stress, and forming a pillow which is elongated along the strike of the basement slope (and thus perpendicular to the paper). The withdrawal of salt upslope leads to overburden subsidence, and the initiation of growth faulting. At a later stage, diapirs may form along the crest of the salt pillow, piercing the overlying growth fault zones, as suggested in sketch (c). This 'up and under' piercement process affecting growth faults has been described in the Houston diapir province, Texas, USA, by Ewing (1984).

The relationship between growth faulting and salt movement in these and other situations appears to be widespread. There is ample evidence of both salt and shale diapirism associated with growth faulting and toe thrusts down the western margin of the African continent, from southern Senegal and Gambia, through Guinea-Bissau, Nigeria, Cameroun and Gabon, to Angola (see Lehner and De Ruiter, 1977). On the other side of the Atlantic, there is a complex interrelationship between gravitational listric faulting and diapiric structures of both salt and shale in the Atlantic marginal basins of Brazil (Dauzacker *et al.*, 1985). Many other examples are to be found in the literature.

Although not an example of salt diapirism, the

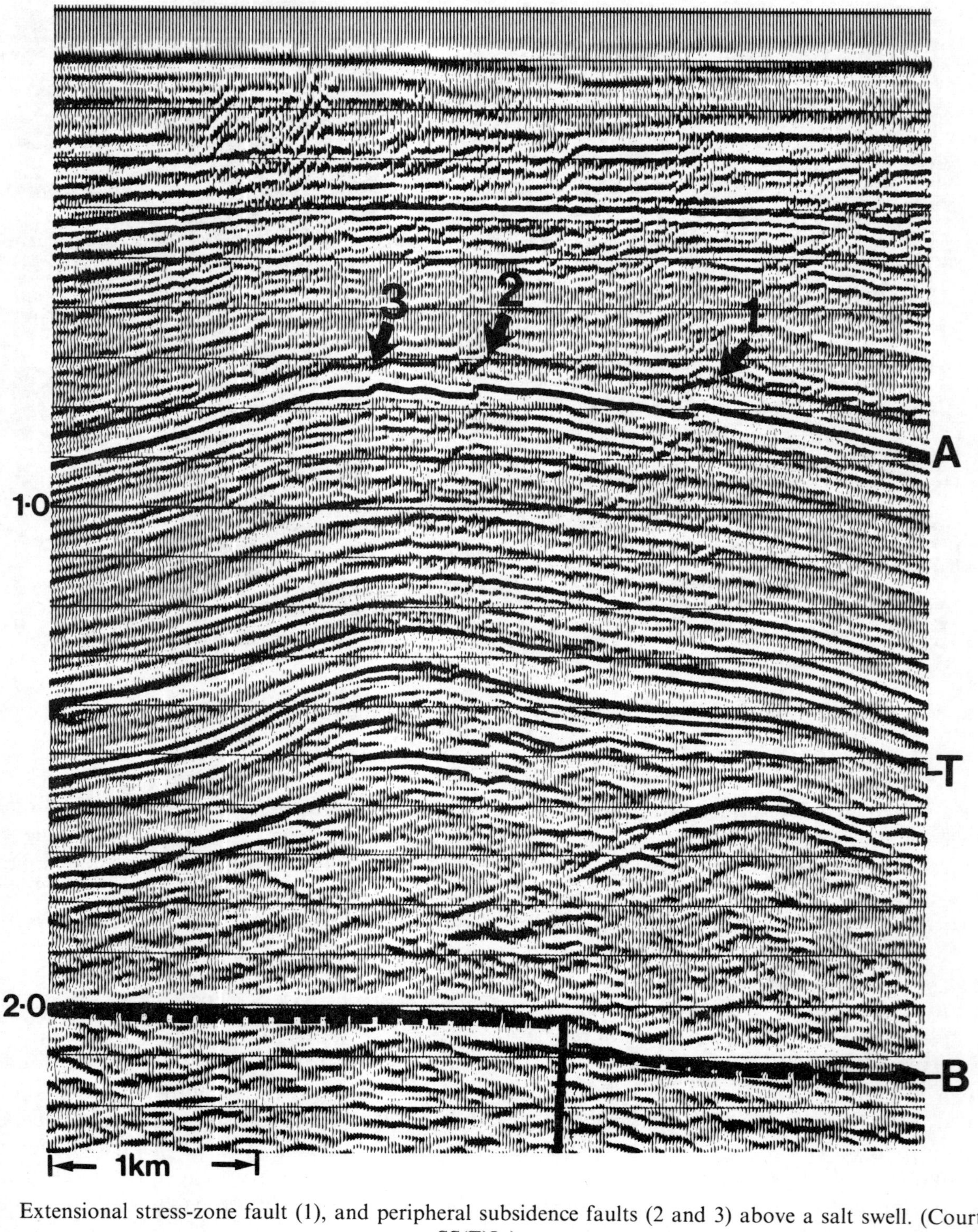

Fig. 5.32 Extensional stress-zone fault (1), and peripheral subsidence faults (2 and 3) above a salt swell. (Courtesy of SS(E)L.)

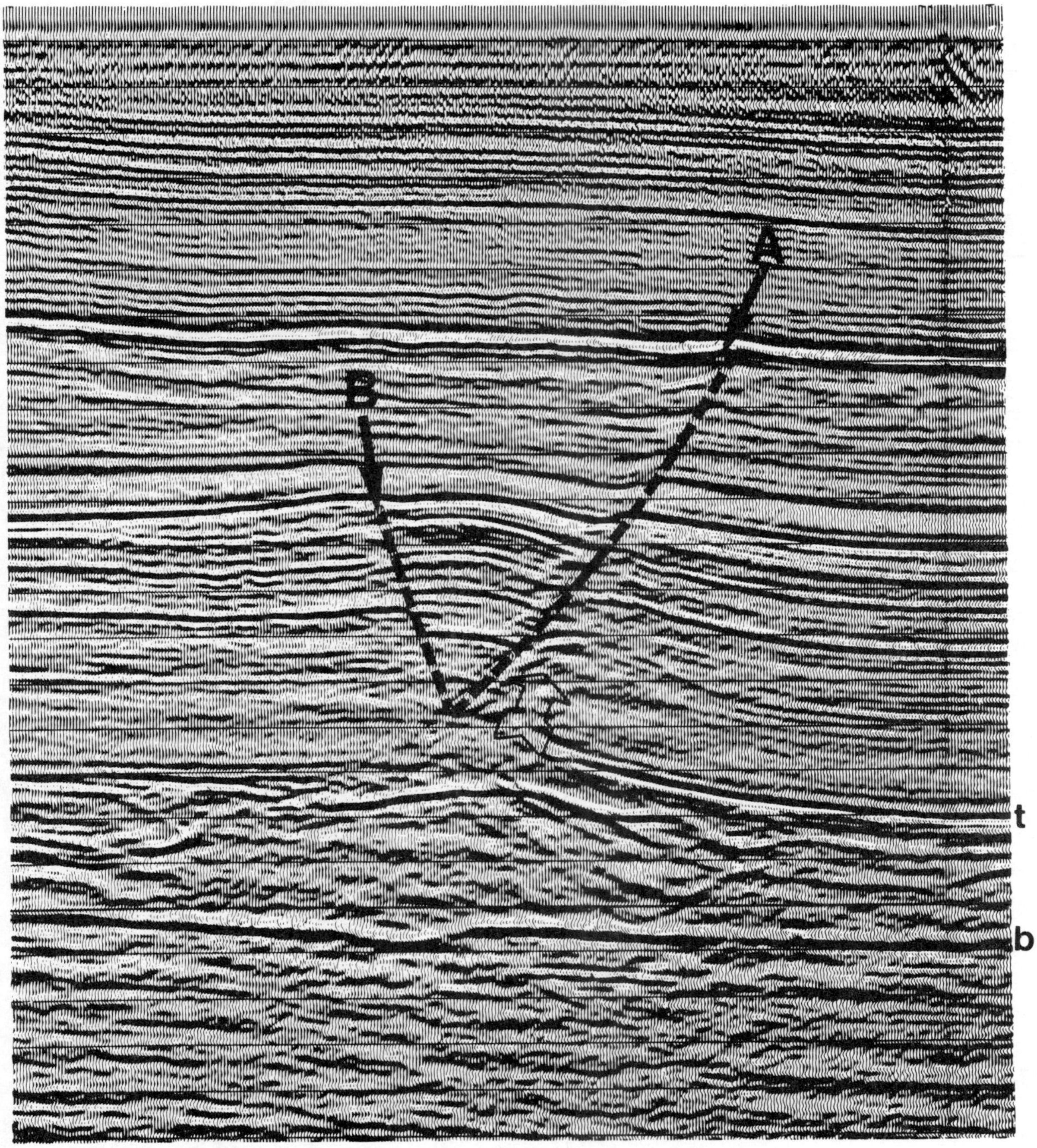

Fig. 5.33 Development of a growth fault (A) and antithetic fault (B) due to uplift/subsidence effects from underlying salt movement. Top salt is at 't', base salt at 'b'. Open arrow shows direction of salt uplift from minor breach in salt top. (Courtesy of SS(E)L.)

circumstances of mobilisation and diapirism in an area of the Nigerian offshore which involves overpressured diapiric shales are of interest in that they present a somewhat familiar aspect when compared with other continental margin examples included in this chapter. The Niger Delta of Nigeria dates from the Palaeogene, with the earliest outbuilding into the Atlantic Ocean taking place in Oligocene times, and with the present shape of the delta developing during the Miocene and later (see Lehner and De Ruiter, 1977).

Overpressured shales result from fluid retention and incomplete compaction. Normal clastic sediments beneath an increasing weight of overburden tend to undergo compaction and de-watering more or less uniformly through the thickness of the unit, with the pore fluids escaping upwards in the section. If this takes

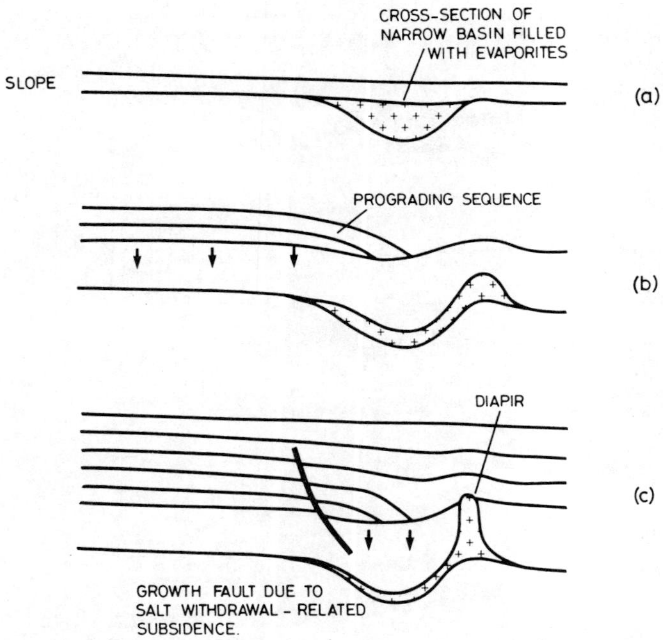

Fig. 5.34 Sketch models showing three stages in the development of a growth fault in the overburden in a continental slope setting, where the salt in a small basin has moved in the downslope direction, resulting in subsidence.

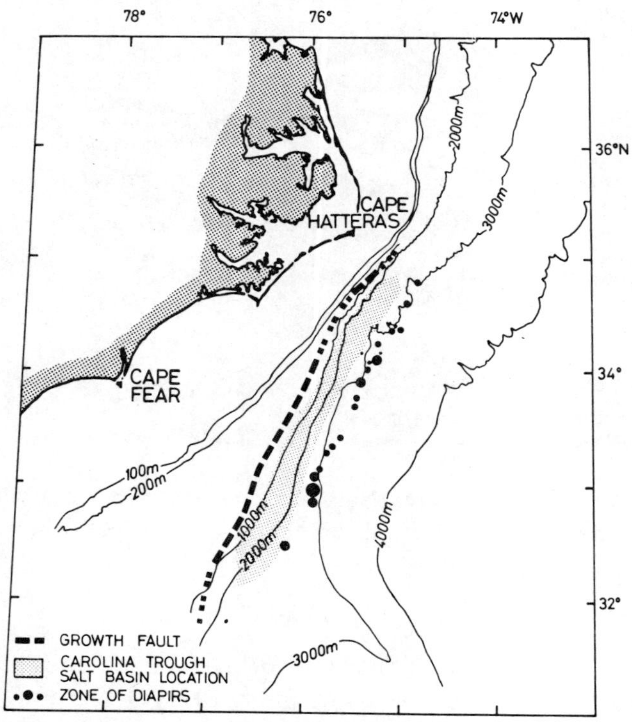

Fig. 5.35 Sketch map, with bathymetry, showing the location of the Carolina Trough salt basin, with a growth fault developed above its landward margin and a zone of salt diapirs along its seaward margin. Compare with Fig. 5.34. (From Cashman and Popenoe, 1985, courtesy of Butterworth Scientific Ltd.)

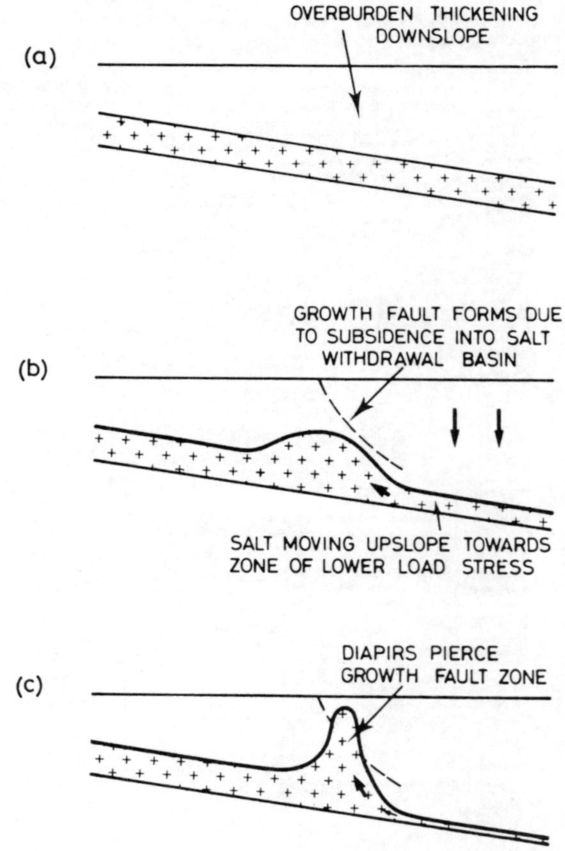

Fig. 5.36 Sketch models of growth fault development related to upslope salt flow and subsidence effects.

place, then eventually the rock matrix is supporting the full weight of the overburden, and any remaining pore fluid takes no part in this support.

With some shales, however, the uppermost part of the unit, consisting as it does of fine argillaceous particles, compacts rapidly to become impermeable and traps pore fluids in the underlying part of the unit. Eventually, the weight of the overburden is being supported to a greater or lesser degree by the remnant pore fluids in the shale, and only partly or not at all by the rock matrix. This prevents complete compaction, leaving the shale with a high fluid content and the presence of pore fluids under very high pressure. In this situation, the shale is in a very mobile condition, and in a field of differential stress will flow readily into zones of lower stress in a similar way to mobile salt rock (although by true interparticulate flow).

Shales of this type have an abnormally low seismic velocity compared with that of fully compacted shales, and can produce velocity anomalies in the seismic section. Sheaths of diapiric shale of this type are often found on the flanks of, and sometimes over the crests of, salt diapirs. Many examples have been found in the Gulf Coast area of the USA. Even where salt is

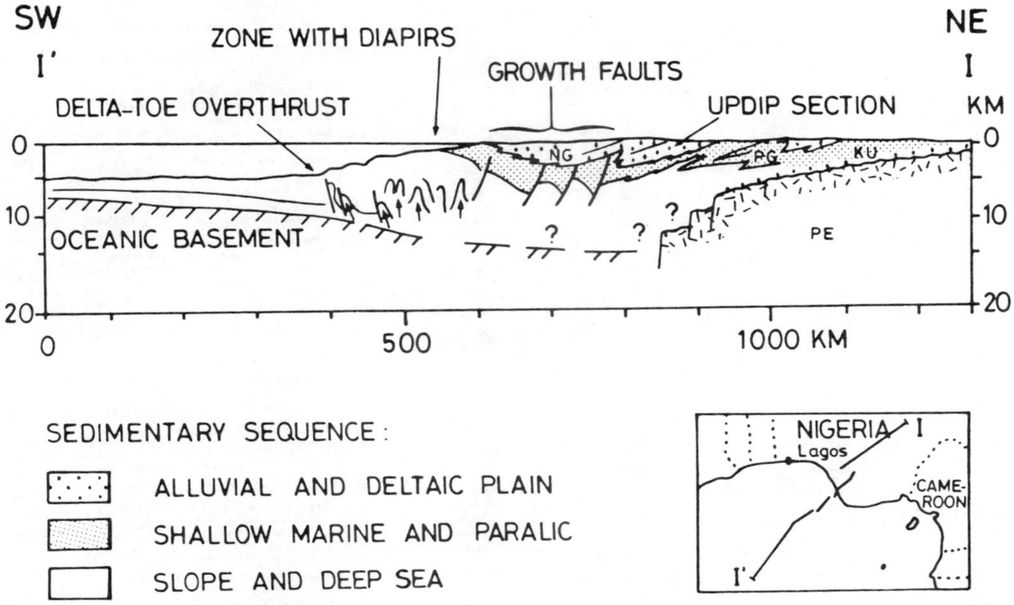

Fig. 5.37 Schematic section across the Niger Delta both off and onshore, showing the zone of diapiric overpressured shales. The small sketch map indicates the location of the section. (From Lehner and De Ruiter, 1977, courtesy of the AAPG.)

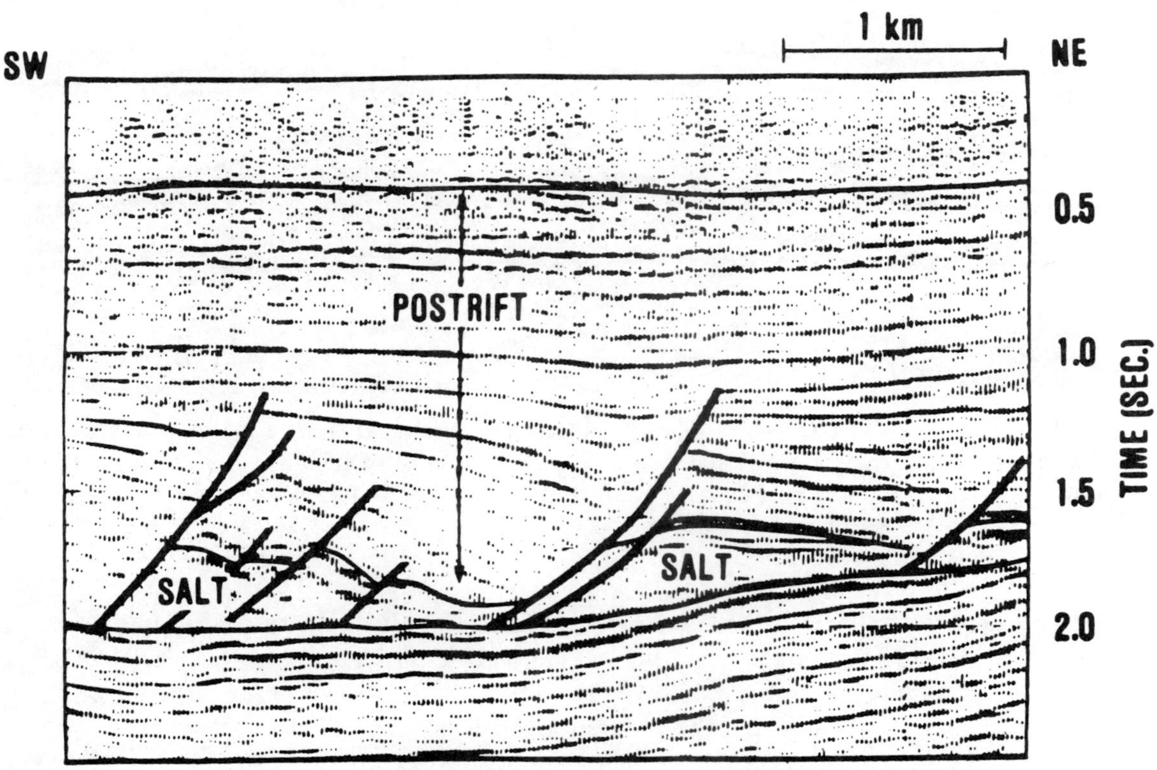

Fig. 5.38 Seismic section, offshore Cabinda, Angola, showing growth faults developed in carbonates due to salt movement and overburden subsidence. Note the salt uplifts on the upthrow side of all the faults. (From Brice *et al.*, 1982, courtesy of the AAPG.)

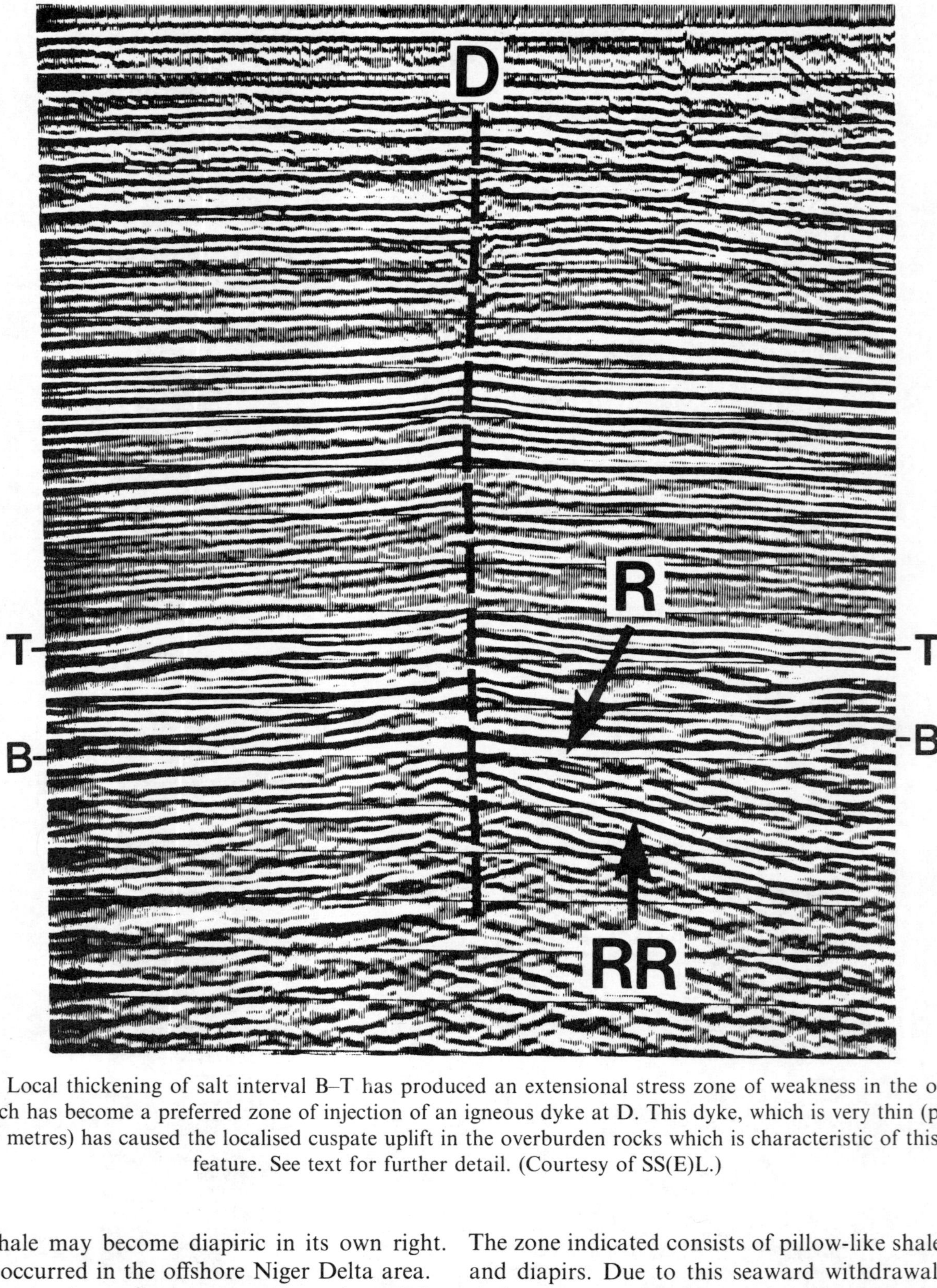

Fig. 5.39 Local thickening of salt interval B–T has produced an extensional stress zone of weakness in the overlying rocks, which has become a preferred zone of injection of an igneous dyke at D. This dyke, which is very thin (probably only a few metres) has caused the localised cuspate uplift in the overburden rocks which is characteristic of this type of feature. See text for further detail. (Courtesy of SS(E)L.)

absent, shale may become diapiric in its own right. This has occurred in the offshore Niger Delta area.

The situation is seen in the schematic section of Fig. 5.37, which is a regional section following the broken line I′–I in the smaller sketch. The thickening sedimentary outbuilding shown has caused the mobilisation and seaward flow of large volumes of overpressured slope shales lying beneath the Tertiary deltaic deposits. The zone indicated consists of pillow-like shale ridges and diapirs. Due to this seaward withdrawal of the shale mass, concomitant subsidence effects in the overburden have led to the zone of growth-faulting indicated. Bituminous shale source rocks are known to exist in the Upper Cretaceous (KU) above the Permian (PE) basement rocks in the NE, but these are probably isolated from the Tertiary deposits by the slope shales, and the source of oil in the numerous fields in the Delta probably lies within the Tertiary (Oligocene/Miocene) section (Lehner and De Ruiter, 1977).

Further to the south, down the west coast of Africa, the growth fault/salt movement relationship is seen again in the offshore area of Cabinda, Angola (see Fig. 5.38). Brice *et al.* (1982) record the situation in which thick halite of Aptian age, with numerous potash intervals interbedded, forms the major part of an evaporite sequence overlain by Albian carbonates and later clastic sediments. Gentle regional subsidence and deposition during the Albian to Eocene period seems to have coincided roughly with the period during which the salt became mobile, and probably developed a tendency to flow to the westward. This led to overburden subsidence patterns and the development of growth faulting in the carbonates, with rollover structures associated, as seen against the major fault zone in the centre of the example in Fig. 5.38. Once again, salt uplift on the upthrow side of the faulting is invariably present.

5.6 OVERBURDEN STRESS AND IGNEOUS ACTIVITY

The movement of underlying salt sets up stresses in the overburden which result in extensional fault and fracture zones. In an area of igneous activity, these zones of weakness can provide preferred routes for the injection of igneous material, particularly in the form of hypabyssal intrusions. Figure 5.39 shows an example from the Southern North Sea where local thickening of Zechstein salt (interval B–T) has resulted in extensional stress in the overlying sequences. A thin igneous dyke has been injected into the zone of weakness which, although it does not extend below the salt, must still have determined the location of the dyke, D. This dyke, from its direction on the map, is probably related to the Mull swarm centred on the Isle of Mull, off the west of Scotland. The unit providing the refracted part of the trajectory of the strong reflected refraction event RR is a basal Zechstein unit with reflection at R. The other reflected refraction seen below RR is related to a sub-Permian unit beneath the base of the Rotliegendes, and shows that the dyke extends below the Zechstein; the reflected refraction above RR is related to an intra-Zechstein unit, which is probably the Plattendolomit. The dyke terminates upwards in the Lower Tertiary, which establishes its age.

5.7 COMMENT

This chapter should not be thought of as closing the subject of the salt/fault relationship. Rather, it should be regarded as a preliminary to the discussion of diapirism, since some of the effects and mechanisms dealt with are of basic importance to the understanding of salt piercement structures in many instances.

CHAPTER 6

Salt Dissolution and Collapse Structures

6.1 INTRODUCTION

The study of structures produced by the collapse of sediments attendant upon the dissolution and removal of underlying salt is largely a recent development. It has arisen partly from observations made during the survey of the deep sea bed by sonar and other means in many areas over the last twenty years, and partly from the increasing availability of good quality seismic data.

From sea bed surveys, for example, circular and sub-circular depressions on the floor of the Red Sea (Schoell *et al.*, 1974; Bäcker *et al.*, 1975), the Orca Basin (a large area of apparent subsidence in the Gulf of Mexico—Trabant and Presley, 1977), and similar types of feature at the sea bed in parts of the Eastern Mediterranean (Belderson *et al.*, 1978, and others), have all been ascribed to collapse associated with leaching of underlying salt in evaporite sequences. A useful review of the literature on this subject is given by Taviani (1984).

As regards the occurrence of deeper-seated dissolution and collapse effects, developments in seismic technology over the last two decades have revealed the widespread nature of dissolution collapse features associated with salt deposits. Lohmann (1972) drew attention to the presence of such features in seismic data of the southern North Sea related to the Zechstein salt interval and, since then, other examples have come to light. Even earlier, and based mainly on the evidence of drilling, numerous features had been described in Canada related to Devonian evaporites which, unlike North Sea examples, did not involve any halokinetic movement of the salt.

Identification of a structure as being due to dissolution collapse requires positive evidence of the leaching of underlying salt. Other processes operate which produce collapse or pseudo-collapse features without the necessity of salt being present. As one example, karst formation due to the leaching of carbonate rocks either at the sea bed or at depth may simulate the evaporite-related processes, and there will be later discussion of what is believed to be a buried karst surface due to areal leaching of the Top Chalk surface in the southern part of the North Sea. Another example is the subsidence effect associated with escape structures such as those reported by Glennie and others in the Rotliegendes sediments in the North Sea, where the deepening Zechstein Sea caused the burst-out of trapped air from dune sands; other possible causes due to dynamic sedimentary processes also exist.

6.2 PROCESSES INVOLVED IN DISSOLUTION AND COLLAPSE

In order that dissolution and collapse may take place, certain basic conditions must be fulfilled. These are that water which is undersaturated with respect to the solute mineral should have means of ingress to bring it into contact with the mineral, and means of egress in order that it can remove the mineral in solution. Thus, a circulatory system is required with a continuous supply of ‘fresh’ water entering the system; and a continuous flow of brine leaving it (‘fresh’, in this context, meaning undersaturated).

This circulatory movement of water is vital, since without it no extensive removal of material can take place. ‘Standing’ water in contact with halite will cause only minimal dissolution of the salt which will cease when the water becomes saturated with respect to sodium chloride (and any other minerals present). According to Baar (1977), salt rocks *in situ* are impermeable at depths greater than 300 m, and therefore a saturated brine will remain in contact with the top of a salt layer at greater depths than this without causing any net dissolution, if the water is static.

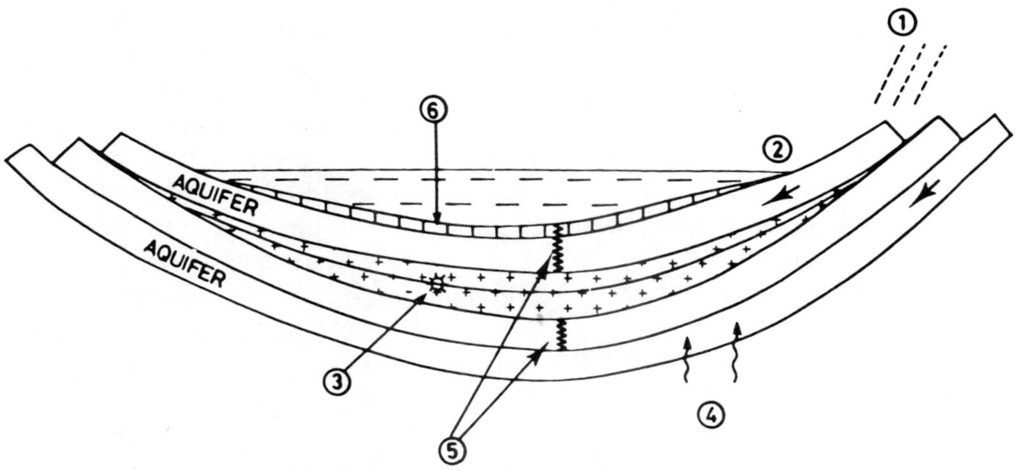

Fig. 6.1 Schematic showing possible sources of undersaturated water.

Several possible sources of undersaturated water exist in different circumstances. Where a basin which includes a salt interval has been uplifted at a later stage in its history (Fig. 6.1), formations above and below the salt may have become aquifers under artesian pressure by entry of meteoric waters at the basin periphery. Sea water itself has an average salinity of 35 000 ppm, of which nearly 80% consists of NaCl. However, hypersaline brines are known, as in the Dead Sea, with salinities of over 250 000 ppm. With this range of variation, and changes of solubility with temperature, sea water brines will normally be undersaturated with respect to NaCl. Thus brine may enter aquifers above and below the salt interval at the basin margin, or through fault systems.

Another possible source of fresh water was pointed out by Borchert and Muir (1964) as resulting from the conversion of gypsum to anhydrite, which occurs when gypsum is subjected to the heat and pressure attendant upon burial. In this conversion, 1 m^3 of gypsum releases 0·486 m^3 of water when replaced by 0·62 m^3 of anhydrite. In a developing basin containing several evaporite cycles, bands of gypsum may exist above and below salt layers. Their subsequent conversion to anhydrite may thus be accompanied by the release, either laterally, or vertically through fractures, of quite enormous quantities of water.

Other possible provenances of relatively fresh water under special geological circumstances are juvenile and connate waters, which might also be available as migrating solvents for at least a restricted period.

The sources mentioned are represented in the simple sketch of Fig. 6.1. The numbers in circles indicate the following sources and features: (1) meteoric water entering the aquifers at the basin margin, (2) brine entering aquifers at the basin margin, (3) water released by the gypsum/anhydrite conversion, (4) entry into the system of juvenile and connate waters, (5) fracture/fault systems, or permeable zones allowing the circulation of water under artesian pressure and brine under gravity control (see later), (6) impermeable ocean floor interval not reaching basin margin; undersaturated brine may still access the aquifers through fractures or at basin margin. Cross symbol is salt, dashed symbol is sea water.

Details of circulatory systems involved where salt dissolution takes place may be extremely complex, and are frequently not determinable from available information. Fracture systems not involving faults of large throw, and higher permeability zones forming parts of a circulatory route, are not normally detectable in seismic data, and little work has been done on this aspect. The most detailed and interesting study carried out recently has been published by Anderson and Kirkland (1980). A basinal aquifer of the Delaware Mountain Group, and the separate Capitan Reef aquifer in the tilted Delaware Basin of West Texas and SE New Mexico, USA, are involved in the system. Salt rock, mainly of the Castile and Salado formations, has been removed in solution, leading to blanket (intraformational) and chimney, or pipe (transformational) brecciation, and collapse structures. A much simplified sketch of the circulatory system as it is interpreted by the above-named authors is shown in Fig. 6.2. Water under artesian pressure, which has entered the basin aquifer of the Delaware Mountain Group (left side of sketch, large arrows) has moved upwards through fractures to contact salt, which has been removed in solution, leading to collapse and brecciation in the overlying basin, these collapse structures becoming progressively younger to the east (right). A western dissolution wedge (W) affecting the

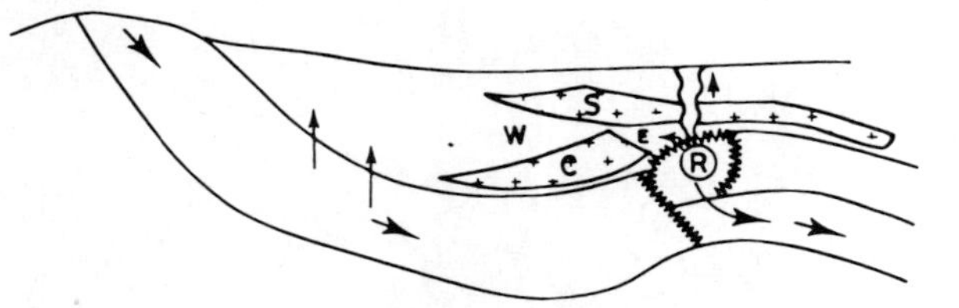

Fig. 6.2 Sketch of dissolution/collapse circulatory system. (Much simplified, after Anderson and Kirkland, 1980, courtesy of the authors and the GSA, from *Geology*.)

bedded salts of the Castile Fm. (C) may be partly due to downdip gravity flow of this water, and partly due to surface drainage, with the resulting brines (of higher density than the incoming water) leaking down into the basinal aquifer. The separate Capitan Reef aquifer (R) has produced an eastern dissolution wedge (E) by lateral flow of fresh water, and overlying collapse structures due to artesian upward flow contacting the Salado Fm. (S), with denser brines percolating downwards, again into the basinal aquifer at the base of the reef. The authors came to the conclusion that

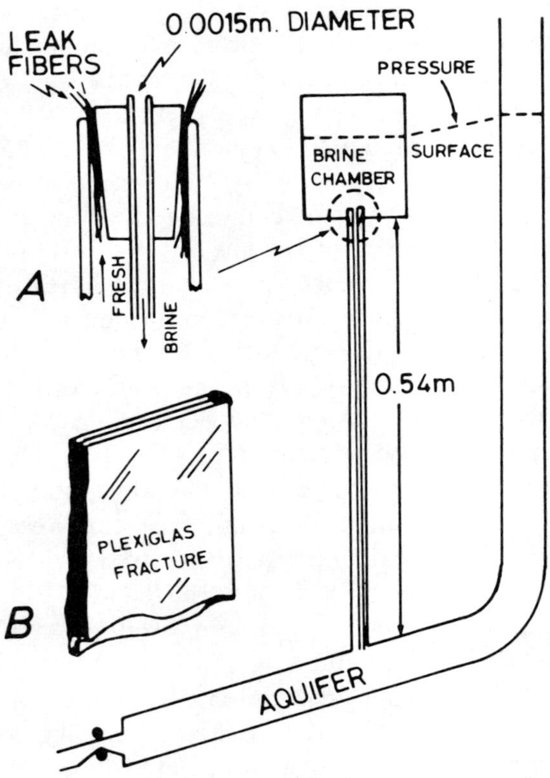

Fig. 6.3 Sketch of apparatus used to demonstrate brine density flow by Anderson and Kirkland (1980). Fresh water under a head of pressure flows from the 'aquifer' up the tube into a chamber containing salt. Brine, in one arrangement (A) flows down a capillary through the centre of the inlet tube. In arrangement B, the exchange is, instead, via two thin plexiglas plates representing a fracture. (From Anderson and Kirkland, 1980, courtesy of the authors, and the GSA, from *Geology*.)

the density of the brine resulting from salt dissolution may provide part of the driving force of circulation due to gravitational effect. This conclusion was supported by laboratory experiments in which fresh water under a head of pressure flowed upwards between two plexiglas plates (representing an artificial 'fracture') separated by 0·001 m or less, into a chamber containing salt. Two-way flow developed readily and continuously with concentration differences between the inflow and outflow of less than 2 g/litre, with the brine flowing downwards through the plexiglas 'fracture' along different paths to the uprising fresh water also passing between the plates (Fig. 6.3). The authors made the point that a complex fracture zone in the real earth would provide ideal upward and downward migratory paths for incoming fresh water and outflowing brine.

The process of brine density flow may explain at least some of the puzzling features seen in certain dissolution/collapse situations. The ultimate destination of the outflowing brines is another problem for which there are probably as many different answers as there are structures. In some instances, where the incoming fresh water is artesian meteoric water, the brine may find its way out of the system at outcrop in other parts of the basin, or be driven upwards into other parts of the system by the head of artesian pressure through fractured and brecciated rocks, and eventually to the surface. In some instances (e.g. Oliver and Cowper, 1983) it is believed possible that it simply moved laterally into permeable subsurface reservoirs, or alternatively vertically downwards through older, fractured rocks and then laterally through sandstones.

When salt is removed by dissolution, the overlying formations tend to collapse, usually due to slow subsidence in concert with the removal of material. It would appear, from drilling evidence, that brecciated units, often resting at their base on the subsalt formation where salt removal has been complete, commonly result from this subsidence. These collapse breccias have been well observed related to the Prairie Formation evaporites of Saskatchewan (De Mille *et al.*, 1964). The lower parts of these breccias consist of rather fine materials some of which, both as fragments and matrix, are likely to be insolubles from the salt interval. Upwards through the formation, the breccia coarsens, with less matrix material, eventually grading into fractured but unbrecciated rock. The breccias in this area are not found where the salt is still present but, significantly, they are widespread as intraformational units in areas adjacent to the present salt location which suggests that the Prairie salt previously existed over a much larger area.

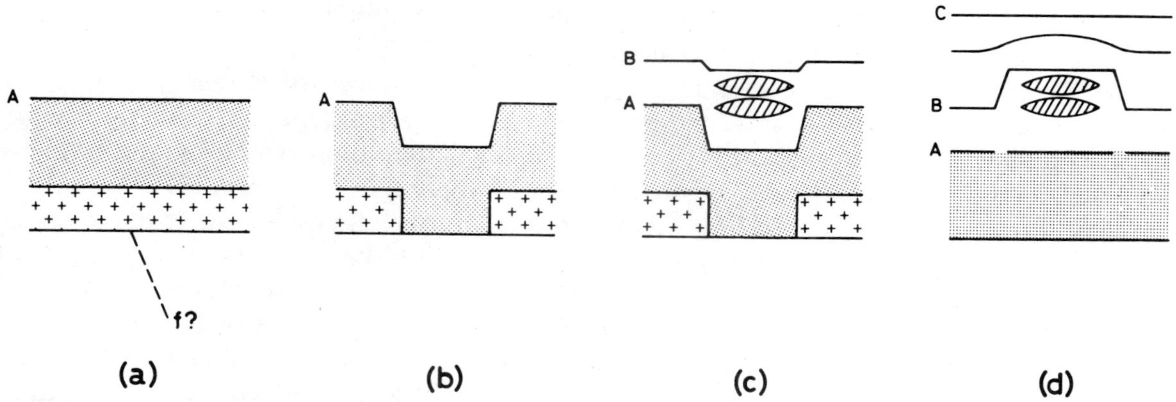

Fig. 6.4 Schematics showing development of a 'sombrero' structure. (a) Movement of a deep fault allows access of fresh water to a salt layer (cross symbol). (b) A localised dissolution feature of sink-hole type results, with subsidence of clastic overburden plug (dot shading). (c) During subsidence, further deposition is taking place, with compensatory thickening (represented by cross-hatched lenses) over the low. (d) After dissolution of all the surrounding salt at a later stage, overburden subsidence leaves the original low as a positive feature.

In the example just quoted, no salt movement of a plastic kind was involved. The Prairie Evaporites have remained *in situ* as deposited, and have not been subjected to any deformation related to salt rock mobility. Where major salt movement has occurred resulting in large-scale fracturing of the overburden, it appears that much coarser breccias than those just mentioned—even up to the scale of megabreccias—may result in any collapse structures which are formed. Examples of dissolution collapse graben which suggest this will be seen later in the chapter.

6.3 COLLAPSE RELATED TO DISSOLUTION OF NON-MOBILE SALT

The area of Saskatchewan mentioned in Section 6.2, and adjacent parts of Canada and the northern USA, form excellent study areas for dissolution effects in non-mobile salt deposits, and much work has been done by Canadian and American geologists. The areal dissolution effects with intraformational breccias occurring were mentioned in the previous section. Other, much more localised effects are also seen which appear to be particularly common in this type of 'static' salt environment, but which seem to be rare in areas where salt mobility has led to overburden deformation.

A feature of these localised dissolution/collapse structures is often the evidence of multi-stage dissolution over a comparatively long geological period. The reasons for the repetitious commencement and cessation of dissolution effects are not clearly established in most cases, usually because the nature of the water circulation system causing it cannot be established. Often the means of ingress of 'fresh' water to the feature is not known. The most plausible hypothesis put forward involves the sporadic movement of deep-seated faulting beneath the formations containing the salt. It is supposed that flow of fresh water up through the fault zone (and any permeable beds intervening between it and the salt layer) is initiated and terminated by movements of the faulting at different periods.

A schematic sketch of one type of structure, known because of its final form as a 'sombrero' structure, is shown in Fig. 6.4. An initially uniform layer of salt is subjected to dissolution in a very localised way, which

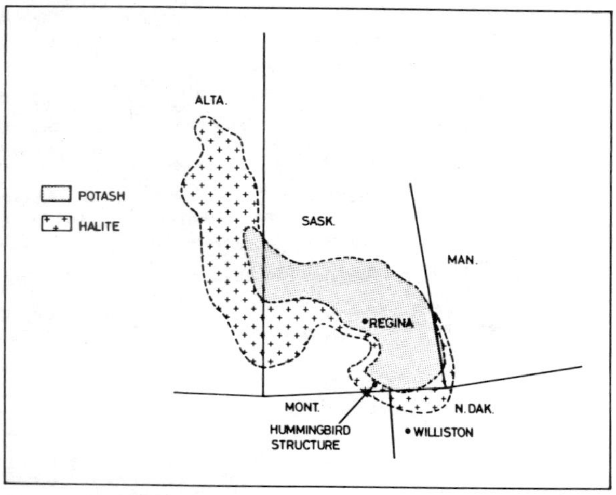

Fig. 6.5 Sketch map showing location of the Hummingbird structure in relation to the halite and potash depositional areas of the Prairie Formation in Saskatchewan. Anhydrite depositional area not shown. (Generalised after De Mille *et al.*, 1964, Smith and Pullen, 1967, with acknowledgements to the CSPG.)

results in the eventual removal of all the salt over a circular or sub-circular area, with concomitant subsidence of a 'plug' of overburden. Due to compensatory thicker deposition in the depression above the feature during continuing sedimentation, at a later stage when the dissolution of the adjacent salt occurs, the overburden around the feature subsides to leave the sediment plug as a positive feature in depth.

The Mid-Devonian Elk Point evaporite basin extends from North Dakota in the USA through Saskatchewan into Canada's Northern Territories (De Mille *et al.*, 1964). Within the Elk Point Group, the Prairie Formation mentioned in Section 6.2 consists of some 670 ft (200 m) of halite with potash beds in its upper part (see sketch map in Fig. 6.5). In southeastern Saskatchewan, the Hummingbird structure is an interesting example of multistage dissolution and collapse which has resulted in a positive feature of the type just discussed; it is of particular interest in that an oilfield has been found associated with it.

The sequence of events postulated (Smith and Pullen, 1967) is that localised leaching of the Prairie salt from below, due to small, intermittent movements of basement faults, aided by the presence of the

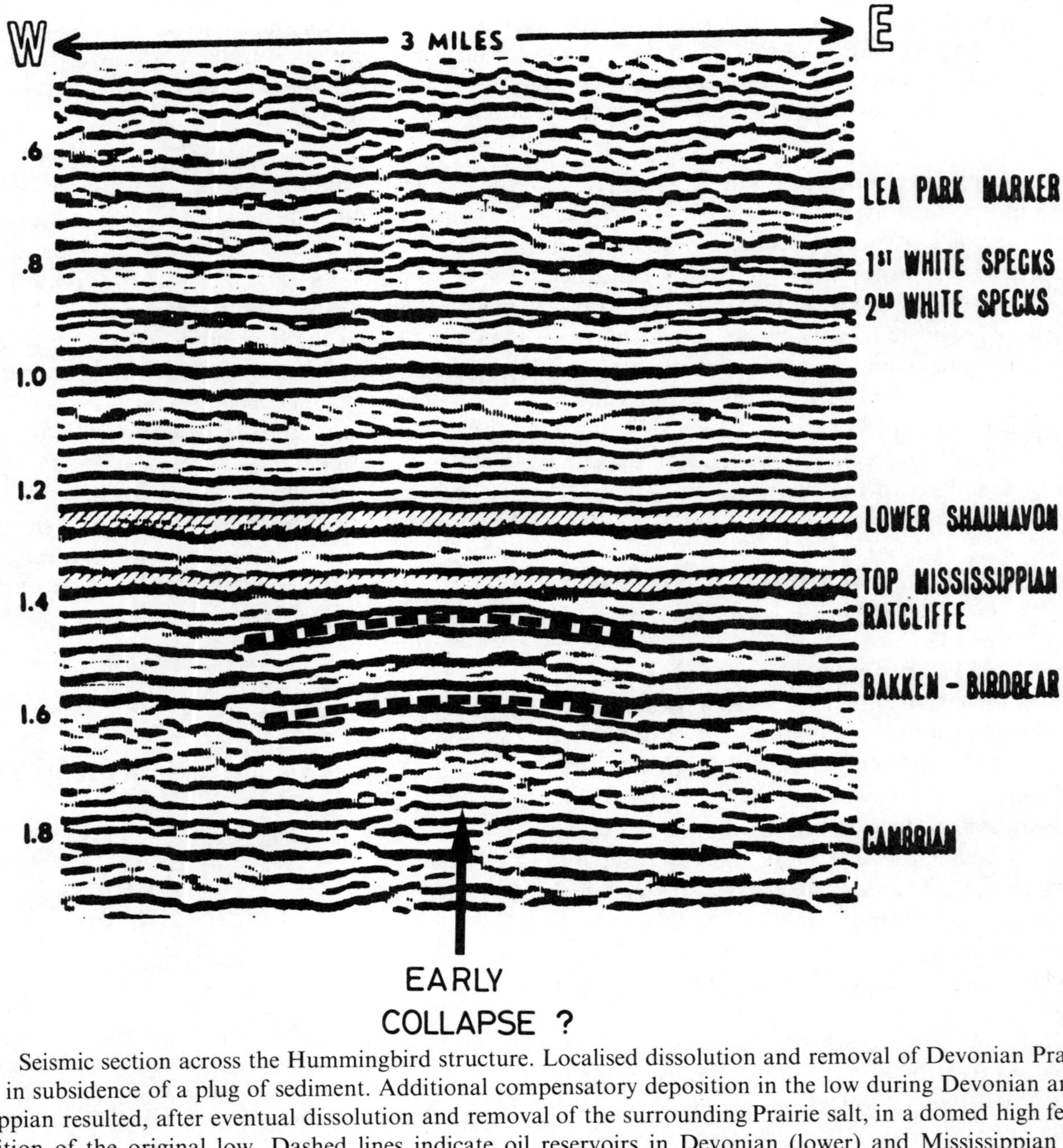

Fig. 6.6 Seismic section across the Hummingbird structure. Localised dissolution and removal of Devonian Prairie salt resulted in subsidence of a plug of sediment. Additional compensatory deposition in the low during Devonian and early Mississippian resulted, after eventual dissolution and removal of the surrounding Prairie salt, in a domed high feature in the position of the original low. Dashed lines indicate oil reservoirs in Devonian (lower) and Mississippian (upper) intervals. Events arrowed may represent earliest collapse over a restricted area before the sink-hole reached its full diameter. (From Smith and Pullen, 1967, slightly modified, with acknowledgements to the CSPG.)

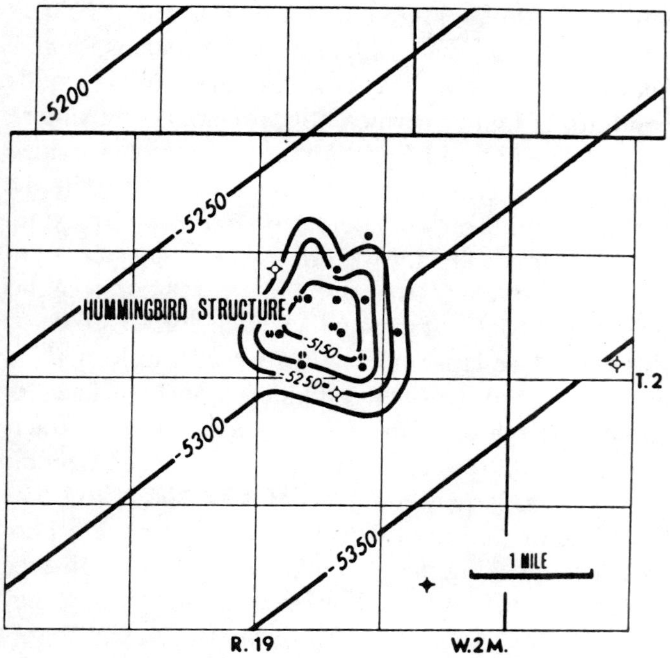

Fig. 6.7 Structure map at the lower (Devonian) reservoir level in the Hummingbird structure, showing the very localised nature of the processes which produced it. (From Smith and Pullen, 1967, with acknowledgements to the CSPG.)

overlying permeable Winnipegosis carbonates upon which the salt rests, allowed formation water access to the salt in a restricted area, resulting in a subcircular zone of dissolution during Middle to Upper Devonian times. The resulting depression was filled partly by overburden collapse, and partly by deposition of an additional 240 feet of Upper Devonian sediments in two or more stages. In early Mississippian time, all the remaining Prairie salt in the depression was removed and there was further collapse and compensatory deposition. After further regional sedimentation, and in Late Jurassic–Early Cretaceous times, all the Prairie salt in the surrounding area was removed by dissolution. This resulted in the subsidence of the surrounding overburden areally, and left the Hummingbird feature as a structural high due to the additional compensatory sedimentation which had occurred. Figure 6.6 shows a seismic section across the structure, and its very localised nature is seen from the structure map at the level of the Upper Devonian pay formation (the lower of two reservoirs) shown in Fig. 6.7.

In such a situation, even when after the localised dissolution and collapse has occurred, the surrounding salt does not undergo areal dissolution but remains effectively as deposited, the compensatory thicker sedimentation in the depression above the salt dissolution 'sink-hole' together with differential compaction can lead to anomalous local thickenings in some units of the overburden sediments. An example has been described by Oliver and Cowper (1983) in the Rumsey area, Alberta, this time associated with dissolution of the Upper Devonian Wabamun salt. In this case, the process resulted in a very small residual low feature on an Upper Cretaceous horizon, but it is not difficult to imagine how, with a different time scale of deposition, a residual high feature may result after differential compaction in such circumstances, even when the surrounding salt has not been removed (Fig. 6.8). See also Fig. 6.9.

Perhaps the most difficult aspect to understand in structures of this type is the reason for their very localised nature. If they are associated with the

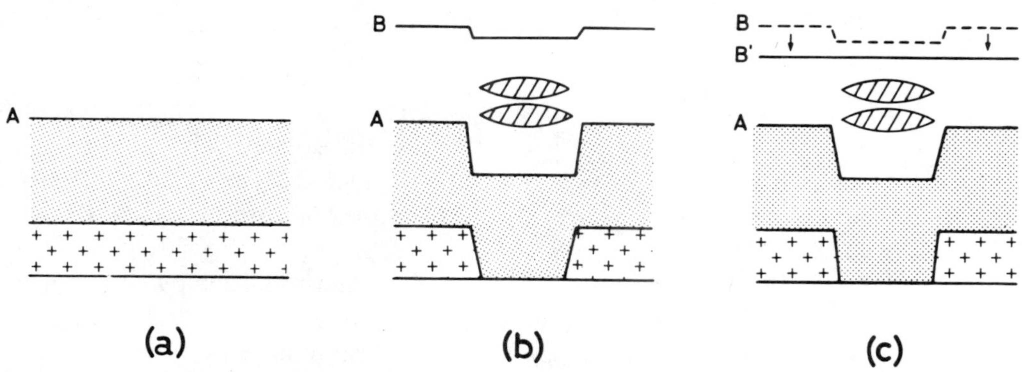

Fig. 6.8 Schematic showing development of another type of localised dissolution and collapse structure, exemplified by the removal of Upper Devonian Wabamun salt in a location in Alberta described by Oliver and Cowper (1983). Developmental stages (a) and (b) follow those of the Hummingbird feature (see Fig. 6.4). Thereafter, however, no general dissolution of the surrounding salt occurs, but because of compensatory thicker deposition above the collapse, the eventual surface depression (B) is of much smaller amplitude than the initial collapse depression beneath, and after compaction of the overburden shales, little or no depression remains (reference surface B′). Variations on this depositional pattern could produce a high feature above the collapse, as in the Hummingbird structure.

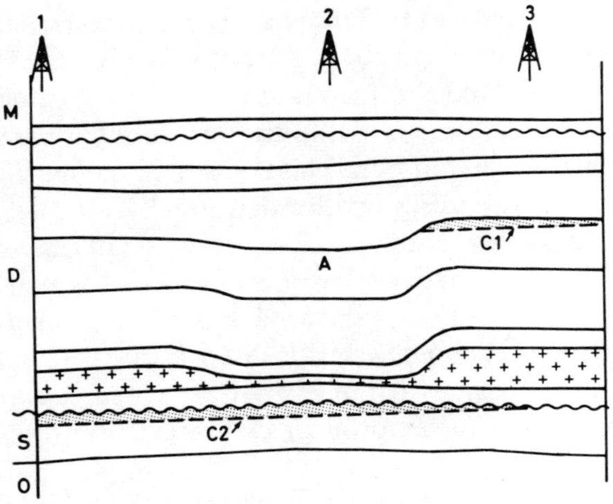

Fig. 6.9 Schematic of another Prairie Evaporite-related structure in Montana—the Outlook Field. Two oil reservoirs have been located, the lower being in a Silurian interval (o/w contact C2), and the upper in a Devonian interval (o/w contact C1). Because of the conformation of overlying strata in relation to the Siluro-Devonian unconformity surface, the C1 oil pool originally lay above the C2 accumulation. Subsequent dissolution of the Prairie Halite (cross symbol) led to collapse of the overburden, and migration of the oil to its present position. The operation of this process was suggested by a drill-stem test in Well 2, which recovered minor oil and gas shows with sulphurous brine in the same stratigraphic interval at A as that in which the C1 oil accumulation is found in Well 3. (Redrawn from Parker, 1967, courtesy of the AAPG.)

movements of faults at depth, it is not easy to appreciate how they come to have the circular, or sub-circular form often shown. Faults are linear features, and any effects of dissolution could reasonably be expected to conform to this by having an elongate configuration, as do some features to be discussed in Section 6.4. One possible explanation, as advanced by

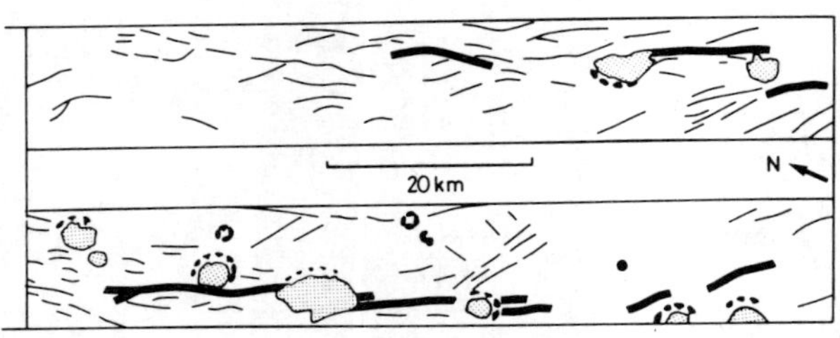

Fig. 6.10 Tracing of long range side-scan sonographs (two-way-scanning, with ship's track down blank central stripe) on the Hellenic Outer Ridge west of Crete in the Mediterranean. Sub-circular, rough patches, some connected by narrow, sinuous belts (heavy lines) are believed to be collapse locations related to leaching of underlying Messinian evaporites. Thin lines are bottom indications of other principal tectonic trends. (Redrawn after Belderson *et al.*, 1978, courtesy of the GSA, from *Geology*.)

Smith and Pullen (1967), could be that they occur at intersections of two different fault systems, where additional and more extensive fracturing may provide conduits of higher permeability for migrating water.

An interesting sidelight on this problem is provided by the investigations of sea bed irregularities in the eastern Mediterranean (Belderson *et al.*, 1978). Long-range sonographs in the area have detected sub-circular to oval craters, as well as elongate, irregular troughs, of rough ground which have been interpreted as areas of collapse breccia incorporating Messinian evaporites and Tertiary sediments above surfaces of submarine salt dissolution (see Fig. 6.10). It is understood that more recent investigations by submersible surveys have largely confirmed this interpretation. The sub-circular and oval features would seem, morphologically speaking, to be similar to the Canadian-type features discussed. It is, however, of interest that they appear to be associated with linear features probably of tectonic nature—perhaps faults, or folds—which does not seem to be the case (from evidence as presented) in the examples in Canada. However, comparisons on a purely morphological basis are hazardous in this, as in many other fields of study; the similarity is mentioned here purely as a matter of interest.

6.4 OTHER TYPES OF DISSOLUTION/ COLLAPSE FEATURE

Apart from the types of localised feature discussed in Section 6.3, others exist, some of which are related especially (although not exclusively) to basins with mobile salt, and others which may occur associated with any salt deposits, mobile or not. These are:

(a) Localised 'sink-hole' features similar to the Canadian examples of Section 6.3 but related to an isolated salt structure.
(b) Linear features related to elongate and salt-wall structures associated with faulting.
(c) Areal dissolution of the salt—the *Salzspiegel* of German geologists—causing widespread, relatively uniform subsidence of the overburden accompanied by blanket brecciation.
(d) Edge dissolution effects at the periphery of a salt basin.

Types (a) and (b) commonly originate due to the faulting attendant upon salt movement, while (c) and (d) occur in any salt basin. The principal 'style' of feature present varies from one salt basin to another, or even within the same basin, depending often on the tectonic setting. In the southern North Sea, for example, type (b) features are the most prevalent within the basin proper, since by far the majority of the salt features present are related to reactivated regional faults, and there are very few salt features present of the form noted in type (a). (It should be pointed out that where such features are located on a seismic line in a widely spaced survey grid, an elongate feature in cross-section will have the appearance of a sink-hole type structure; it is not known whether the examples shown, for instance, in Lohmann's (1972) paper as sink-hole type features were definitely established as such.)

The four types of feature mentioned are often found in various combinations in a single basin. In a general way, the Canadian examples of Section 6.3 can be related to types (a) and (c), with the rider that the type (a) features came about by a process not related to salt uplift. Also, type (c) and (d) features are present in many, perhaps most, salt basins, since they do not relate to situations brought about by salt movement. Similarly, features of types (a) and (b) can be envisaged as occurring in areas where salt uplift and overburden fracturing have resulted from tectonic movement, rather than plastic mobility of the salt.

In a salt basin where plastic salt movement has taken place, however, the details of the movements involved become most important for the interpretation of collapse structures. In some locations, major salt thickening and uplift results in large-scale fracturing of the overburden, allowing free migration of undersaturated water and resulting in enormous collapse features in the overburden. In other circumstances, quite minor late-stage movements of salt can result in complex and subtle collapse features in the overlying sequences which often appear to have no connection with the deformation in the salt interval until subjected to very close scrutiny.

It is also of importance in such areas to distinguish

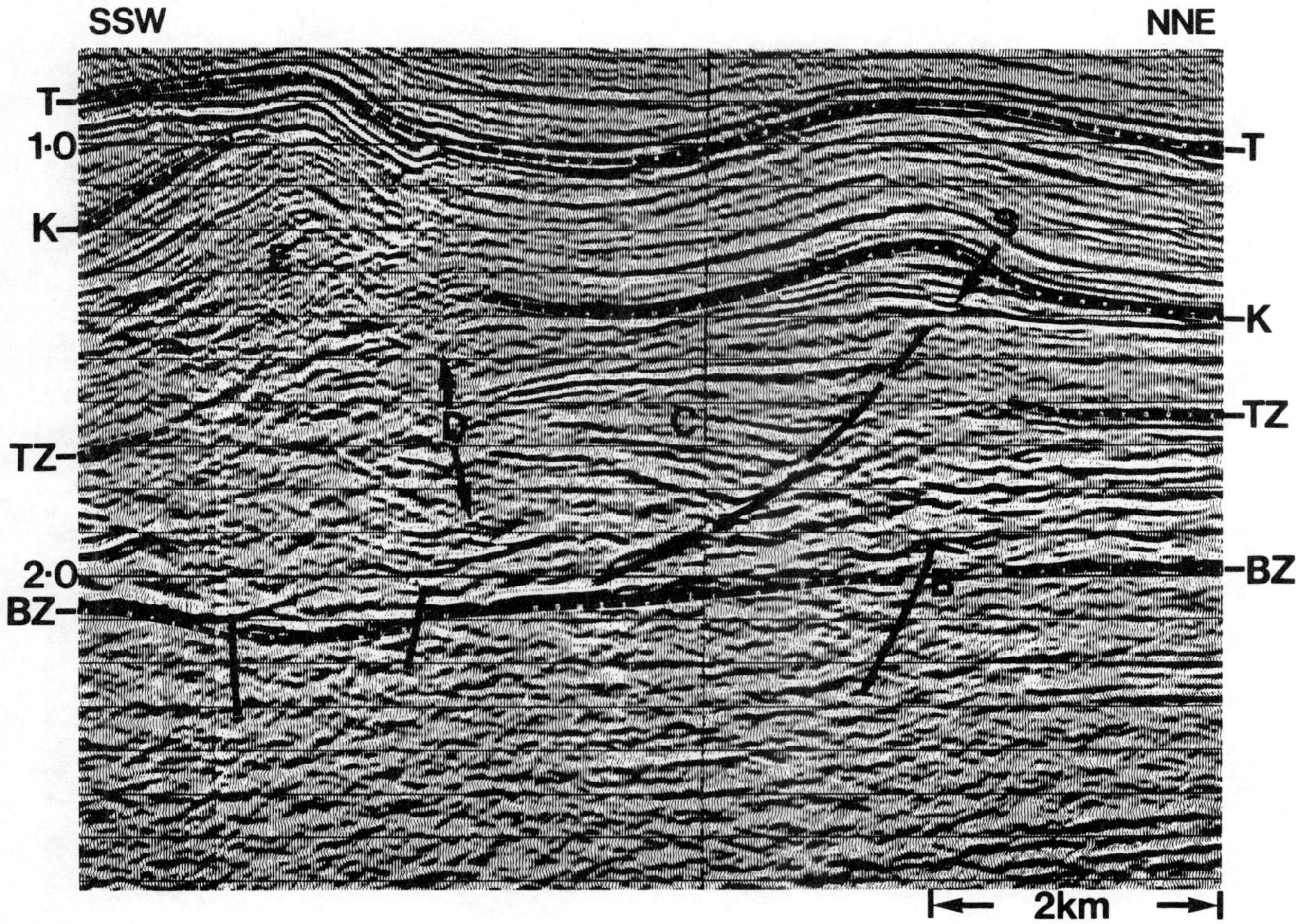

Fig. 6.11 Migrated seismic section showing subsidence of a wedge of overburden (C) related, not to dissolution/collapse, but to flow of underlying salt laterally to the left, and upwards (D and E) in the section. Salt interval within BZ–TZ on left. The BZ–TZ interval on the NNE side of the feature consists mainly of marginal (shelf) facies. (Courtesy of SS(E)L.)

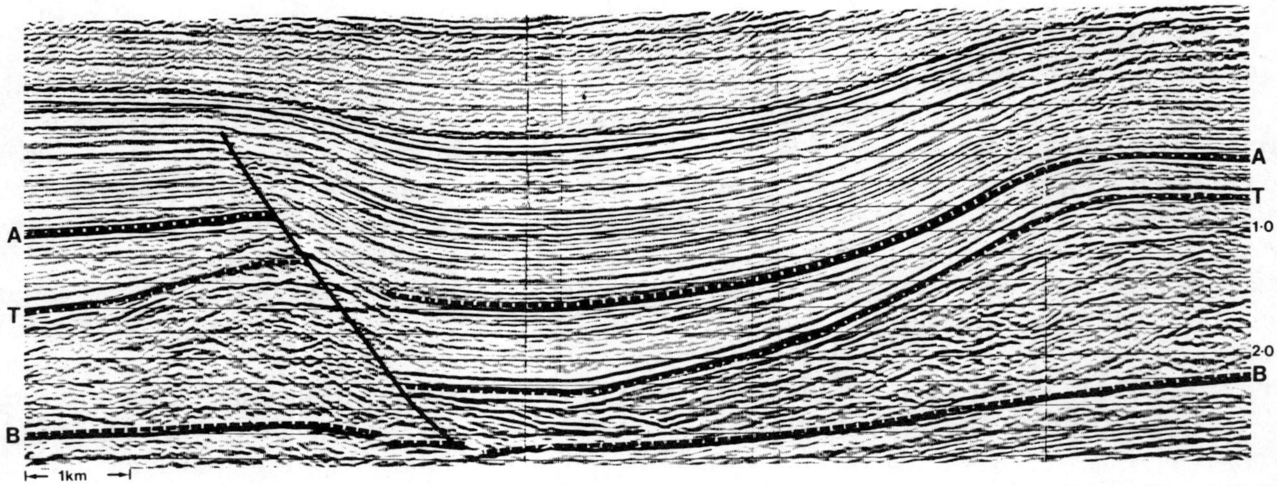

Fig. 6.12 Migrated seismic section. Another example of subsidence due to salt flow. Salt interval is within B–T. The half-graben structure is due to flow of salt into the pillow on the right. The top of the salt (event T) is intact (apart from at the fault location), indicating that no dissolution/collapse effects are present. The thickening of the salt into the fault zone on the upthrow side probably indicates salt flow up the fault zone due to differential loading—a mechanism which is considered in more detail in the chapter on faulting and salt movement. (Courtesy of SS(E)L.)

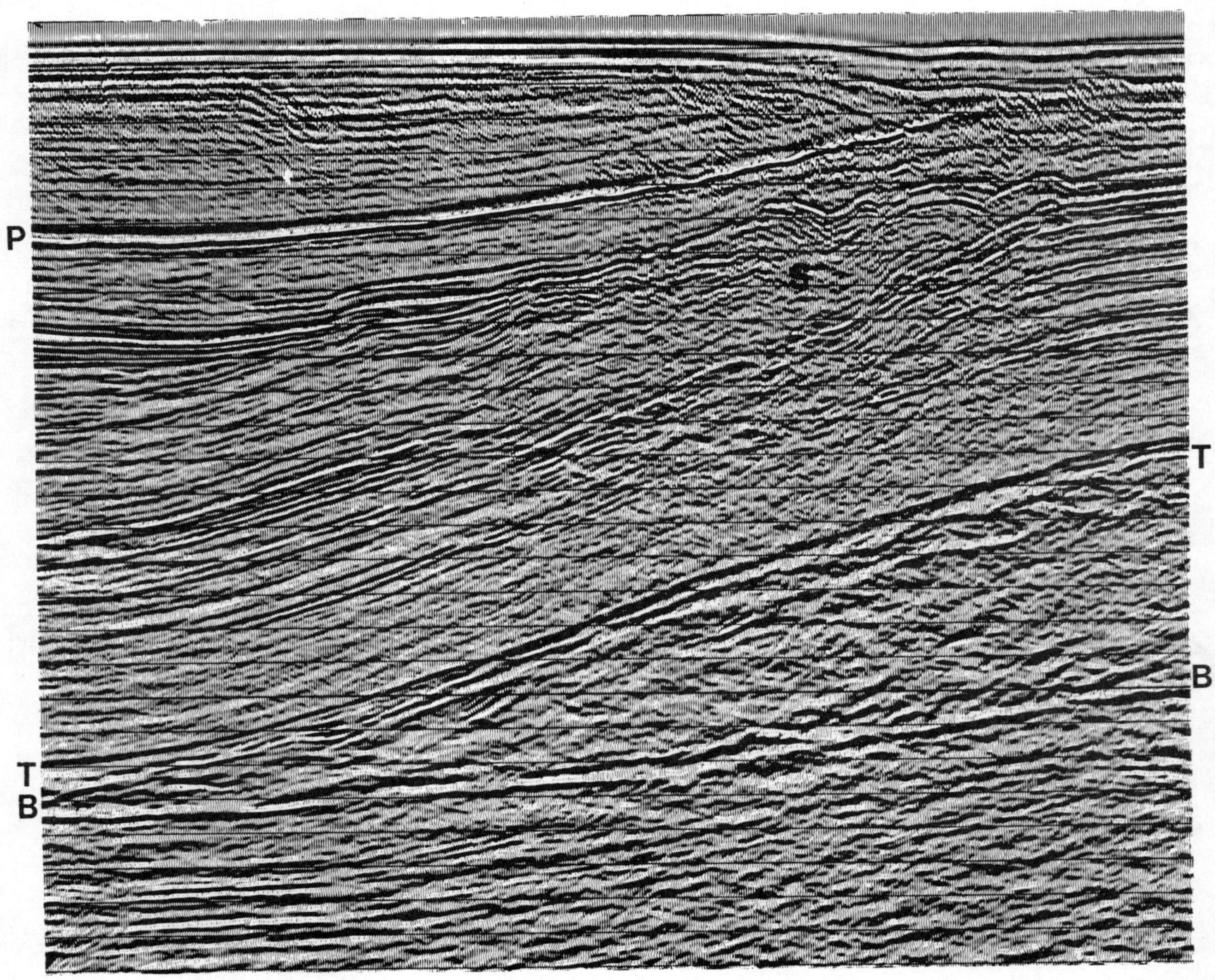

Fig. 6.13 Migrated seismic section. Mass movement—another collapse effect not related to dissolution. Undercompacted sediments have slumped, as indicated by the chaotic events around S in the section. The flank of the salt pillow (interval B–T) has exceeded the angle of rest of Mesozoic sediments below the Base Tertiary (P). The more indurated Triassic sediments immediately above T show no signs of slumping. (Courtesy of SS(E)L.)

between subsidence/collapse features due to dissolution of the underlying salt, and similar features due to withdrawal of salt laterally by flowage. In Fig. 6.11 the wedge of overburden marked C is a subsidence feature which is believed to be due entirely to the withdrawal of previously underlying salt laterally (towards the left of the section), and also vertically (D) into a diapir which was later removed in the erosional period represented by the unconformity K. (Sometimes effects due both to salt withdrawal and to dissolution are combined in the same structure, and only a very careful analysis of the history of the salt movement will enable effects to be attributed to the two causative processes.) Figure 6.12 is another example.

Another type of collapse sometimes seen in areas of salt movement, and which is not related to salt dissolution, is slumping. This can occur in the overburden on the flanks of a rising salt feature, where the overburden sediments are, in parts of the section, relatively unconsolidated. Figure 6.13 shows an example where the steepest part of the flank of a salt pillow has exceeded the angle of rest for a major part of the Mesozoic sediments above; contorted slumping and slide features can clearly be seen, particularly in the higher parts of the section almost up to the base of the Tertiary. The Triassic section immediately above the salt interval appears to be essentially intact, indicating that the sediments are well compacted and indurated due to greater depth of burial and minimal residual pore fluids. Greater permeability and high formation water content have doubtless influenced the instability of the overlying sediments. Further examples of mass movement phenomena will be shown in later chapters.

6.4.1 Fault and Fracture Related Collapse

Access of undersaturated water to salt will be most readily achieved where a fracture system exists in the rocks above or below the salt interval, as for instance where a joint system is present. Such fracturing can also be caused by tectonic activity, or movement of the salt. Slight flexing by either type of movement may be sufficient to cause microfracturing, while greater stress will produce macrofractures and faults. Evidence of faulting will often be seen on the seismic section. On the other hand, stresses which produce jointing, microfractures, or faults of small throw (including strike-slip faults with little or no dip-slip component) will not leave evidence which can be resolved by normal seismic methods, in most cases. This may explain the difficulties mentioned previously in identifying the causes of some structures; possible migratory routes for water cannot be determined with certainty, since these require only fractures, and not necessarily faults of resolvable throw on the seismic section, to operate as a circulatory system.

One of the commonest causes of dissolution and collapse in a mobile salt basin stems from stress-zone fracturing and faulting above rising salt features. This subject was mentioned briefly in Chapter 2 (see Figs. 2.11 and 2.12), and will be dealt with further in the chapter on faulting and salt movement later in the book. Here the associated dissolution effects will be discussed by reference to examples of two more collapse structures of this type shown in Figs. 6.14 and 6.15. In the first of these, a large collapse graben has formed along the top of an elongate salt structure (seen here in cross-section), which is a large pillow of slightly asymmetric form that has invaded the overburden diapirically due to the stress-zone faulting. Pillow formation began in the pre-unconformity (U) period, and continued in the post-unconformity period. The salt was still rising in the post-C (middle unconformity) period contemporaneously with the major dissolution/collapse episode. Effects at the shallow unconformity (A) suggest that, although the collapse effects had ceased by this time, the salt body was still rising slightly at and above this level on the section, possibly indicating that the continuity of the 'plumbing' system was interrupted, or that available water had become concentrated to saturation with salt. Figure 6.15 shows an example where an elongate salt pillow (seen in cross-section) had developed a less strongly marked collapse graben before the major erosional phase represented by the unconformity U. The arrowed event possibly represents a shard of shattered anhydrite/carbonate sequence from the salt roof which has been 'rafted', or carried up within a minor salt invasion of the overburden. The shape of the unconformity indicates, again, later upward movement of the salt. The post-unconformity collapse effects may be due partly to stress fracturing, and partly to the proximity of the unconformity surface to residual invading salt in the graben, allowing the downward percolation of undersaturated water or brine through the unconformity. The apparent onlapping events above the unconformity on the right are strong, shallow multiple reflections; the fainter, dipping events on the left probably indicate the true, concordant dips above the unconformity. A minor point of interest on this section is the strongly dipping,

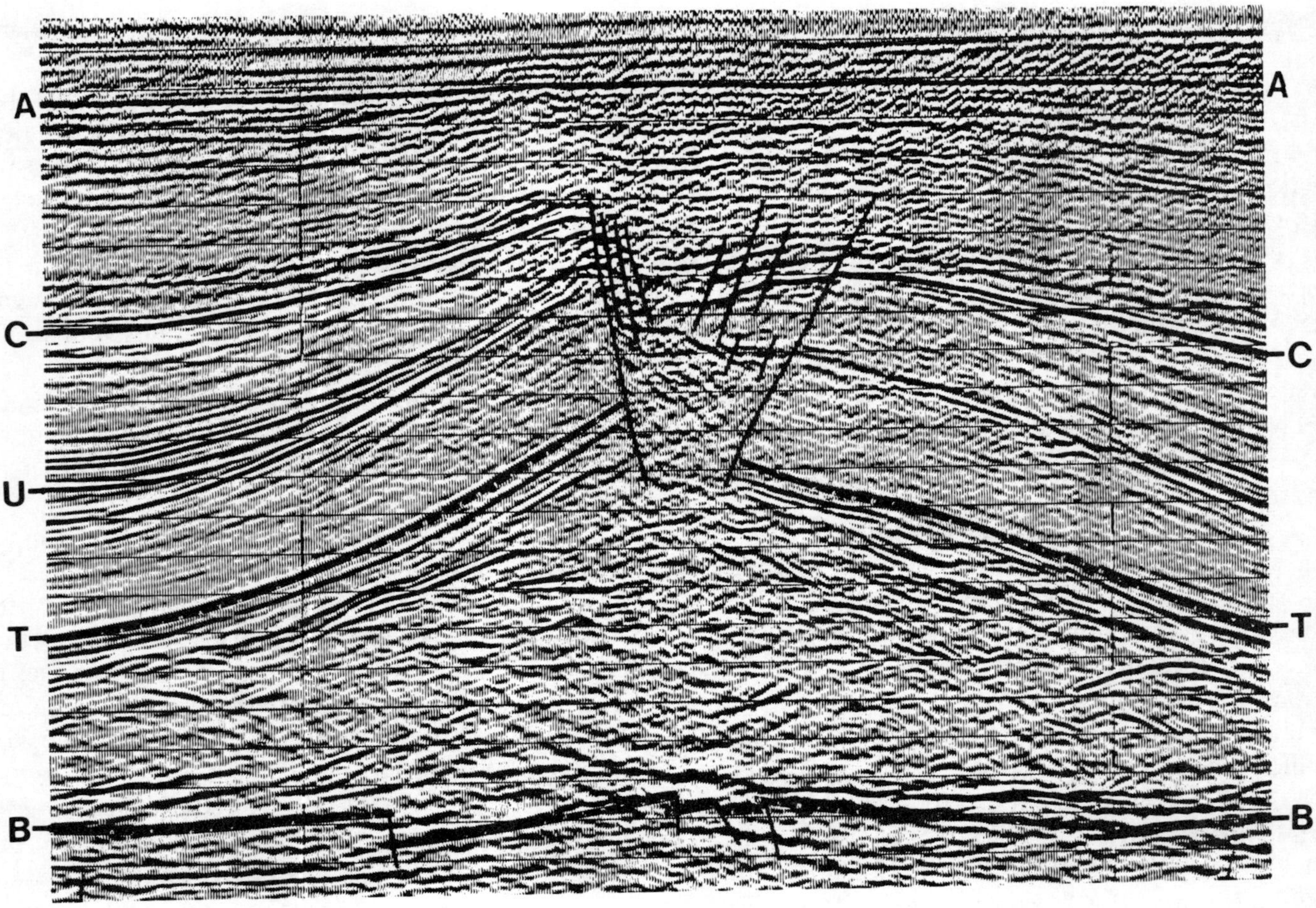

Fig. 6.14 Migrated section. A large dissolution/collapse graben at the crest of a linear salt pillow (seen here in cross-section). Salt interval is within B–T. U, Late Cimmerian unconformity; C, Base Tertiary; A, intra-Tertiary unconformity. (Courtesy of SS(E)L.)

truncated stratal reflection seen at the unconformity at the right-hand end of the section. The same feature is seen at the unconformity over the left flank of the salt feature. At both locations, the truncated event is producing a small feature at the unconformity surface, and must represent a strongly erosion-resistant lithological unit, probably a limestone, in the Mesozoic section.

A similar type of feature, but seen at a much more advanced stage than in the previous examples, is shown in Fig. 6.16. Here the dips of events within the TZ–CU interval on each side of the major collapse structure in the centre of the section give an indication of the early history of the salt movement.

Salt within the BZ–TZ interval thickened into a swell beneath the present location of the collapse graben A. It is probable that movement of the strike-slip fault seen at BZ level below the structure, combined with lateral movement within the salt interval, induced the vertically-upwards movement of the salt here. Stress-zone fracturing took place at a later stage, and salt injection of the faulting resulted in an asymmetric diapiric invasion of the overburden by the salt which lifted the overburden on the right of the structure higher than that on the left, to form a 'trapdoor' diapir. It should be noted that the structure is an elongate linear type, being seen in cross-section in this illustration.

From the configuration of events in and adjacent to the collapse graben in its lower part, it seems likely that the rise of the salt swell/pillow took place at about the time of the erosional phase represented by the unconformity, CU, and that salt probably extruded at the contemporary surface. Very soon after this, dissolution of the salt commenced, and a large volume of salt was removed from the wedge-shaped zone below A. The steeply-dipping, strong events to the left of the letter A are of interest in that they appear to pass continuously into the horizontal events to the left of the graben. To a lesser extent, the same effect is seen to the right of A also. There is just a possibility that these events represent stratal reflections from a

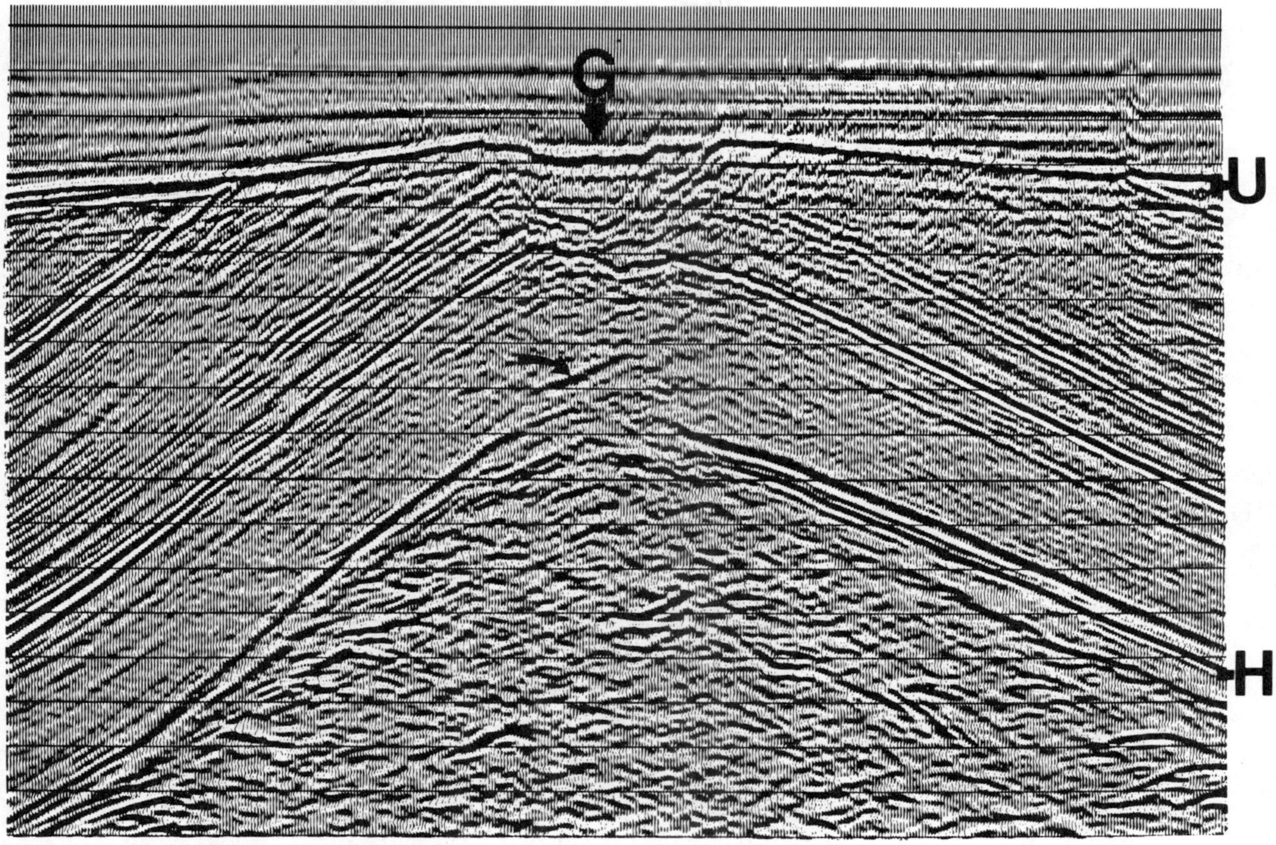

Fig. 6.15 Migrated section. Another dissolution/collapse graben, G, at the crest of an elongate salt pillow, seen in cross-section. H, top of salt; U, unconformity. The arrowed event is probably a rafted segment of the anhydrite/carbonate lithological units capping the salt, which has been carried up in a minor diapiric invasion of the overburden. The dissolution of this invading salt through stress-zone fractures created by the salt pillow has led to the collapse, both in pre- and post-unconformity times. (Courtesy of SS(E)L.)

'flap' of sediments which were laid down over an extruded salt surface, and which subsided to form a kind of 'lining' to the flanks of the graben when the salt was removed by dissolution. If the dip of these events seems large for the sediments to have withstood the effects of gravity, it should be remembered that the vertical exaggeration of the section is about 2·5:1, and the actual dip is much less. It is realised that this is a highly speculative suggestion. However, extruded salt *can* exist at the surface (e.g. in present day Iran), perhaps long enough to be covered, in a very arid environment, by terrigenous deposits.

The later history of the feature appears to be one of continued deposition and subsidence of the superincumbent load of sediments during further dissolution of the underlying salt, as far up as the shallow intra-Tertiary unconformity near the top of the section, resulting in the much wider and shallower collapse seen above A. The small feature and faulting present at the level of the intra-Tertiary unconformity above the high right-hand side of the graben suggests that upward salt movement was continuing on this side of the feature coevally with the main subsidence effects. This continued upward movement probably produced the almost horizontal thrust fault at S.

In the same illustration, the smaller collapse feature at B is also of great interest as an example of a subtle and complex type. A small synform shape occurs at the Base Tertiary/Top Chalk boundary BT which seems to indicate that the basal Tertiary sediments have collapsed into the Top Chalk surface. Vertically below, e.g. at CU level and deeper, there are signs of rather wider and shallower graben-like collapse effects. Superficially, there appears to be no connection between these features and the salt interval (BZ–TZ) below.

A number of features of this type are present in this general area in the southern Zechstein Basin of the North Sea. They are produced by linear collapse channels (seen here in cross-section), which are narrow, but very long and persistent for such small features. One of them has been traced for over 100 km

Fig. 6.16 Major (A) and minor (B) collapse features, discussed in the text. S is a thrust produced by salt uplift deformation which has led to bed repetition as seen by the events in the CU zone. BZ–TZ is the salt interval. TB is the Top Bacton horizon. CU, unconformity. BT, Base Tertiary. (Courtesy of SS(E)L.)

whilst retaining an almost uniform width of about 0·5 km. A feature of such persistence clearly requires some explanation.

It has been observed that they are parallel to the main tectonic trend in the area, which is also imposed on the major salt structures, that they normally occur at locations where the underlying Zechstein salt is locally *relatively* thin (the local thinning of the salt is often very subtle, but exists), and that following from this, the salt thickens laterally away from this minimum in both directions. As a result, there is commonly a distinct dip change in the post-Zechstein bedding reflections so that a shallow V-shaped overburden feature is formed in contrast to the shallow U-shape of the top salt locally.

In the light of these facts, is there then, in spite of immediate appearances, some connection between these minor collapse channels and the movement of the underlying salt?

An interpretation of these channels has been suggested in a paper (Jenyon, 1984a), and is as follows. Due to late, very minor salt movements, synclinal stress-zoning has occurred in the post-Zechstein sequences at these locations, with the compressed fold-core being approximately at the Base Tertiary/Top Chalk boundary. The schematic in Fig. 6.17 illustrates the situation envisaged. The compressional stress acting at this boundary appears to have caused severe fracturing of the top of the Chalk. The compressional stress decreases vertically downwards probably quite rapidly to a surface of null stress, below which a zone of increasing extensional stress is present. It appears that undersaturated water has percolated downwards through the fracture system resulting, and dissolved the disrupted top of the Chalk to produce a sink channel into which the basal Tertiary sediments have collapsed. It should be noted that in the schematic diagram the radius of the zone of greatest compressional stress, the position of the trace of the null stress surface, and the location and width of the zone of extensional stress are suggestions only, since precise indications cannot be given from the seismic data alone.

It seems that vertically below the sink channel, the zone of extensional stress has caused micro- and macro-fracturing. Further dissolution—in this case, of several thin Middle/Upper Triassic evaporite units and also possibly some thin carbonate intervals—appears to have occurred from the observed evidence on the seismic sections.

Another point of interest is the short time interval within which the collapse effects die out in the overlying

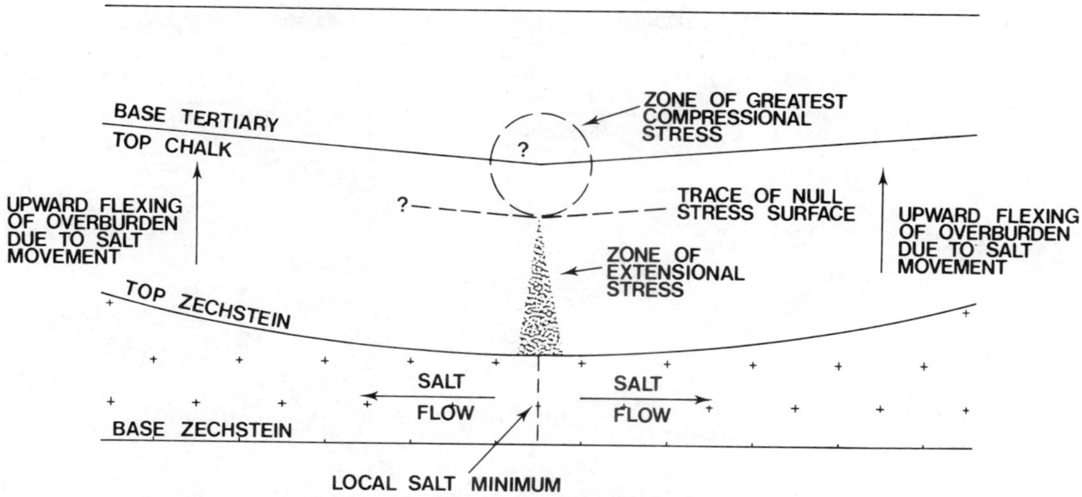

Fig. 6.17 Schematic showing an interpretation of the minor collapse features discussed in the text.

basal Tertiary stratal reflections, usually within a TWT interval of 100–200 ms. This suggests that the minor localised salt movements which probably caused these features can be dated to within quite narrow limits as having occurred shortly after the deposition of the basal Tertiary beds.

In Fig. 6.18 is shown at B another well-developed feature of this type; in this instance, the example is of an unmigrated seismic section, which leaves visible the strong hyperbolas which are being produced by the basal shape of the collapse at the Tertiary/Chalk boundary. The zone of extensional stress below B has led to the development of several wider but time-restricted collapse events, three of which are indicated by arrows. Within this part of the section, three thin Triassic salt intervals are known to exist—the Röt, Muschelkalk and Keuper halites—and it is also possible that thin carbonate intervals may have been

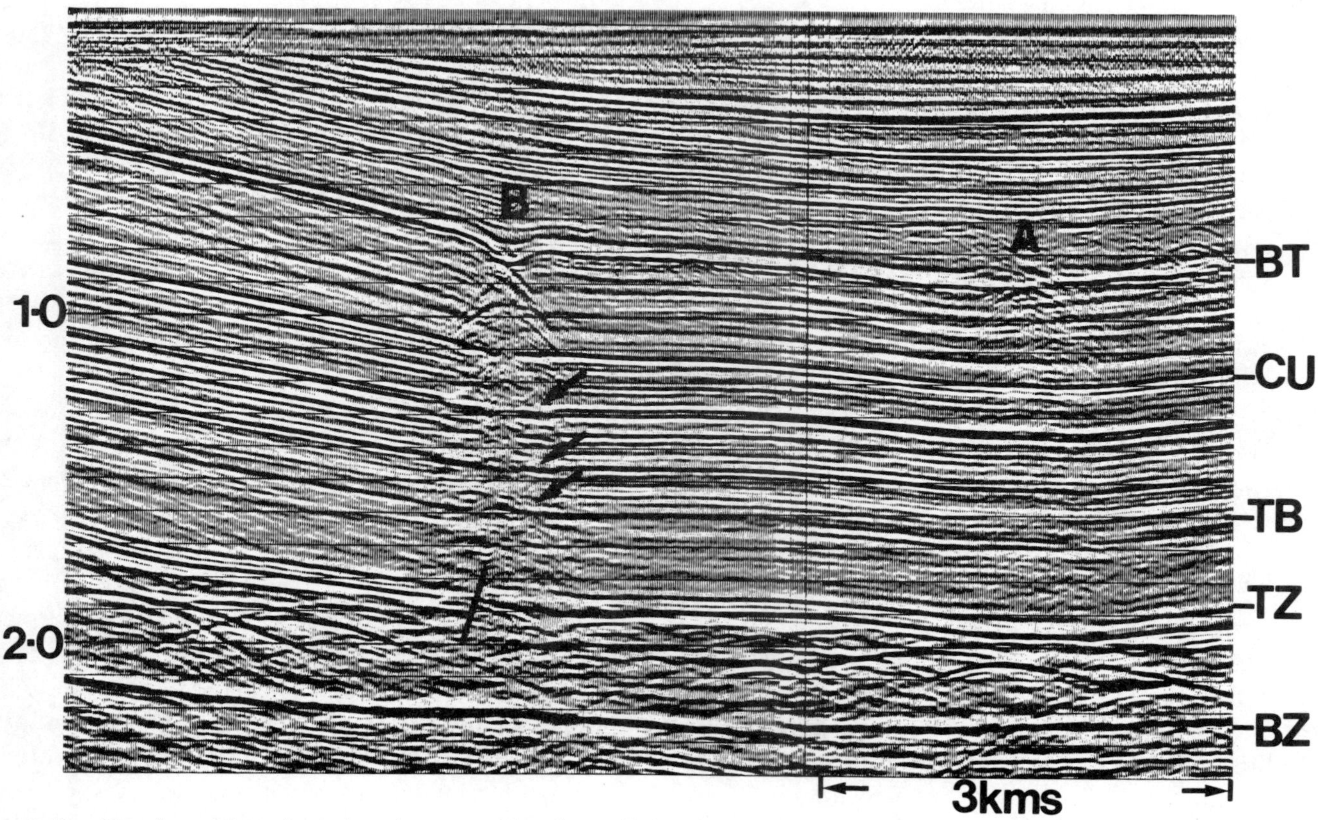

Fig. 6.18 Unmigrated section. Another example of a well-developed minor collapse feature (B), and an incipient feature of the same type (A). See text for explanation. Key is as for Fig. 6.16. (Courtesy of SS(E)L.)

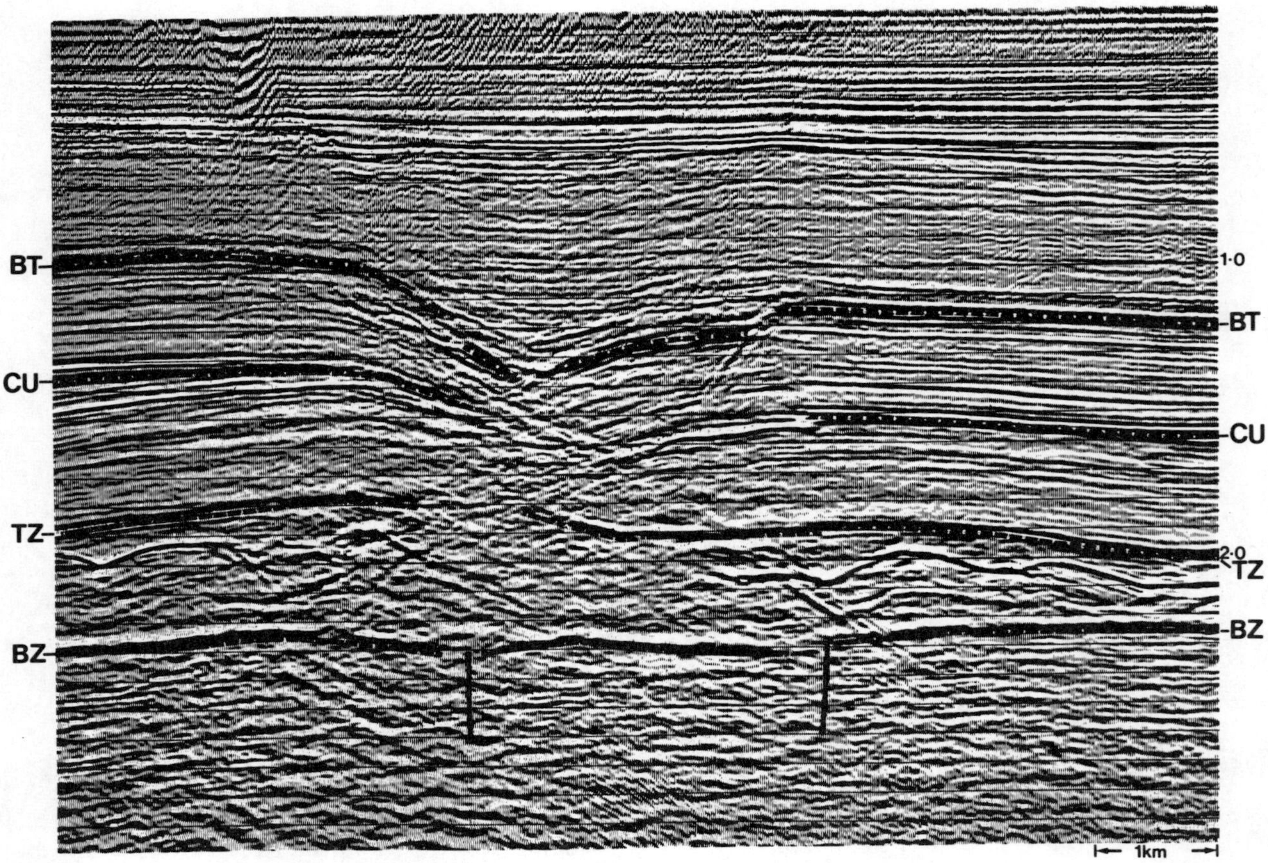

Fig. 6.19 Large linear collapse structure, seen in cross-section. This may be a case where surface extrusion or near-surface spread of salt has occurred. Key: BZ–TZ, salt interval; CU, Late Cimmerian Unconformity; BT, Base Tertiary. (Courtesy of SS(E)L.)

affected by dissolution. The extensional stress is also likely to be the cause of the normal fault at the top of the BZ–TZ interval which includes the main Zechstein salts.

Indicated at A in this same example is an incipient feature of the same type, where the zone of compression has begun to develop with thickening and disruption of the event at BT level which marks the top of the Chalk. As yet, the extensional stress zone below has not developed. Note particularly the subtle but visible localised thinning of the BZ–TZ salt interval below both A and B, bearing in mind that it is the upward flexing of the overburden sequences in both directions laterally away from salt minima such as these which is causing the synclinal stress-zoning in the overlying sequences.

Features of this type may be of importance in hydrocarbon exploration, since the breaching of an otherwise impermeable seal formation may result, providing an upward migratory route for hydrocarbons into the shallower section. This may be either good or bad news depending on the reservoir-trap-seal situation.

In Fig. 6.19 is shown a section across a large linear collapse feature measuring some 3·5 km in width at BT level. Perpendicular to the section, it runs for some 20 km, eventually passing into a salt wall diapiric structure. This passage from one type of structure to another is not uncommon in some areas. The development of structures of this type is often extremely complex, involving multiple dissolution and salt flow episodes.

In this instance, considerable asymmetry in development is evident from the differing thicknesses of the salt layer (BZ–TZ) on each side, and the presence of events truncating beneath the CU unconformity on the left, but not on the right of the section. This indicates that a salt uplift occurred initially on the left of the structure during pre-CU Mesozoic times, possibly related to reactivation of faulting at BZ level. Piercement seems to have occurred during the CU period, probably with some lateral invasion of the TZ–CU overburden interval, and almost certainly with extrusion of salt at the contemporary surface.

The most intriguing aspect of this structure is the much greater collapse effects at the BT level than

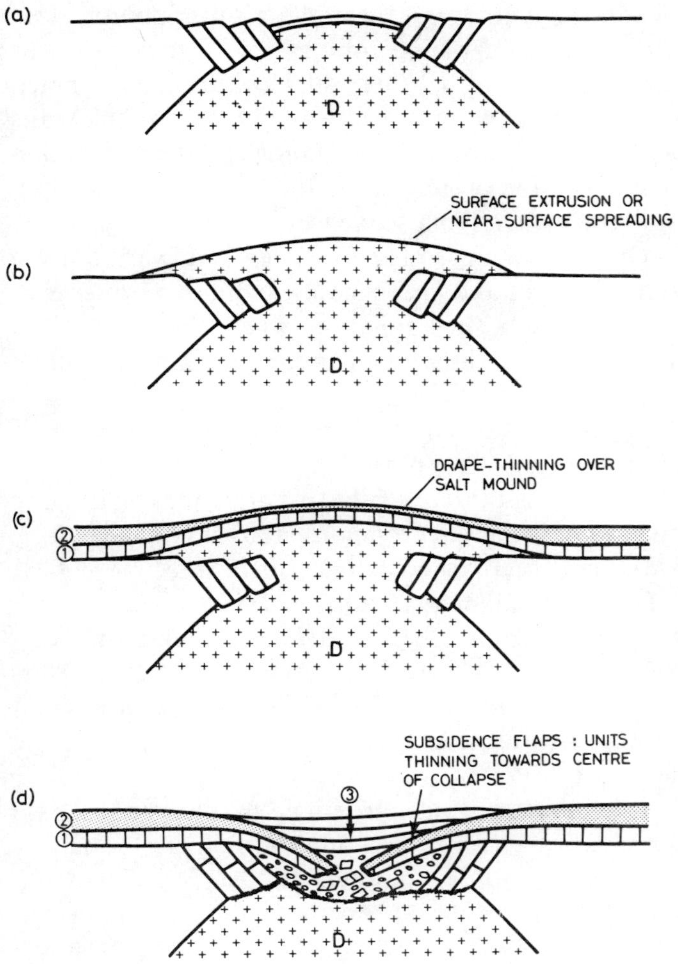

Fig. 6.20 Diagrammatic suggestion of possible effects of extrusion or near-surface spreading of rising salt mass. (See text.)

beneath. One explanation for this can be arrived at by assuming that a surface mound of salt was maintained, in spite of erosion/dissolution, by continued salt flow during the CU erosional phase and for some time afterwards during the deposition of the post-CU sediments. In this case, a substantial thickness of Cretaceous (CU–BT) sediments would have been laid down over the surface of the salt mound. At some time early in the Tertiary, salt flow failed to compensate for dissolution, the balance was disrupted, and the major collapse effects seen in the section ensued. The volume of salt in the mound may have been greater on the left-hand side, and/or the post-salt sediments thinned markedly over the mound; either effect could explain the strong convergence of the CU–BT interval. Something similar, but of less extent, may also have occurred in the TZ–CU interval. In this respect, the structural features may be analogous to those in Fig. 6.16, but on a larger scale.

The process suggested is sketched diagrammatically in Fig. 6.20, where in (a) a salt diapir (cross symbol) approaches the surface, with some fracturing of the overburden. In (b), extrusion of the salt at the contemporary depositional surface takes place. Sketch (c) shows a later stage where post-salt sediments have been deposited over the surface salt mound, and continued salt flow keeps pace with any erosion/dissolution effects which may occur. In sketch (d), salt flow has ceased to compensate for these effects, and dissolution has taken place leading to graben formation in the pre-salt section and collapse of the post-salt deposits (units 1 and 2), which form a species of 'lining' to the graben feature. Normal syndepositional drape effects cause the thinning of post-salt units 1 and 2 in (c) over the salt mound, and it is this thinning which, after collapse, gives the effect of convergence of the post-salt units in towards the centre of the graben feature. After stage (d), renewed deposition infills the collapse feature at '3'.

The process outlined is a hypothetical suggestion only, and is not supported by any corroborative evidence. The principal difficulty involved is the concept of a surface salt mound being maintained during continued sedimentation, but this is not considered to be beyond the bounds of credibility. Salt mounds exist at present at the surface in very arid areas such as Iran, and it is conceivable that they could be covered by terrigenous sediments. On the other hand, where indurated bedrock is covered by superficial deposits such as clays which may be undercompacted but still impermeable, the rising salt could be injected between the bedrock and overlying material, spreading out to form a mound and doming up the overlying superficial deposits which may protect the salt from dissolution if in the marine environment. Spreading overhangs seen associated with salt diapirs have been interpreted at several locations in this way. Later, this impermeable seal may break down, or undersaturated water may find some other route to the salt.

6.4.2 Areal Dissolution Effects

Widespread superficial dissolution of a salt layer has been recognised in the subsurface in a number of areas. Lohmann (1972) referred to a horizontal surface of leaching of a salt layer by the German geologists' term *Salzspiegel* (salt mirror), and noted that it could be caused, for example, by formation water movement in a sandstone aquifer overlying a salt layer and effecting dissolution over a wide area. Another cause of widespread dissolution is where a surface of unconformity has cut down to within a short vertical distance of a salt interval, and any sequences between the unconformity and the salt layer are permeable. Provided that the necessary circulatory system exists, then undersaturated water percolating down through the surface of unconformity could dissolve and remove the upper surface of the salt progressively.

From the seismic viewpoint, such a situation may be very difficult to recognise in the data, since if carried to completion the process results in the absence of an interval from the stratigraphic section, rather than the presence of any identifiable features, where a uniform salt layer has been removed.

An areal leaching process is probably responsible for the presence of the dissolution and collapse breccias over wide areas adjacent to the present Prairie Salt deposits mentioned in Section 6.2. These, however, are known to exist from drilling results rather than seismic data. It would be impossible to identify such relatively thin intervals of breccia on a normal seismic section. Slow, widespread subsidence of the overburden, even where associated with some brecciation of the lowest units, is unlikely to produce any major faulting or other diagnostic effects.

On the other hand, where a salt interval is not uniform in the lateral sense, features may become identifiable which can strongly suggest that an areal dissolution process has been active. Lateral variations in the salt thickness may be due to the original depositional environment, to later salt mobility, or to the presence of buried topographic features such as carbonate buildups. Where salt is locally contaminated with large concentrations of impurities, or thin salts are interbedded with other lithologies, later dissolution may be of a very irregular nature, leading to localised collapse of overlying sequences.

One type of environment where lateral variations in salt thickness, and interbedding with other lithologies, may occur is a shelf situation, where thin halites and marginal facies lithologies occur together as infilling between reef structures.

A migrated seismic section from the North Sea is seen in Fig. 6.21, from an area where the Zechstein shelf facies is developed. In this area, marginal facies rocks consisting of carbonates and anhydrites mainly, with thin interbedded halites, underlie some 8000 ft (2500 m) of sediments, mainly of Tertiary age. Beneath the Tertiary, thin Chalk and Lower Cretaceous sediments lie on the Late Cimmerian unconformity surface, while below this surface there is a thin interval of Triassic together with, in places, a thin Jurassic sequence.

It is envisaged that erosional cutdown of the Late Cimmerian unconformity into the thin Mesozoic section, and its close approach to the Top Zechstein, allowed downward percolation of undersaturated water, which led to leaching of the Zechstein evaporites. Collapse of the Mesozoic sequence, together with remnant Zechstein evaporite and carbonate insolubles,

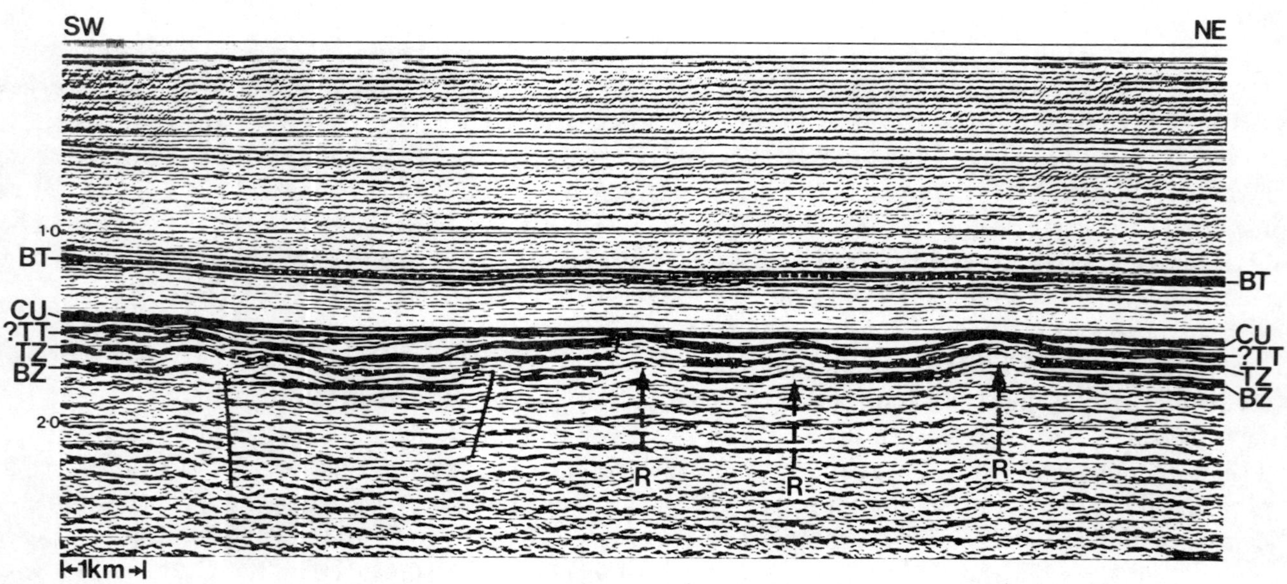

Fig. 6.21 Areal dissolution effects. Collapse structures due to unconformity (CU) cut-down close to evaporite interval BZ–TZ. Leaching of salt from the latter has caused the collapse features in the TZ–CU interval, with drape over features marked R, believed to be carbonate buildups of Zechstein Z1 or Z2 age. Note the pull-up anomalies. ?TT believed to be Top Trias. (Courtesy of SS(E)L.)

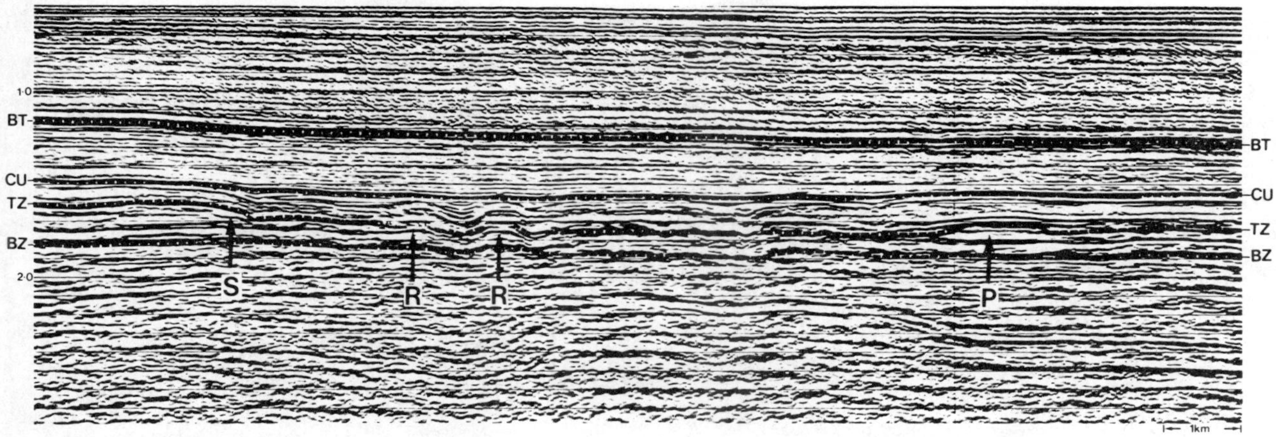

Fig. 6.22 Areal dissolution effects. Remnant features left after leaching of salt due to cut-down of unconformity CU. S, salt slope; R, carbonate buildups, possibly of atoll form; P, pod of undissolved salt, or evaporite insolubles. Note the collapse and drape effects. Key as for Fig. 6.19. (Courtesy of SS(E)L.)

occurred, resulting in the very irregular pattern of sinks and channels seen in the example. It is estimated that the Zechstein sequence probably had an original TWT interval of over 200 ms, which has been reduced in some locations to 50 ms or less by the dissolution and collapse effects. Seismic data alone are inadequate to resolve the detail in the resulting *melange* of evaporite insolubles, anhydrite, and carbonates of varying density/porosity characteristics. Within this jumble of lithologies plus any reef structures which may be present, direct hydrocarbon indications—'bright spots and flat spots' are sometimes seen in this area, and probably indicate the presence of gas in more porous reef carbonates (Jenyon and Taylor, 1983).

Another example from the same general area is shown in Fig. 6.22, where collapse features over a wide area can be seen. These are associated with a salt slope (to be discussed in Section 6.4.3), and other features which are believed to be carbonate buildups—one of them possibly of atoll form. Note the irregularities at the base of the Zechstein interval (BZ) in places, which are due to velocity anomalies related to the strongly contrasting acoustic impedances between various lithologies and structures (salt, anhydrite, carbonates of various densities in reefs and in the infill, and collapsed Mesozoic material) which are present. The atoll-like structure, other examples of which can be seen in scores of locations in the area, is of interest in that it has what appear to be flat-spot gas/liquid contact events within the two highs. These highs are separated by a collapse structure thought to be the result of the leaching-out of infill material from the central 'lagoon'.

Reference was made in Section 6.1 to areal dissolution not related to the presence of a salt layer, but to karstification of a carbonate surface. An example of what is probably such a surface at the top of the Chalk, now buried beneath 3500 ft (1000 m) of Tertiary clastics in the southern part of the North Sea, is shown in Fig. 6.23. This example of a migrated section shows the Top Chalk seismic marker to consist of a prominent black peak (A) followed by a much weaker, irregular peak (B). Despite the apparently weak nature of the latter, it is extremely persistent and can be followed over a wide area; in places it is defined by a series of minute diffraction hyperbolas. There are frequent irregularities affecting both A and B which strongly suggest local collapse features, and two of these are indicated in the figure.

The equally persistent peak C has been correlated from well log data with the top of a marked change in lithology within the Chalk, where a highly glauconitic interval over 100 ft (30 m) thick occurs having an interval velocity 1000 ft/s (300 m/s) lower than that of the Chalk above and below. Peak C shows no signs of the irregularities present at the levels of peaks A and B.

The interpretation of these events is that peak A is the response to the Top Chalk surface, and peak B the response to the base of a zone of karstic solution/collapse effects which may coincide with either a change in the Chalk lithology less well marked than that at C, or the base of a zone of superficial fracturing at the Chalk surface (see later).

Nearby well log information supports this interpretation. Acoustic and caliper log data indicate a zone of about 130 ft (40 m) below the Chalk surface with characteristics of a mixture of collapsed Tertiary shales and Chalk rubble. The sketch map in Fig. 6.24 shows

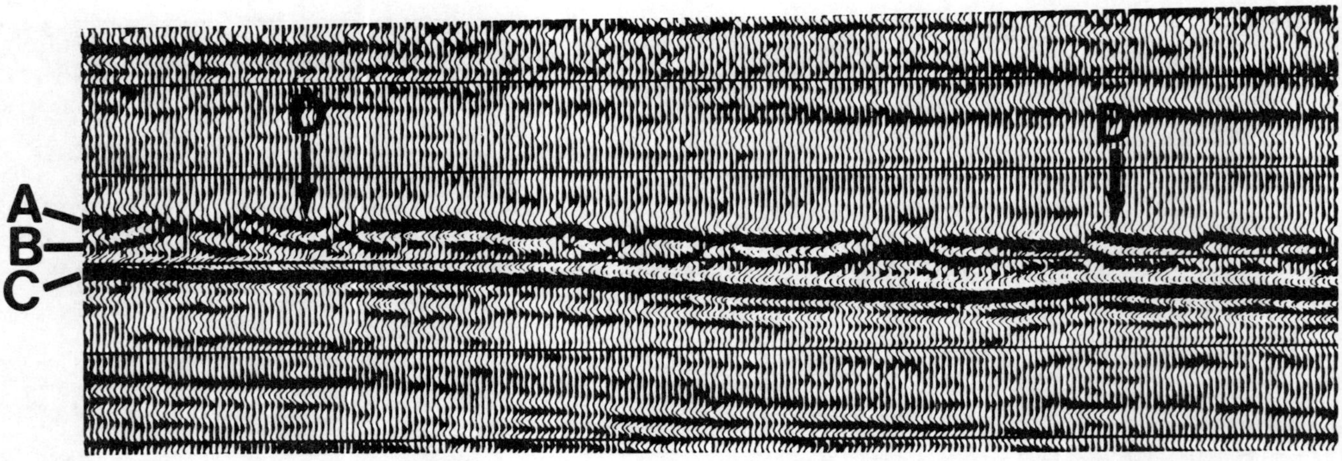

Fig. 6.23 Karstification of Top Chalk surface. D,D, two of several collapse effects to be seen. See text for discussion of A, B and C. (Courtesy of SS(E)L.)

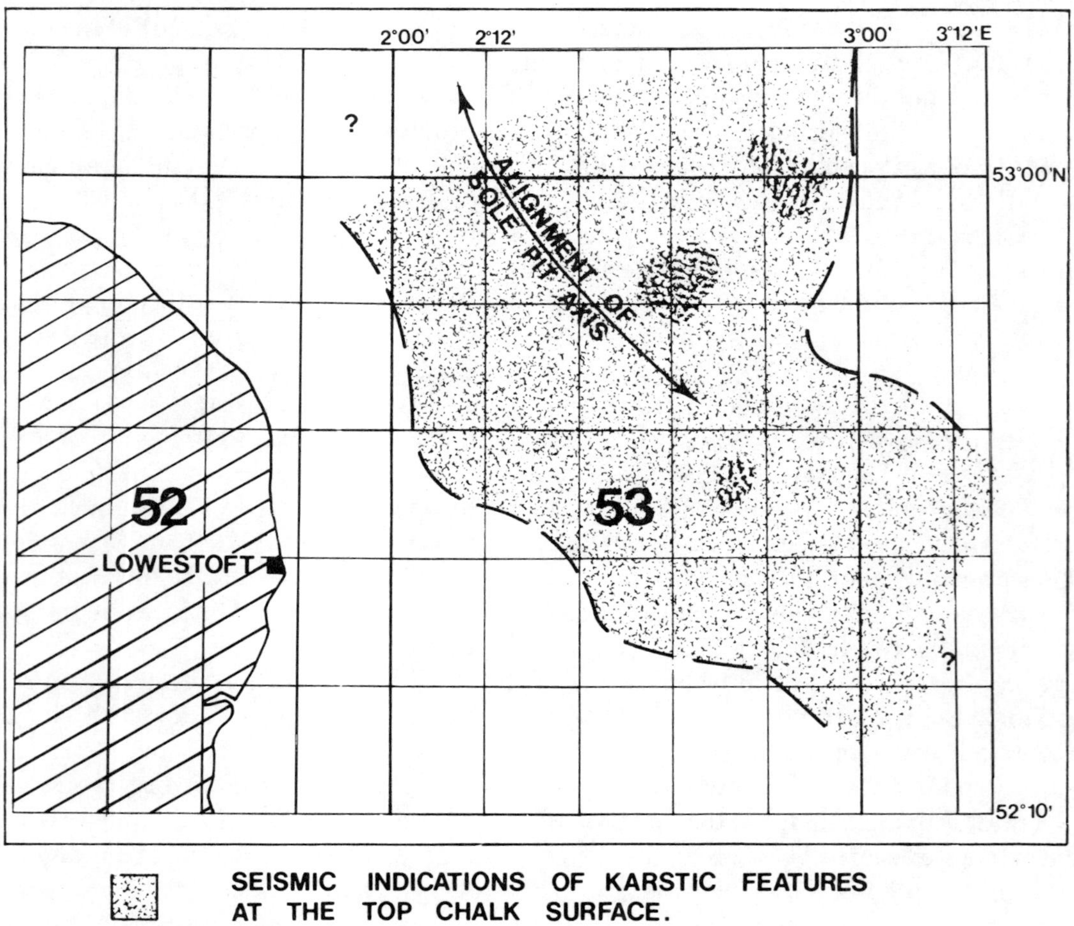

Fig. 6.24 Sketch map showing area affected by karstification off East Anglia in the southern North Sea. (Courtesy of SS(E)L.)

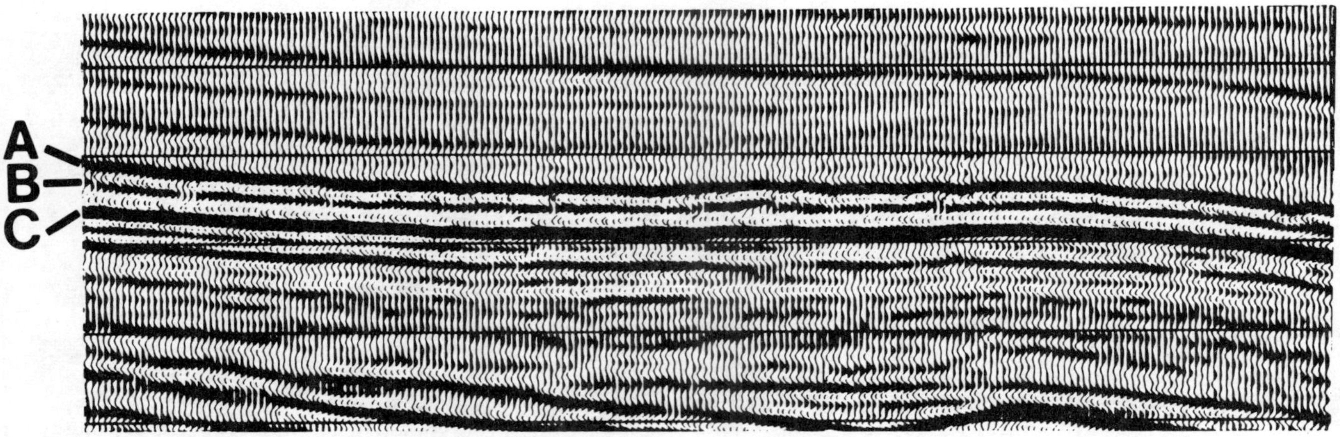

Fig. 6.25 Section showing the undisturbed character of the Top Chalk seismic marker outside the area of karstification. (Courtesy of SS(E)L.)

the extent of the karstification of the Chalk surface in the immediate area, and in Fig. 6.25 an example from a seismic line only 10 km away from the previous seismic example, but this time just outside the limit of the karst area, shows the unaffected nature of the peaks A and B. The presence in the seismic data of buried topography features at Top Chalk level as indicated in the sketch map, and the location of the area in alignment with the Sole Pit inversion axis, suggests the reason for the observed effects. Uplift occurred along the inversion axis, with the Chalk being brought to the surface under either subareal or shallow water conditions; the affected area probably represents the region of maximum curvature of the

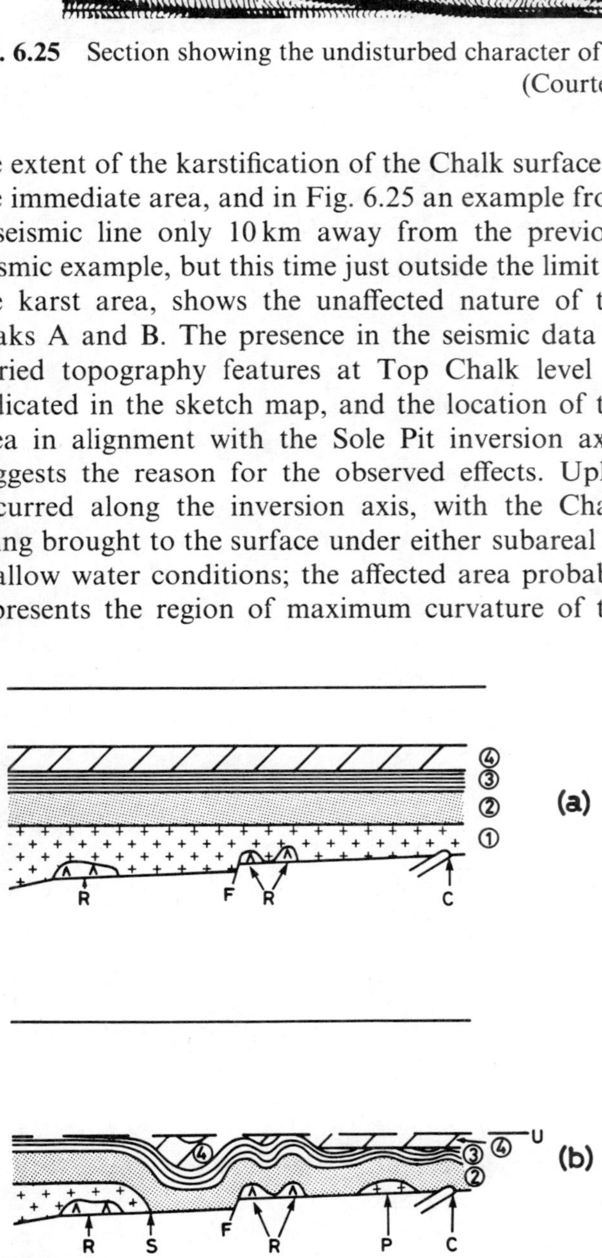

Fig. 6.26 Schematic of types of remnant feature left by areal dissolution of a salt layer. There is overlap between the effects seen here and those produced by edge-dissolution. Compare this with Fig. 6.22.

uplift in the overburden, with the resultant stress causing superficial fracturing of the top of the Chalk and leading to the dissolution and karstification.

It seems clear from this example that the process as described results in extremely superficial and subtle effects in the seismic data which are unlikely to be confused with phenomena related to salt dissolution and collapse structures.

In the process of areal dissolution of salt, where the latter is removed over a large area, certain types of remnant feature may be seen at the erstwhile base salt level which result in drape structures in the subsided overburden. The schematic in Fig. 6.26 illustrates the most frequently seen of these. The model used is the fairly typical one of a faulted carbonate ramp (a) before and (b) after areal dissolution has occurred. In (a), carbonate buildups (R) which have been buried by the salt layer (1), and a cuesta feature (C) due to a more erosion-resistant unit in the truncated, dipping series below the unconformity marking the salt base are shown. The salt layer lies beneath an overburden of clastics (2–4). In the lower sketch (b), an erosional phase has resulted in the cut-down of an unconformity into the overlying clastic sequence, and during the cut-down process, water has percolated down through the unconformity surface and the permeable clastic units, effecting dissolution of the salt layer over a wide area. Clastic bed 2 now lies directly on the pre-salt unconformity over the greater part of the model section, although some salt, terminating in a dissolution slope, S, still remains at the left-hand end of the section. (This is to illustrate that such a salt slope may result from areal dissolution as well as purely edge-dissolution by intraformational water moving down from the basin periphery, as will be discussed in the next section.) To the right of the salt slope, the remnant

Fig. 6.27 Marked drape effects produced by a remnant feature (R) after edge or areal dissolution of salt layer (B–T). In right-hand half of section, all salt has been removed, and only a thin basal sequence of carbonate/anhydrite lithologies remains, the top and base of which cannot be resolved (B/T, right). On the evidence of the primary reflections below the base salt here (which are the events showing strong dip to right), R may be an erosional cuesta at the unconformity surface. (Courtesy of SS(E)L.)

carbonate buildups at R have resulted in drape features in the subsided clastics above, as have also the remnant pod P, which may be of undissolved salt or other evaporite insolubles, and the erosion-resistant unit at C. Late stage dissolution may also cause subsidence effects at and above the unconformity U.

This model supposes that the dissolution is due to water percolating downwards from the unconformity U. However, if clastic unit 2 were an aquifer, a similar sequence of events would result, as it would also if the subsalt unit were also an aquifer carrying undersaturated water. In the latter case, an inverted *Salzspiegel* would result, the surface of dissolution of the salt being the basal, rather than the upper, surface.

It can be seen from the type situation model shown that any remnant structures present after removal of the salt will, because of the drape effects produced in the overburden above them, advertise their presence with some clarity. Edmunds (1980) noted this fact, commenting that the process '. . . significantly affects the mapability of underlying reef accumulations and . . . mapping of prospective trends', quoting the Middle Devonian Keg Reef trend of northern Alberta, and the Silurian Niagara Reef trend of Michigan as examples.

The point is well illustrated in Fig. 6.27, which shows such a remnant feature at R, with very marked drape effects in the overlying sequences. The situation here is probably of a combination areal/edge-dissolution process, with the thick salt interval B–T on the left terminating in a dissolution slope in the centre of the section, and a wide swathe of collapse effects in the overburden above and to the right of the salt slope. The horizon indicated as B/T represents thin basal carbonate/anhydrite lithologies, the top and bottom of which cannot be resolved here, but which were the floor of the salt layer before dissolution took place. The nature of the remnant feature R cannot be determined with certainty in this instance. Dips in the pre-salt sequence truncating at the unconformity are strongly to the right, and it is possible that this is an erosion-resistant cuesta. Without further information, however, the possibilities that it is either a small pinnacle/patch reef or an insoluble remnant evaporite pod cannot be ruled out. It does serve to illustrate, however, with what clarity such remnant features stand out in the data after the removal of the salt layer.

6.4.2.1 Upward persistence of drape

The marked effect on the overburden of quite small irregularities at the original base salt level, as seen in Fig. 6.27, is due to the condition of compaction and induration of the subsided overburden units.

Normal synsedimentary drape over a small irregularity tends to die out quite rapidly upwards in the sedimentary section (see Fig. 6.28, upper sketch) due to effects of erosion and winnowing of material from the high and re-deposition in the adjacent lows.

In the areal dissolution/collapse context, however, the situation is quite different. The overburden, which eventually subsides on to the remnant feature, is already at least partly, if not completely, compacted and indurated; the contemporary depositional surface may be many hundreds of metres above the subsiding basal overburden beds. Upon complete removal of the salt layer, therefore, the shape of the remnant irregularity is transmitted upwards without change through the overlying stratigraphic units, as shown in Fig. 6.28 (lower sketch), which is an idealised sketch showing a small lenticular irregularity A resting on the original salt base level.

The salt layer which was originally interposed between the small irregularity A and the basal overburden unit B has been removed by dissolution. Unit B has subsided, taking up the antiform shape of A. Units C and D take up exactly the same shape. The geometry of the situation demands, however, that the zone W between the inflection points on each side of the drape structure (i.e. the points of inflection between the horizontal bedding and the arc of the drape structure) widens with increasing upward distance from A. At some later time, horizon E will still be showing a curve concentric with A, but over a much wider arc than is present at A. This could be of importance in a trap situation, as it could lead to increasing areal extent of a closure upwards.

This situation is also of importance in that drape structures in a subsided interval may give an exaggerated impression of the size of the initiating irregularity, e.g. a very small feature such as a patch reef may produce a large drape structure in the subsided overburden; an erroneous idea of the size of the initiating irregularity may lead an interpreter of seismic data to discount the possibility of the presence of patch reefs. Very small features of this type may actually be below the limit of seismic resolution at a given depth on the cross-section, only their effects on the overburden being visible.

6.4.3 Salt Edge Dissolution Effects

Where a salt layer terminates at a feather-edge towards the margin of a basin, undersaturated water (either meteoric or otherwise) moving down from the basin periphery may attack the feather-edge of the salt, removing it as brine and causing the salt edge to retreat down the basinal dip as a salt slope, the amplitude of which increases with increasing thickness (Fig. 6.29) of salt as it retreats from the location of the feather-edge.

Part of a migrated section from near the western margin of the southern Zechstein Basin of the North

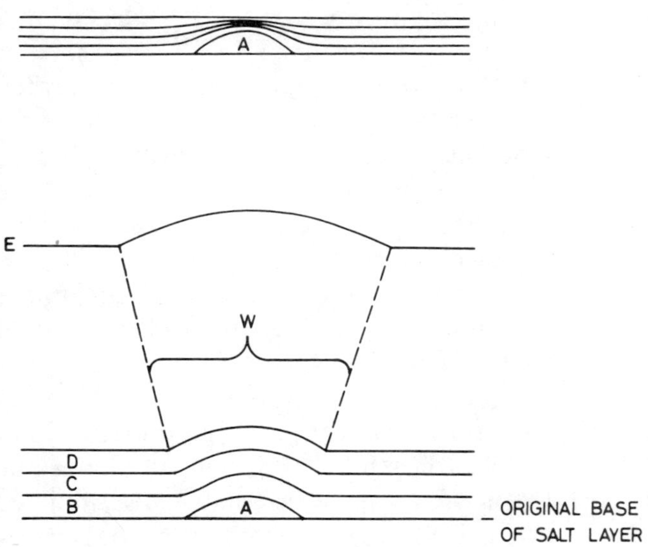

Fig. 6.28 (Upper) Syndepositional drape over a small lenticular irregularity, A. (Lower) Subsidence drape over a similar feature.

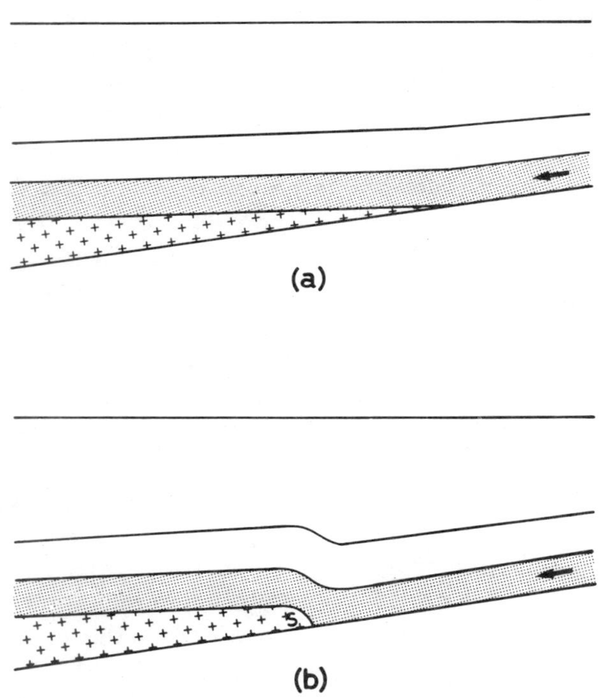

Fig. 6.29 Schematic of edge-dissolution process. Water circulating through aquifer (dot-shaded) in (a) leaches feather-edge of salt (cross symbol). Salt retreats down basinal dip (b), evolving into a slope (S) of increasing amplitude, leaving behind it progressively increasing subsidence effects in the overburden. Amplitude of salt slope may or may not reach total thickness of salt layer. A similar effect may develop by areal dissolution due to water percolating down through an unconformity.

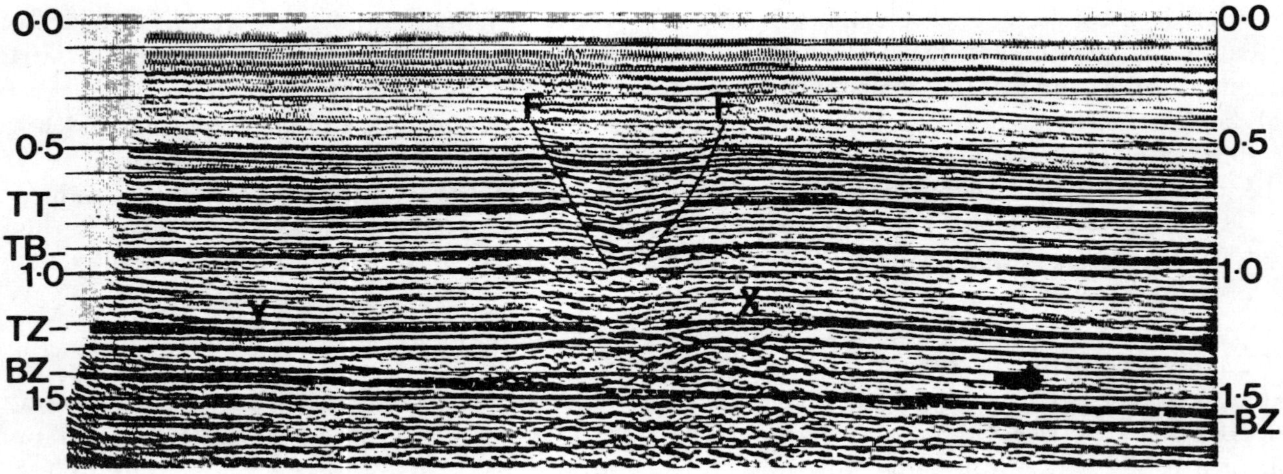

Fig. 6.30 An example of edge-dissolution. Key: BZ–TZ, interval containing thin salts interbedded with other lithologies near a basin margin; TB, intra-Triassic horizon; TT, Top Trias. See text for explanation of other letters. (Courtesy of SS(E)L.)

Sea is shown in Fig. 6.30. The arrow within the Zechstein interval (BZ–TZ) indicates the relative direction of the basin centre, and the basinal dip at Base Zechstein level is continuous across the section. At Top Zechstein level and above, however, there is an interruption in the basinal dip, a reversal having taken place from point X to point Y at TZ level. In the overlying sequences to the left of X there is a collapse graben (F–F) affecting the major part of the post-Zechstein section; this graben will be considered later.

The effects here are interpreted as being caused by edge-dissolution of the Zechstein halites. Water undersaturated with respect to NaCl is envisaged as having entered the formations updip, at or near the basin margin, and having moved intraformationally down dip until in contact with the feather-edge of the salt within the Zechstein interval. Continuing dissolution has caused the salt feather-edge to retreat down dip, and the vertex of the incompletely-collapsed diffraction hyperbola below and slightly to the left of X probably marks its current position. The removal of the salt has caused the gentle subsidence of the overlying sequences that has resulted in the dip reversal. Although no well data are available in the immediate vicinity, regional studies (e.g. Taylor and Colter, 1975) show that there is likely to be a compressed sequence of the Zechstein present from the Z1 to Z5 evaporite cycles, with thin to moderately thick beds of Z2, Z3 and Z4 halites separated by carbonate/anhydrite sequences of varying thickness. The dissolution process in such a group would be complex. It appears likely that updip from the edge of the salt in this example, a thinning wedge of halite insolubles together with remnant interbedded carbonates and anhydrites would be present. This would explain the gentle nature of the dip reversal over the major part of the feature between points X and Y.

In hydrocarbon exploration, this type of situation is of potential importance, since the critical (i.e. contra-basinal) dip reversal produced in sediments updip from the basin centre could provide an important series of traps in the linear high feature (following the salt edge round the basin margin) which results. Given suitable reservoir and seal lithologies, together with source intervals deeper in the basin, major producing fields can and do occur. Langstroth (1971) noted the close association of many of the producing oilfields and the 'zero edge' of the Devonian Prairie Salt on the east flank of the Williston Basin in North Dakota, resulting from '. . . drape over salt edges or salt ridges . . .'.

The interesting collapse graben mentioned earlier, which is located to the left of and above X in the figure, is not believed to be the result of the salt-edge dissolution. It may itself have caused some localised dissolution of the salt and collapse of the overlying sequences either before the onset of the salt edge dissolution process, or at an early stage of the latter. The graben faulting is believed to be the result of movements associated with an important tectonic axis (Market Weighton) located to the west of this area. It is likely that there was only one fault originally at the graben location, and that salt dissolution along the strike of the fault, with attendant collapse, initiated an antithetic normal fault to produce the graben.

The simplified schematic of Fig. 6.31(a) shows the basic situation, with the dissolution slope S of the salt layer having retreated down dip, and resulting in an

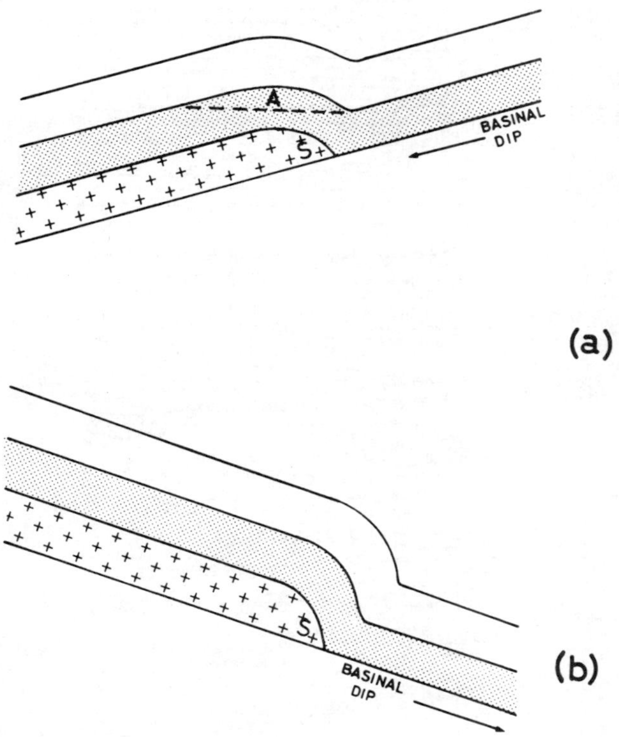

Fig. 6.31 Effect on hydrocarbon trapping potential of basinal tilting.

antiform flexure in the overburden due to progressive collapse, with a potential hydrocarbon trapping situation above the spill-point of the flexure, at A in the cross-section.

In many areas, a basin has been subjected to later tectonic movement, and basinal dip may have been reduced or even reversed. Figure 6.31(b) shows such a situation schematically and clearly the hydrocarbon trapping situation shown in the upper illustration has been removed. Any hydrocarbons present originally will have migrated up dip from such a location. Figure 6.32 shows an example of this unfavourable type of configuration, although in less exaggerated form. Due to tectonic tilting of the area, the collapse flexure above the salt slope S no longer constitutes a trap for hydrocarbons. The basin centre in this instance is off section to the left, and it can be seen that, although the contrabasinal dip is shown by the overburden to the right of S, the basinal dip in the overburden to the left of S no longer occurs, and so a simple monoclinal flexure remains which does not constitute a trap configuration.

A more favourable scenario is illustrated in Fig. 6.33, where the basinal dip still remains. Because of this, both contrabasinal dip to the right of the salt slope, and basinal dip to its left, is shown by the overburden sequence (although the latter is not well seen within the confines of the section in this example). Although such a situation is favourable, it may not necessarily result in a hydrocarbon trap, of course. Assuming the presence of source intervals downdip, favourable geometry *along* the linear feature seen here in cross-section is required for closure, as are also the presence of suitable reservoir and seal formations.

An unusual situation involving an edge dissolution salt slope is shown in Fig. 6.34, where a thick salt interval within BZ–TZ has been deposited across an old fault escarpment (F), with a slump surface (E) due to mass movement down the fault which shows that the fault existed either from the very earliest time of

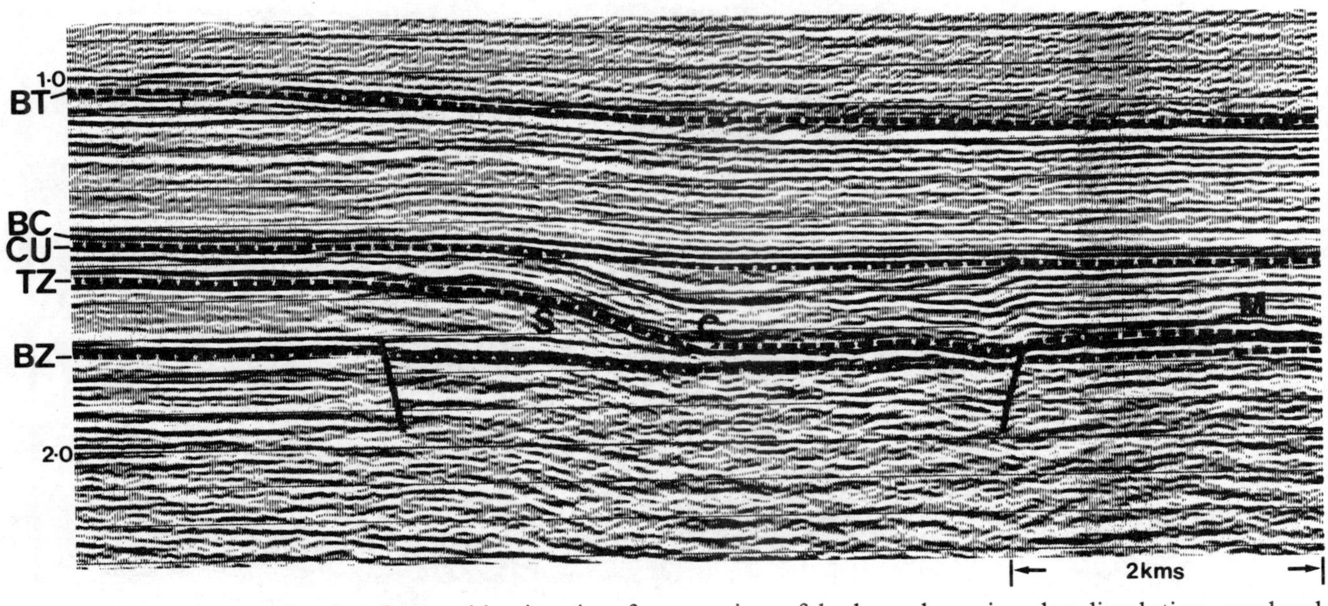

Fig. 6.32 Seismic example of unfavourable situation for trapping of hydrocarbons in edge-dissolution overburden flexure. (Courtesy of SS(E)L.)

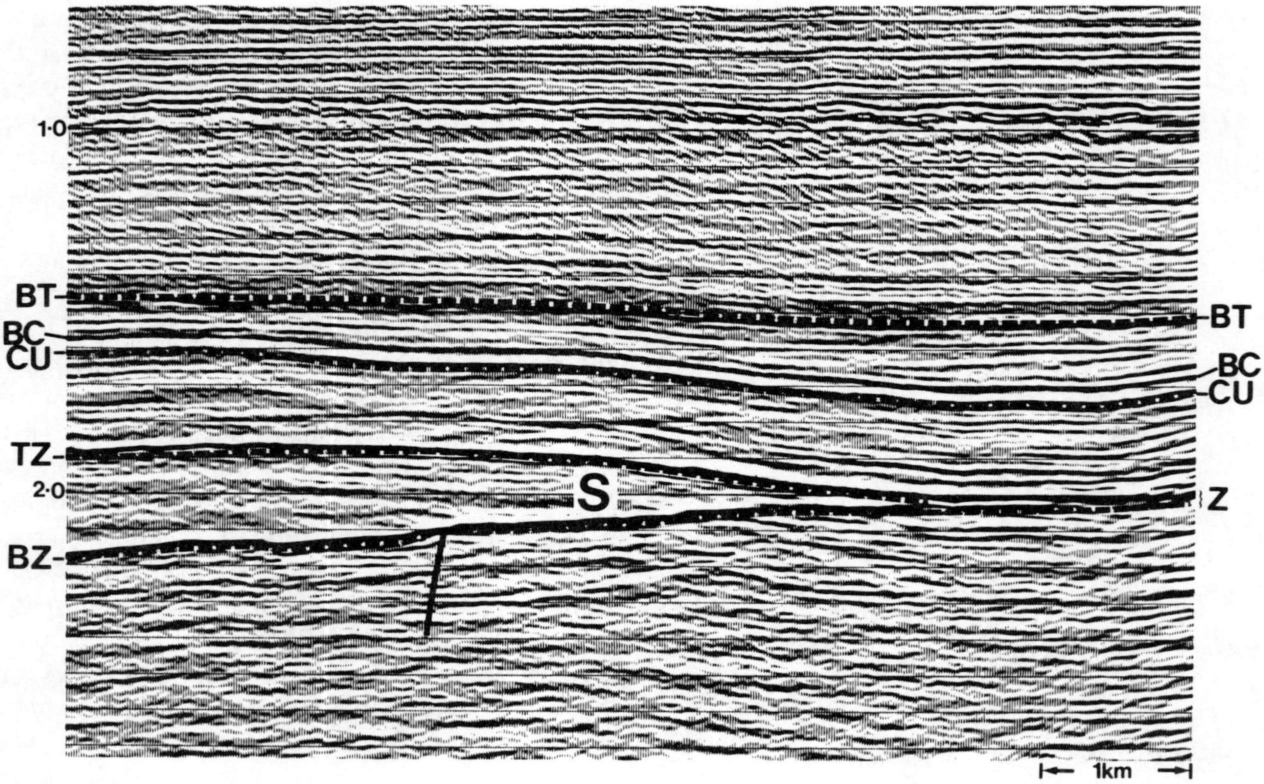

Fig. 6.33 Seismic example of more favourable situation. Key: S, salt slope; BZ–TZ, salt interval; CU, unconformity; BC, Base Chalk; BT, Base Tertiary. (Courtesy of SS(E)L.)

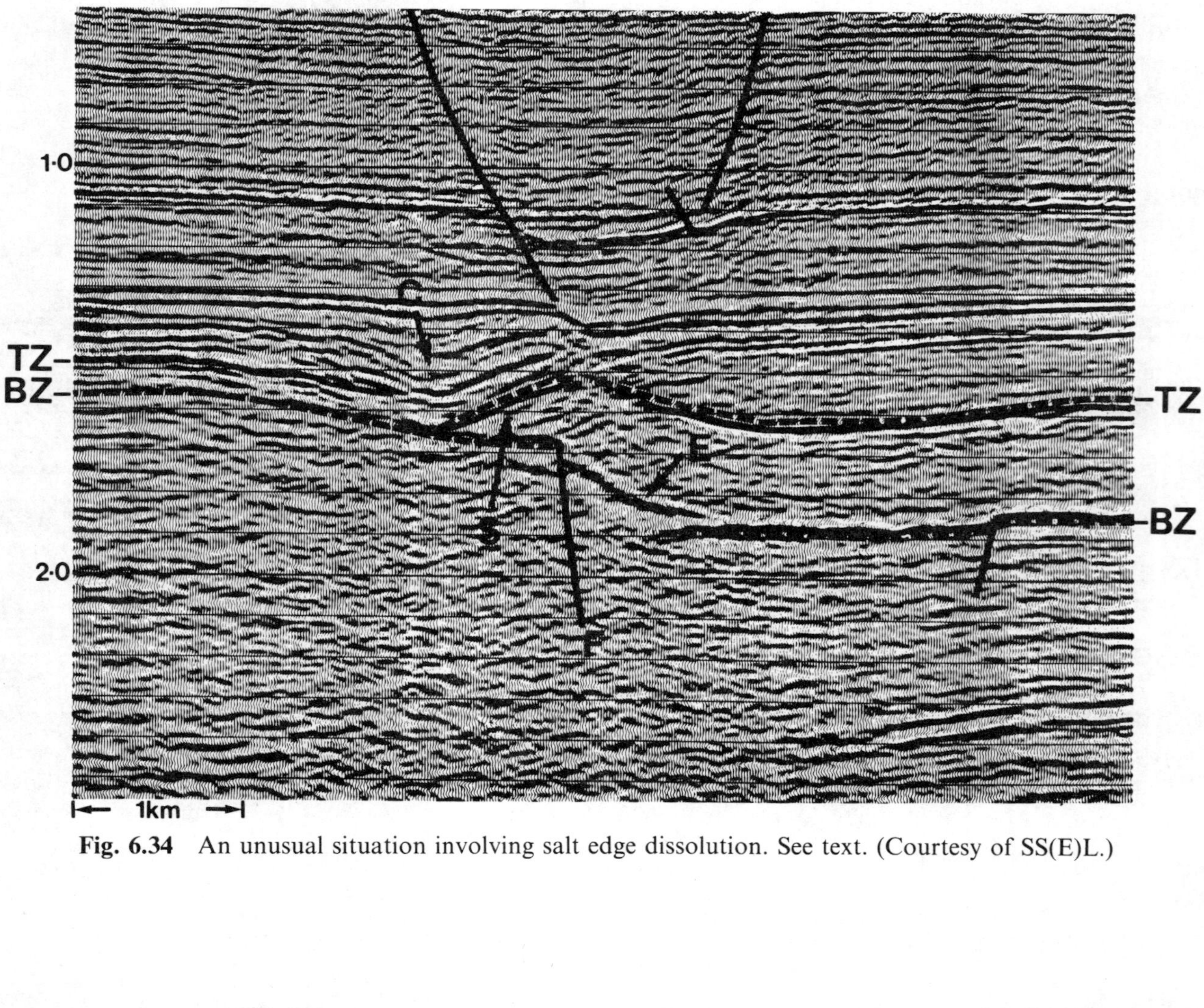

Fig. 6.34 An unusual situation involving salt edge dissolution. See text. (Courtesy of SS(E)L.)

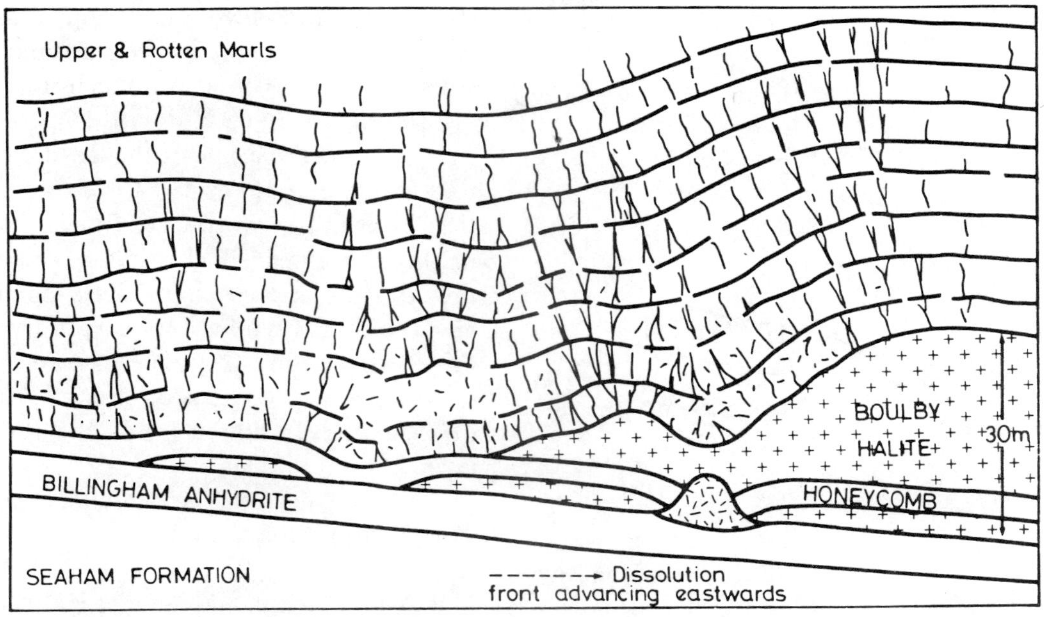

Fig. 6.35 Diagrammatic sketch of a salt-edge dissolution slope and accompanying collapse effects in the overlying sequence at a surface outcrop on land in NE England. Although comparatively small, this example shows remarkable similarities to the much larger feature seen in Fig. 6.34, where a similar slope at S and collapse effects at C can be seen. (Reproduced by courtesy of Dr D. B. Smith and JAPEC (UK) from Smith, 1985.)

salt deposition, or before it. On the upthrow side of the fault, at the extreme left of the example, only marginal lithologies are present in the BZ–TZ interval on this platform area, but the salt dissolution slope, S, and the collapse C in the overburden between the salt top and an unconformity indicate that previously the salt layer on the upthrow side of the fault extended to a feather-edge further to the left than its present position.

An interesting comparison can be made between the situation in Fig. 6.34, and that in Fig. 6.35. The latter is a diagrammatic sketch of a surface outcrop on land in the NE of England made by Dr D. B. Smith from observation, and shows a dissolution slope in the Boulby Halite (the EZ3 halite). The dissolution front has advanced eastwards—from left to right—and the progressive removal of the salt rock has led to the subsidence and fracturing of the overlying marls. Although on a much smaller scale, the similarity between the situation and effects shown by this sketch, and the structure on the 'upthrow' side of the Fig. 6.34 fault, is striking.

As suggested in the previous section, and evidenced by Figs. 6.27 and 6.32, there may be considerable overlap between areal and edge-dissolution processes in the zone around the margin of a salt basin. In Fig. 6.32 particularly, the overburden collapse which begins approximately at the location marked M on the right of the figure, seems to have been a process of quite uniform subsidence, certainly between M and the right-hand fault at BZ level. This is more typical of areal dissolution than the progressively-increasing subsidence effects to be expected in a purely edge-dissolution process. With the close approach of the unconformity CU to the top of the salt interval, TZ, in the example, this would support the argument for an areal dissolution process having resulted in the observed collapse effects, and in the salt slope here. On the other hand, in the example of Fig. 6.30, there is no evidence that any process other than intraformational water movement and purely edge-dissolution was effective.

It has been assumed so far that the salt simply feather-edges at the basin margin to produce the effects seen. In some cases, however, the salt changes facies from basinal to marginal, with the salt interdigitating with carbonate and anhydrite lithologies, for example. In this situation, the hypothetical case of areal or edge-dissolution can be considered, which would perhaps lead to the feature illustrated schematically in Fig. 6.36. Undersaturated water is supposed to effect dissolution of the salt across the zone of facies change, resulting in the collapse of the overburden sequences as seen in cross-section in the figure. Although no unequivocal real-earth example is known, it seems quite a feasible situation, and is of potential interest because of the possible hydrocarbon trap situations which could emerge. Should the basin be unaffected

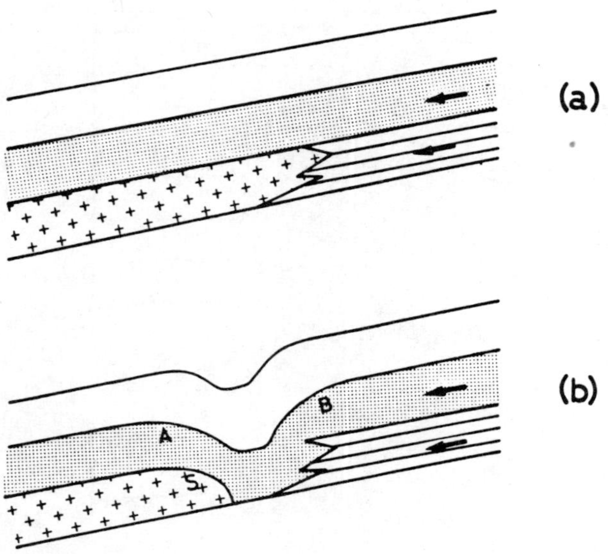

Fig. 6.36 Schematic of edge-dissolution situation, hypothetical only, where salt does not feather-edge but interdigitates with other lithologies at a basin margin. See text for discussion.

by later tectonic activity, then the collapse flexure at A could provide the normal trap configuration to be expected. However, should tilting of the basin occur (in a clockwise direction in this example), where the normal trap configuration would disappear, the collapse flexure at B may still provide the requisite geometry.

6.4.4 Stratigraphic Dating from Collapse Features

A knowledge of the processes involved in various types of dissolution/collapse features often enables the interpreter to identify the period over which the dissolution effects occurred. In Fig. 6.30 it is possible to follow the history of the edge dissolution process, since at TZ level the subsidence low in the overburden is at Y, but gradually migrates basinwards (to the right) with decreasing time on the section. In Fig. 6.32, the areal and edge-dissolution processes began in the early Mesozoic (just above TZ) at the right-hand end of the section, reaching a maximum during the Lower Cretaceous (between CU and the Base Chalk BC—the greater part of the Lower Cretaceous having been removed by the Cimmerian erosion phase). Some of the sediments in the deep collapse trough at C are Mesozoic deposits which have been removed from adjacent areas by the CU erosion.

Occasionally, detailed stratigraphic correlation may be assisted by a study of dissolution/collapse effects. In Fig. 6.37, a complex minor feature of similar origin to the type described earlier and illustrated in Figs. 6.16 to 6.18 occurs. In this instance, the synclinal stress zoning has been produced not by a late minor salt thinning as such, but by the upward flexing of the overburden to one side of the feature only. This is due to a thickening of the salt to form a small swell at S, the effects in the overburden indicating the late nature of this movement. The result has been, as described previously, compressional fracturing of the Top Chalk at its boundary with the Base Tertiary (BT) and below this in the feature C a zone of extensional fracturing down to a small extensional normal fault at the top of the Zechstein salt layer (T). B is the base of the salt layer.

It is known in this area that between the Triassic Top Bacton horizon (TB) and the Cimmerian Unconformity (CU), three thin Triassic halites occur—the Röt, Muschelkalk, and Keuper Halites. The Röt halite lies almost immediately above the TB horizon, and the associated collapse can be seen in the section. Two other strong collapse effects below the CU level are evident.

Although there is no well in the immediate vicinity, from other wells it is known that (where they are not affected by erosion) the three halites maintain relatively constant intervals. The two other horizons marked M and K are tentatively identified from the collapse effects seen as the Muschelkalk and Keuper Halites, and the time intervals between these, and the interval between the effect at M and that associated with the TB horizon for the Röt Halite check acceptably with the relevant time intervals determined in the nearest well. Thus, an understanding of the specific collapse process here has enabled the stratigraphic correlation of the Triassic halites to be made with reasonable certainty just from the seismic data.

6.5 DISSOLUTION AND COLLAPSE IN THE CAMPOS BASIN

Many salt areas have examples of dissolution effects, and those discussed so far have been limited to the North American and European areas. Another area exhibiting these features is the Campos Basin, Brazil, which lies partly on- and partly offshore to the NE of Rio de Janeiro. In the offshore area, a number of oilfields have been discovered in a zone sub-parallel to the coastline and approximately following the 200 m isobath (see the sketch map in Fig. 6.38). Bacoccoli *et al.* (1980) have described the area offshore as having probable Precambrian basement rocks overlain by

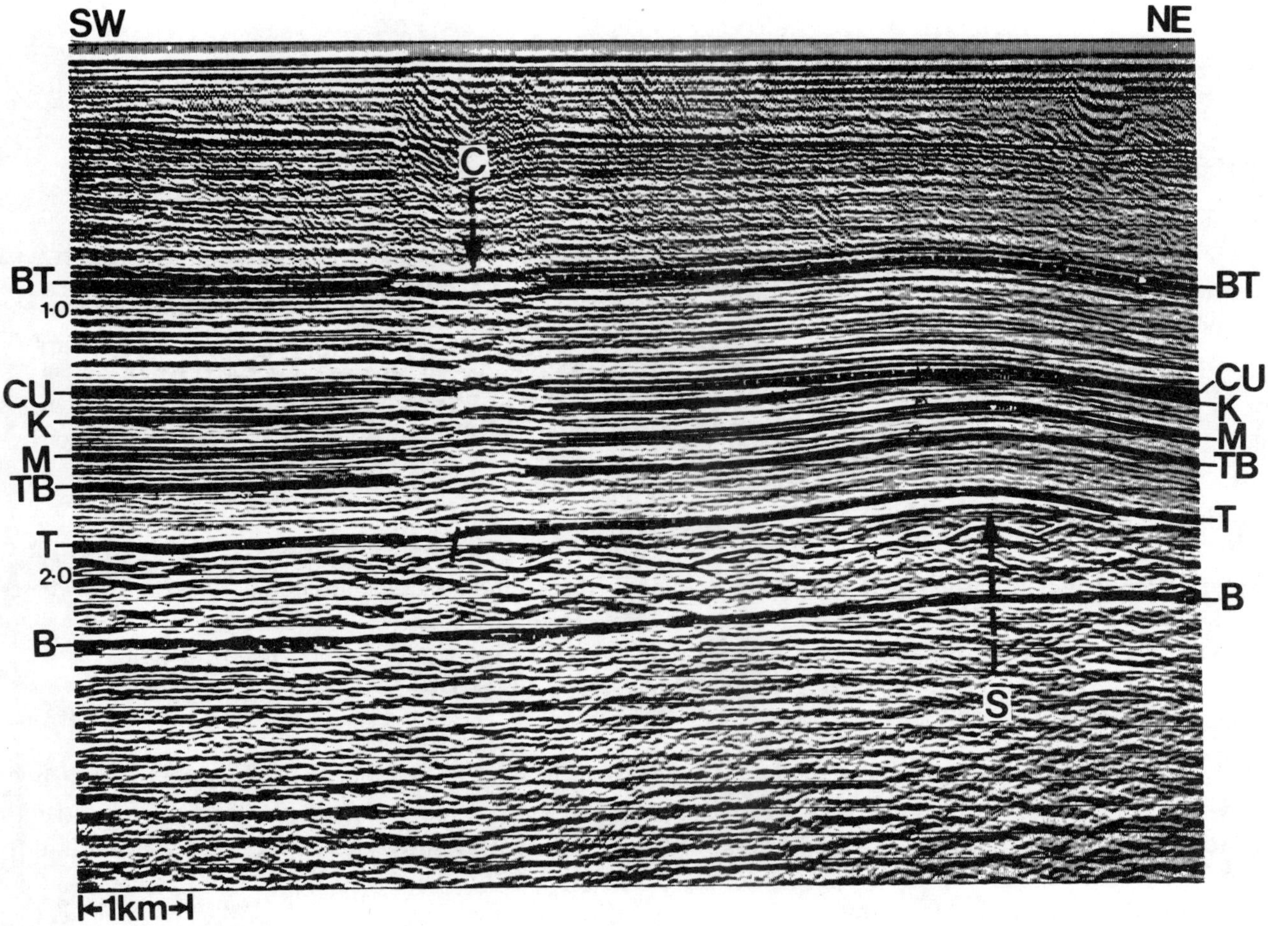

Fig. 6.37 Example of how dissolution/collapse effects may be of help in stratigraphic correlation. See text for key and explanation. (Courtesy of SS(E)L.)

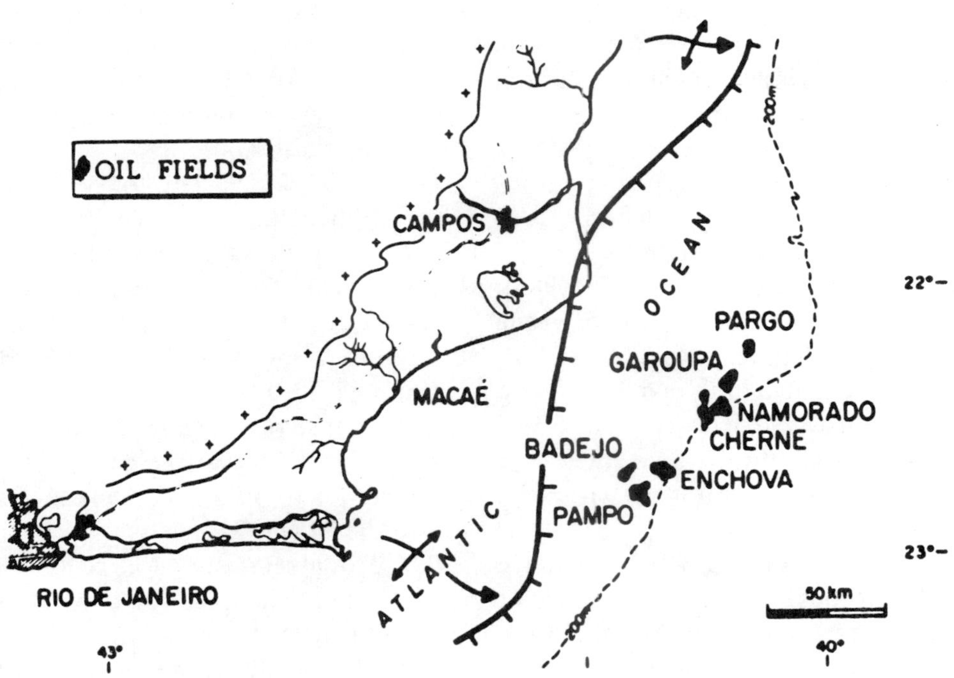

Fig. 6.38 Sketch map of offshore area of the Campos Basin, Brazil, showing the zone of oilfields, and a down-to-basin fault which in places is hinge-like with little or no throw. (From Bacoccoli *et al.*, 1980, courtesy of the AAPG.)

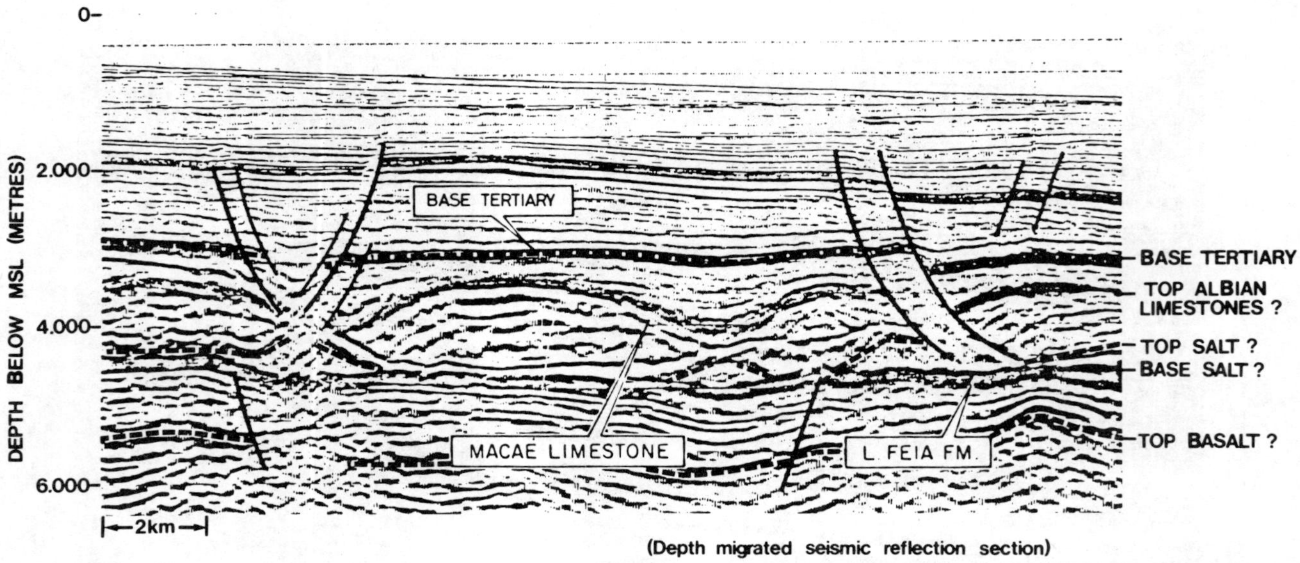

Fig. 6.39 Depth-migrated seismic section of a line passing to the south of the Namorado and Cherne oilfields seen in Fig. 6.38, the line direction being from WNW (left) to ESE (right). Features due to dissolution collapse and salt withdrawal (probably in part involving outflow at the depositional surface) can be seen. (From Bacoccoli *et al.*, 1980, slightly modified; courtesy of the AAPG.)

basalts ranging in age from early Cretaceous to early Tertiary in different areas. Immediately above the early Cretaceous basalts, a syntectonic continental rift stage sequence of early Cretaceous (Aptian) occurs—the Lagoa Feia Formation. The upper part of this sequence includes evaporites, specifically thin anhydrites and thick halite (up to several hundreds of metres encountered in some wells). Above this, an interval of massive Albian carbonates (the Macaé Limestones) was deposited. Hydrocarbon reservoirs occur in turbidite sand bodies interbedded with basinal shales in the Upper Macaé sequence overlying the Macaé Limestones.

Tectonics associated with the rifting are very complex, this area being one of two major deflections in the zone of rifting, caused by the structural grain of the Precambrian basement (Szatmari *et al.*, 1985). The NNE-trending down-to-basin fault shown in the sketch map (Fig. 6.38), which is very much simplified, is in fact more a hinge-line than a fault in many places, with only small downthrows to the east. In some areas, however, it shows larger throws associated with NW–SE trending strike-slip faults (not shown in the sketch) which traverse the zone of oil-fields seen.

Movement of the Aptian evaporites was most active during the Upper Cretaceous, with the occurrence of small salt pillows, together with growth faulting. Bacoccoli *et al.* believe that much salt escaped through the depositional surface during the erosional period represented by the Top Cretaceous/Base Tertiary unconformity, leaving 'salt scars' in the overburden, inversions of relief, and dissolution collapse structures.

Figure 6.39 shows a seismic example of a line to the south of the Namorado and Cherne Fields, the line direction being from the WNW (left) to ESE (right). A growth fault zone near the ESE end of the section, with an uplift on the upthrow side which is probably due to the flow mechanism discussed in Chapter 5 (faulting, erosional levelling with development of load stress differential, and flow of salt from downthrow to upthrow side). Marked thickening and rollover is present in the overburden sediments on the downthrow side of the fault zone.

Near the WNW end of the section, a marked dissolution collapse graben is present, showing the 'double cone' effect mentioned by Lohmann (1972) as being typical of such features in the North Sea area. Between this structure and the growth fault there is a pronounced low which may be due partly to salt withdrawal into the fault zone, and partly to dissolution collapse (note again the 'double cone' effect). It is of earlier date than the collapse graben to the left, showing onlap infill of the late Cretaceous shales above the Macaé Limestone.

The probable escape of large amounts of salt at the surface, and the lack of any substantial diapiric structures in the immediate area, suggest that the overburden at the time of commencement of salt movement was relatively thin. It did, however, consist in part of massive carbonates, which may have been sufficient to produce a density inversion at the salt top. The general situation suggests the probability that updip lateral salt movement initiated subsidence and growth faulting in places, and salt uplifts elsewhere—related perhaps to faults affecting the base—and maybe the whole—of the salt interval. The growth fault zone in this example is superficial, soling-out at the base of the salt interval, but the uplift and collapse seen at the WNW end of the section is probably related to the deep fault which occurs here.

CHAPTER 7

Diapirism—1

7.1 INTRODUCTION

This chapter will be concerned with the nature, origin and mode of occurrence of salt piercement structures. Among the topics covered will be the morphology of some typical forms of diapir, and their distribution in the evaporite basin setting; the concept of diapirism as a process in which at least three more or less distinct stages can be recognised will be explored, and consideration will be given to the important interactive effects between diapirs and post-evaporite depositional patterns. In these and other matters it will be found that the chapter draws together many threads running through the earlier parts of the book.

Because of the present state of knowledge of salt tectonics, the subject is not without its controversial aspects, and attention will be drawn to a number of problems and inconsistencies which are encountered in current concepts of salt movement and diapirism.

The great majority of examples shown, and cases cited, are drawn from Western European and North American sources simply because it is in these two areas that the bulk of work has been effected in salt tectonics, and the most voluminous literature has arisen, over the past century. The availability of illustrative seismic data is unevenly distributed in the historical perspective; in the case of many salt structures found at earlier times in the search for oil and gas, the discovery, exploration and mapping were based on drilling results without any extensive use of subsurface geophysical methods. In many instances, published results at earlier periods included line-drawn depth profiles based on drilling data, or even simple schematic diagrams as the sole cross-sections available. It is only during the past two decades that large volumes of high-resolution seismic data have been produced and (subject to the limitations of confidentiality) become available for illustrative purposes. This period coincides with the massive expansion of effort in marine hydrocarbon exploration, and so it is not surprising that many examples shown are of offshore salt features.

7.2 FORMS OF DIAPIRISM

According to Odé (1968), it was Mrazec (1915) who first introduced the term 'diapir' in studies of Romanian salt structures. The word is derived from Greek, meaning 'to pierce', and is now understood by geologists to refer to structures formed when one rock material which is in a plastic state pierces overlying rocks trans-formationally.

Salt rock is only one of several lithological types which may be involved as the mobile material in piercement structures. Other evaporites such as gypsum and the bittern salts of potassium and magnesium—either alone or as intergrowths with halite—are observed to have flowed under certain circumstances.

Other sediments, including shale (particularly if overpressured), clay and mud also form diapirs. Clay or mud diapirs frequently occur in the deltaic environment, as in the delta of the River Niger, and in the Mississippi Delta area (e.g. the 'mudlumps' referred to in Chapter 5). So-called 'mud volcanoes' which occur in different parts of the world, e.g. in Mexico, are in fact diapiric features. Shale diapirs are present in many areas, such as offshore Niger Delta, and a sheath of diapiric shale is frequently encountered on the flanks of, and sometimes over the crest of, a salt diapir. Many examples of shale sheaths have been found associated with salt diapirs in the Gulf Coast area of the USA.

Igneous intrusive bodies—particularly of granite

and gneiss—are also often diapiric in form; Ramberg (1981), in his centrifuged model studies of gravitational effects in crustal deformation, uses basically the same model materials to study both salt-type features and igneous intrusive-type structures, and the morphological aspects of real-earth sedimentary diapiric structures and igneous intrusive bodies are often very similar.

It is frequently not immediately apparent from seismic data alone what type of mobile material is producing diapiric structures. Velocity effects, including the observation of bounding acoustic impedance contrasts (both magnitude and sign), may be of assistance. The presence or absence of velocity anomalies below the diapiric body, the presence and type of disruption at the crest, including evidence of any typical dissolution collapse effects, and the configuration (or absence) of any primary reflection events within the suspect interval are other observations which may provide diagnostic clues. Lohmann (1980) provides a useful discussion on methods of differentiating, in particular, between salt and shale diapirs.

Of the various forms of diapirism, that involving salt rock as the mobile material has commanded the most attention and resulted in the largest body of literature over the past century. This is hardly surprising when it has been estimated that over 90% of the world's major oil and gas fields are associated with salt deposits. Halbouty (1967) estimates that some 80% of the oil and gas fields in the Gulf Coast province have been geologically influenced by salt movement.

The relationship between hydrocarbons and salt cannot be regarded as in any way mysterious. Apart from source rock potential of carbonates in evaporite sequences, thick salt deposits are generally found in a subsiding basinal setting. In the same setting, but in a different time-frame and palaeoenvironment, organic-rich marine shales occur. Deltaic deposits may underlie the basin, as in the Southern North Sea, providing other sources which will contain terrigenous organic matter. In the same time-frame, the evaporites may interdigitate marginally with marine shales. Subsequent salt movement, if occurring at a propitious time in relation to maturation and migration of hydrocarbons, will provide excellent traps in overburden sediments. Salt and other evaporites can also act as efficient seal formations, preventing upward migration and escape of underlying hydrocarbon accumulations. Considering the immense importance and value of hydrocarbons, which are fundamental to the existence of modern civilisation, the sustained interest in evaporites in general and salt tectonics in particular is hardly surprising.

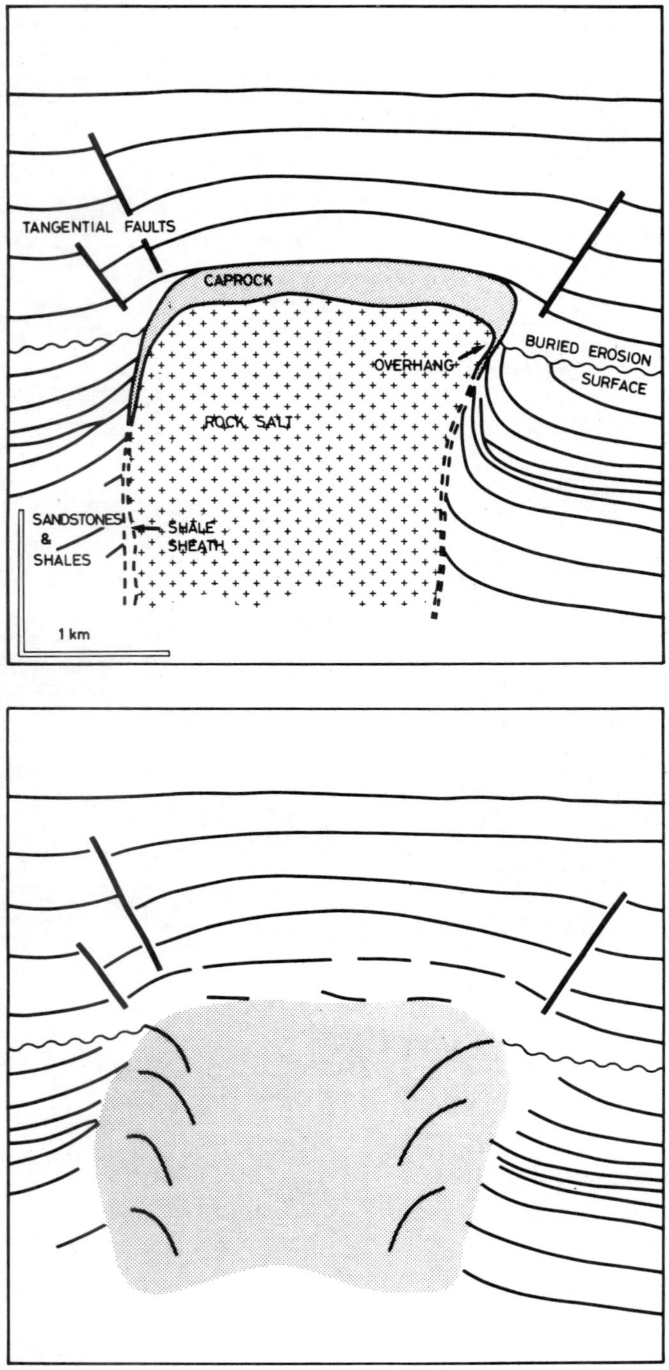

Fig. 7.1 Diagrammatic cross-section of a salt diapir (upper sketch) characteristic of many structures in the US Gulf Coast area. Due to limitations of the seismic method, a seismic section across the feature would somewhat resemble the lower sketch, with no representations of the diapir flanks or adjacent segments of truncated units where the dips exceed some critical value. By means of special studies, and reprocessing of small amounts of data, it is now possible to recover the steep dips of flanks and truncations not seen in sections produced by standard processing. (Upper sketch from NBS Monograph 167, by kind permission of the National Bureau of Standards, US Dept. of Commerce.)

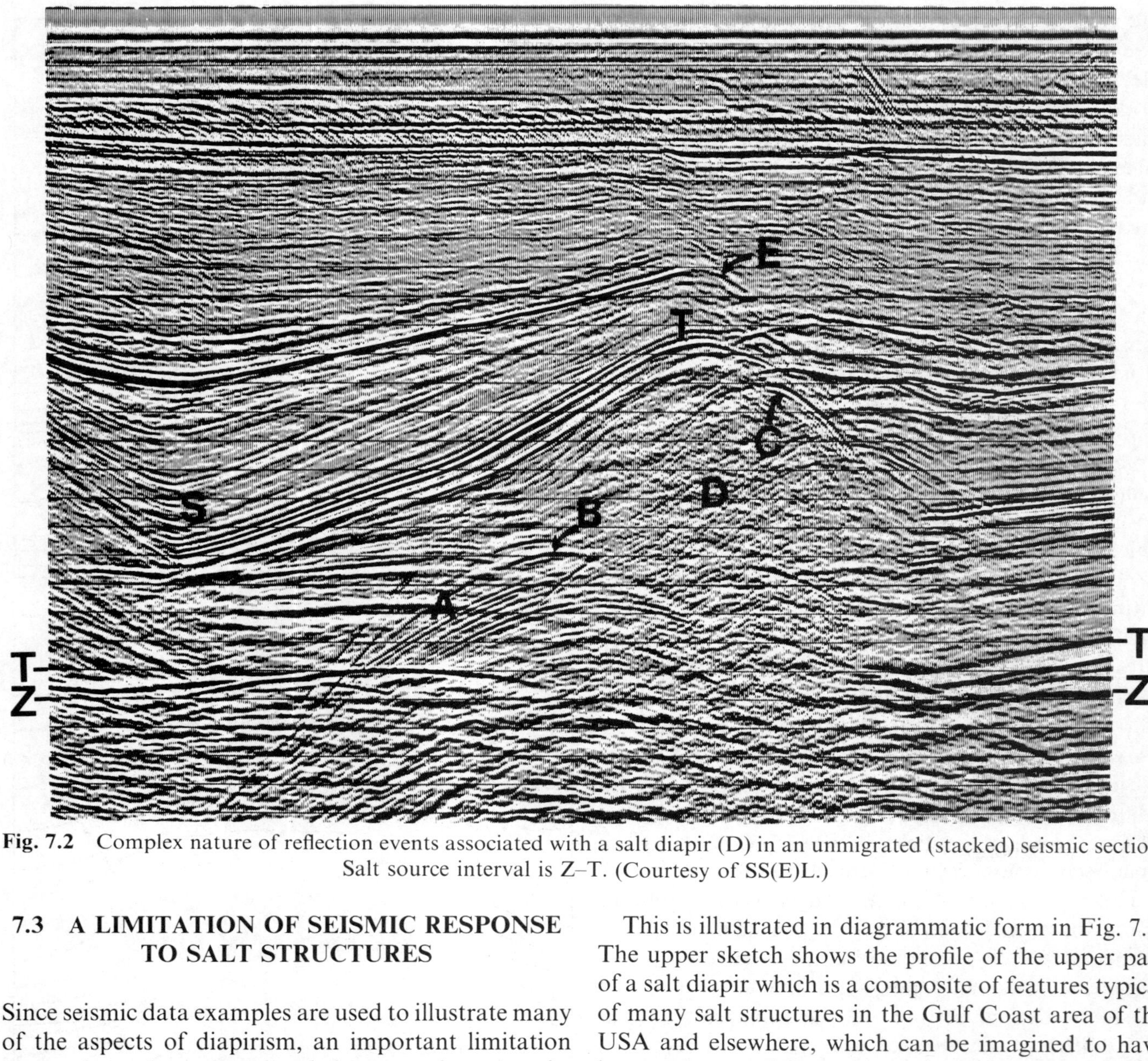

Fig. 7.2 Complex nature of reflection events associated with a salt diapir (D) in an unmigrated (stacked) seismic section. Salt source interval is Z–T. (Courtesy of SS(E)L.)

7.3 A LIMITATION OF SEISMIC RESPONSE TO SALT STRUCTURES

Since seismic data examples are used to illustrate many of the aspects of diapirism, an important limitation imposed on these data by their processing must be realised.

The stacking routine in a standard seismic processing route cannot deal with reflection events which are dipping very steeply. Where a salt feature with vertical, or very steep, flanks is present, having steeply dipping or even overturned overburden units truncating against it, none of the very steep events will be seen in the final section display.

The exact limiting figure for dip on any horizon depends on the time and velocity to that horizon, but typically an event from an interface in the real earth dipping at more than 45° could not be recovered by the correction routine effected prior to the stacking process, and so would not be plotted on the seismic section.

This is illustrated in diagrammatic form in Fig. 7.1. The upper sketch shows the profile of the upper part of a salt diapir which is a composite of features typical of many salt structures in the Gulf Coast area of the USA and elsewhere, which can be imagined to have been constructed from detailed drilling evidence. The lower sketch shows the type of response to this structure which might be expected in a migrated seismic section (although the diffraction curves seen would be more typically present in a stacked section). The dot-shaded area would be a chaotic noise zone, which might include incoherent background seismic noise, fragmentary multiple reflections, side events, reflections produced by velocity 'pull-up' anomalies acting on sub-salt units, partial or complete diffraction patterns (some as indicated) from both within and outside the plane of section originating at truncated bed-ends, and shape-response hyperbolas from deformed beds adjacent to, above, and sometimes within, the salt.

A real seismic example is shown in Fig. 7.2, which

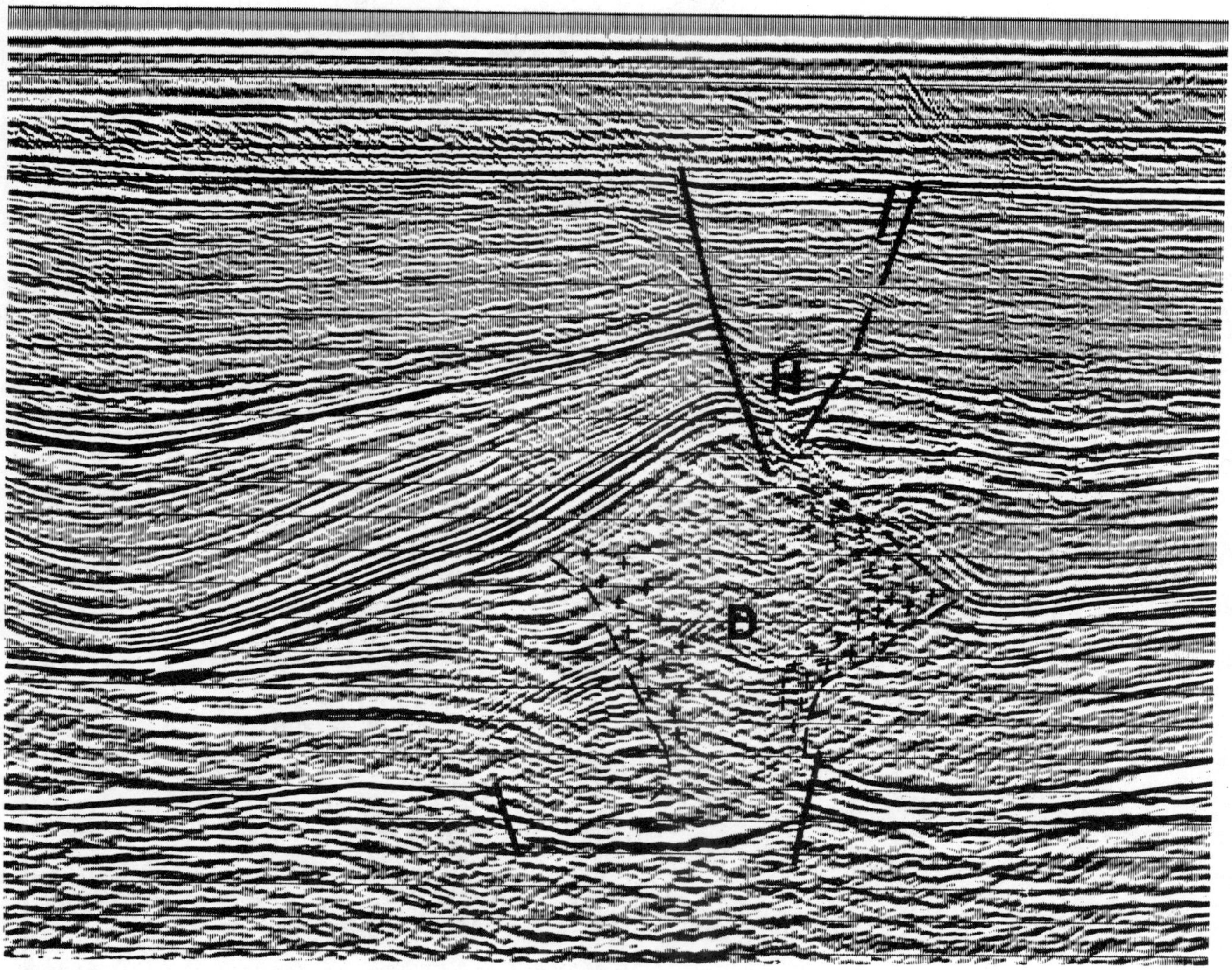

Fig. 7.3 Migrated version of the section example of Fig. 7.2, with estimated boundaries of the diapiric salt marked (cross symbol), and faults defining the graben, H. (Courtesy of SS(E)L.)

is a stacked section across a salt diapir, D, and should be compared with a migrated version of the same data shown in Fig. 7.3.

In Fig. 7.2, a marked partial diffraction pattern is seen at A, while at B and above, several strong, slightly antiform events occur which at first sight belong to the diffraction pattern at A, but in reality have a different origin. At C, a diffraction curve which has the same phase throughout is present, whilst at E, a partial diffraction curve can be seen. The location T marks the local crest of the deformed overburden anticline, and S is an adjacent syncline.

Referring to Fig. 7.3, the migration has not completely collapsed the diffraction pattern A which suggests that it is originating from outside the plane of section. The shape responses of B can be seen to originate in some shallow synform drag-folds produced in truncated overburden units by frictional drag from the rising salt. The event C is probably a shape-response hyperbola related to the sharp synform lower termination of the dissolution collapse graben H in Fig. 7.3. The event E in Fig. 7.2 is observed to be a truncation diffraction at the left-hand bounding fault of the graben. The migration process has restored the real-earth shape of the syncline S, and the overburden flank up to T, quite well.

From a study of the coherent noise events and the effects of the migration, estimated positions for the flanks of the diapir have been marked by dashed lines in Fig. 7.3, with cross symbols for the salt at the boundaries. The resulting shape of the diapir is, it will be agreed, far from obvious at first sight.

There will be some further discussion of this limitation in Chapter 9. However, the reason for the difference between salt diapirs as seen in seismic data and line-drawn profiles constructed from drilling and other information should be borne in mind. It is also remarked that current work shows that, by special

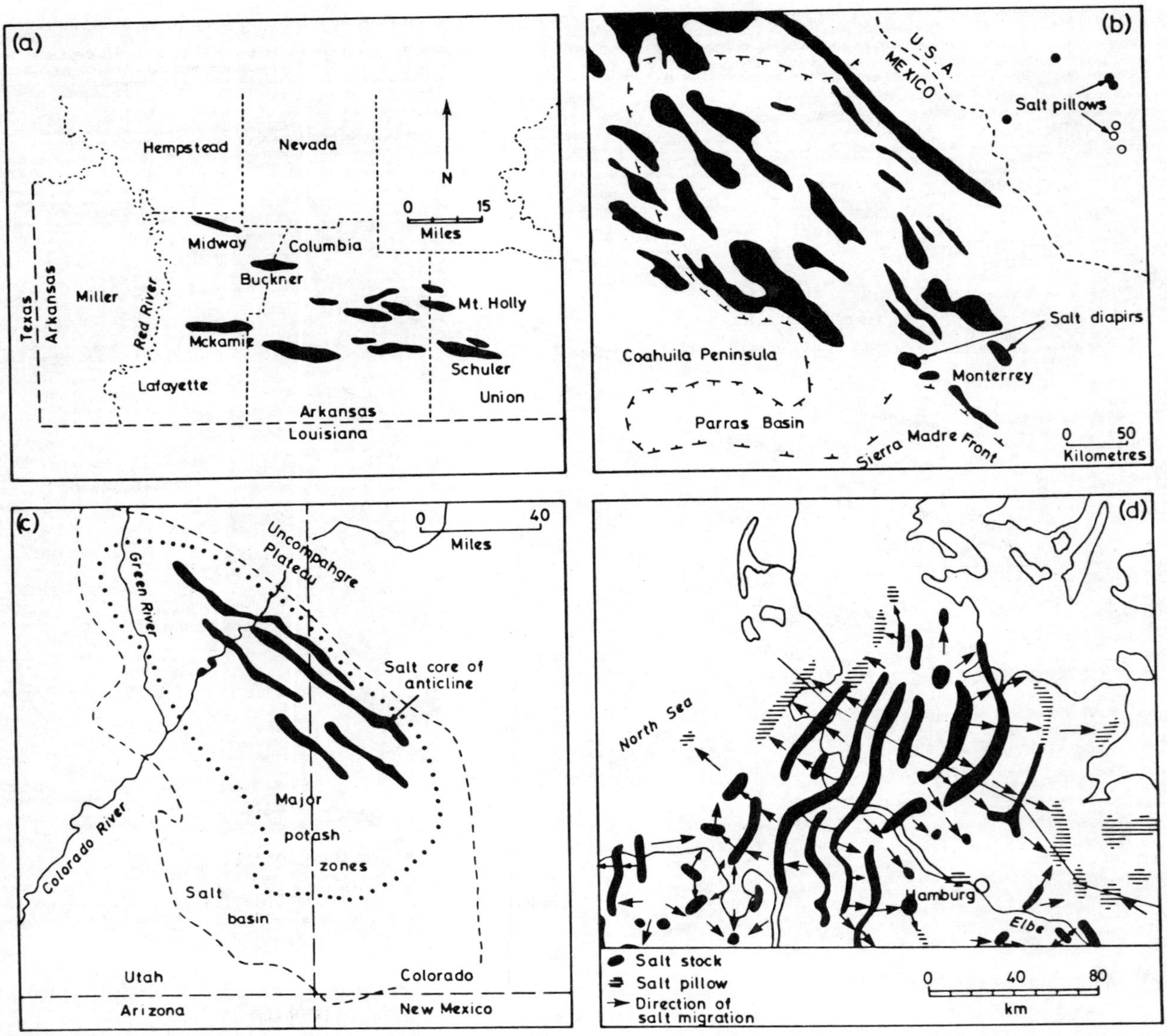

Fig. 7.4 Elongate nature of actual or suspected salt structures in (a) Southern Arkansas, (b) Mexico, (c) Paradox Basin, Utah and Colorado, (d) Northwest of Hamburg, Germany. (Sketch maps redrawn after Murray (1968); courtesy of the AAPG.)

reprocessing of small amounts of data, the steep dips mentioned here can be recovered, if this is necessary for the interpretation in specific cases.

7.4 SALT MOVEMENT IN THE BASINAL CONTEXT

Most salt movement can be explained in terms of either halotectonic or halokinetic processes (these terms having been introduced by Trusheim, 1960, and earlier papers). Halotectonism involves the application of (usually) tangential compressive stress to a stratigraphic section containing a salt rock layer. Halokinesis is an isostatic process involving the gravitationally unstable situation which exists when a denser overburden rests on a more buoyant salt layer. In a halotectonic situation, the bulk density of the overburden may be, but is not necessarily, greater than the salt rock density, since the driving force of movement is an external one. In halokinesis, there *must* be a density inversion at the salt layer, since it is this internal instability of the system which provides the driving force for salt movement.

In many areas, halotectonic and halokinetic processes are closely associated; it seems likely that the former process may often be important in the initiatory and early stages of salt movement, with the latter process becoming increasingly important in controlling the intermediate and later stages. Where salt structures are elongated parallel to important tectonic trends, as in the southern North Sea (Fig. 5.1) and

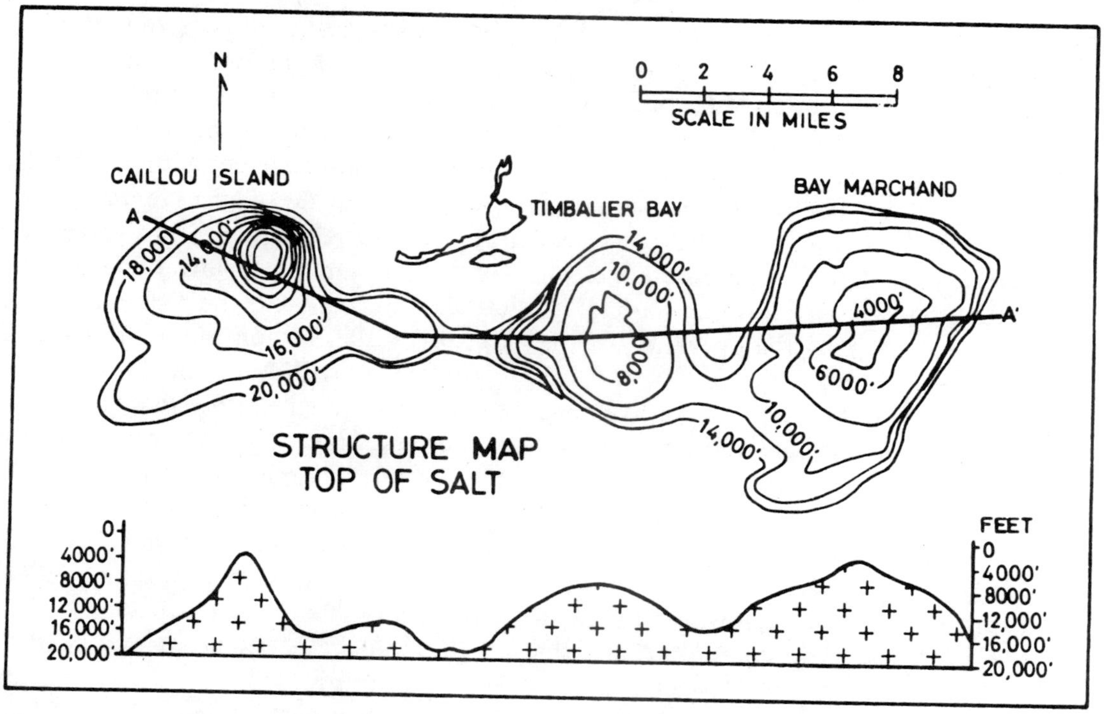

Fig. 7.5 Sketch contour map and longitudinal vertical section of the Bay Marchand salt ridge, offshore Louisiana. (Redrawn after Frey and Grimes (1970), with acknowledgements to the AAPG.)

elsewhere (Fig. 7.4), there is obviously a strong implication that the initiation of salt movement and diapirism was halotectonic.

When a salt rock layer is buried, and conditions of temperature and pressure cause it to become potentially mobile, a number of factors influence the form taken by any subsequent movement. Two important factors on a regional scale are the configuration of the salt basin and the pattern of deposition in the overburden. To illustrate this, two simple models were considered in Chapter 4.

The first of these is illustrated in Fig. 4.7, where a salt layer (cross symbol) in sketch (a) is covered by a relatively uniform thickness of overburden clastics (cross-hatched) towards the margin of a basin, the centre of which is off towards the right. If a wedge of sediment—deltaic or otherwise—progrades into the basin at a later stage (sketch (b)), the additional depositional loading will cause potentially mobile salt to flow from beneath the zone of higher load stress towards the zone of lower loading ahead of the advancing sedimentary wedge. A vertical component of movement may develop due to compaction depression of the section below the sedimentary wedge (analogous to the situation discussed in relation to the Mississippi mudlumps in Chapter 5), in which case a salt ridge will form just basinward of the margin of thickened overburden.

The salt ridge is not likely to rise in a regular way; due to unevenness along the crest, some parts of the ridge will 'lead' other parts upwards, and the locations which are highest will receive salt flow preferentially to adjacent locations as a result of the differential head of overburden pressure at such locations. Individual features will develop, as suggested by the cartoons of Fig. 4.8, although perhaps not in the regular manner shown. A classic example of this type of development is the Bay Marchand salt ridge just off the coast of Louisiana, USA, as discussed by Frey and Grimes (1970). A sketch map and profile form Fig. 7.5. The salt ridge, which is at least 28 miles long and may extend eastwards for another 30 miles or more, is believed by the authors of the paper cited to be due to a prograding wedge of deltaic and related sediments of Late Mesozoic/Early Tertiary age which advanced southwards into the Gulf area, causing deeply buried underlying salt to flow southwards ahead of the depositional wedge, to form the ridge. The salt structures are believed to have risen as intrusive spines from the crest of the ridge, as suggested previously. (For a discussion of Gulf salt tectonics, see Bouma *et al.*, 1978.)

The same general mechanism is suggested by Seni and Jackson (1983) as having operated in the salt deposits of a Jurassic failed marginal rift basin north of the Gulf of Mexico, in East Texas. Activated initially by the deposition of a carbonate wedge, and then by differential loading due to prograding terrigenous clastics, the salt flowed into ridges and troughs perpendicular to the basinal dip; subsequently, diapirs rose as spines from the salt ridges.

It is of interest to note that the development of spine-like diapiric features from salt ridges was also suggested by Trusheim (1960) in his classic paper on the Zechstein structures of Northern Germany (Fig. 4.9), although without invoking the operation of the mechanism just discussed.

The second mechanism to be considered is sketched in Fig. 4.5, where a simple basinal situation is shown with the thickest overburden in the basin centre. The salt layer shown here is of uniform thickness, but more usually would be thicker in the centre of the basin. If the salt becomes mobile, it will move away centrifugally from the basin centre towards the margins due to the load stress differential. Also, the deepest salt in the basin centre will be at a higher temperature, and under greater pressure—hence, less viscous than salt further towards the margins. Trusheim quotes examples of basins in which, at the present day, there is a marked thinning of salt in the central part of the basin, with thickening of salt towards the periphery. He also notes that in a subsiding, flat-floored basin, the elastic limit of the salt will be exceeded simultaneously over large areas, whereas in simple saucer-shaped basins, the first movement of salt should be in the central part.

Taken to its limit, the latter situation will lead to the central area of a basin being free from salt. As noted in Chapter 4, there are two areas (at least) in the centre of the Southern Basin of the British North Sea where salt has withdrawn completely towards the margins; a seismic section across the short axis of one of these forms Fig. 4.6. Immediately surrounding this elongate salt-free area, there is a ring of salt pillows and, at one end, a salt wall diapir. This illustrates the fact that a regular thickening of the salt towards the basin margin may not occur; basinal irregularities such as faulting or the application of tectonic stress may intervene. There is evidence, however, in the Southern North Sea Basin that some salt movement effects did reach the margins in places—as discussed under basin-edge diapirism in Chapter 8.

The salt-free zones just mentioned represent breaches in a salt layer which may be sealing underlying hydrocarbon accumulations. This may be good or bad news, depending on the timing of the maturation and migration of any hydrocarbons present. If the timing is right, these zones may act as important alternatives to faulting as migratory routes for the hydrocarbons. If it is wrong, they may represent routes by which the oil or gas escape to the surface. Much will depend on the formation of faulting and updip pinch-out traps together with overburden anticlines as a result of the subsidence/uplift effects related to the salt withdrawal; equally, the presence of good seal formations related to any traps which form will be critical. In Fig. 4.6, the well-marked graben in the centre of the section shows the location at which complete salt withdrawal first occurred, and the potential in adjacent subsurface locations for pinch-out and turtle-type traps due to the progressive withdrawal of the salt and the related subsidence is clear.

There is the appearance of a paradox in this matter. If, according to Trusheim and others, the largest salt structures in the simple basinal model occur at the basin centre, how does this square with the concept of flow outwards to the margin, with large areas in the basin centre denuded of salt? At some time after lateral salt flow has commenced, there should be little or no salt left at the basin centre, with any large salt structures forming towards the margin of the basin.

It seems likely that where the large structures are indeed found at the basin centre, the reason must be related to the cause of onset or initiation of vertical pillow and diapir formation, as opposed to purely lateral movement of the salt. The initiation of such vertical movements, from whatever cause, may take place early in the period of lateral flow, in which case the large structures will certainly be located towards the basin centre, since the thick salt there provides the greatest supply for vertical movement. There is the possibility that one specific cause of salt movement—major fault movements in the basin floor—may provide both the 'triggering' mechanism to set the salt in motion, and at the same time the preferred locations for pillow and diapir development.

This is speculative, since some believe that, given a deep enough basinal shape, thick salt and adequate overburden loading, salt movement may be self-initiating. However, neither of these hypotheses is 'scientific' in the Popperian sense. All that can be done is to examine the evidence in as many salt basins as possible, and draw whatever conclusions seem most reasonable. Both of these processes, and various others, may explain salt movement patterns in different basins. The two simple models discussed are examples only—each basin must be examined to determine the relationship between subsidence, depositional patterns and salt movement.

7.5 DIAPIR MORPHOLOGY

As noted, it is not usually easy to determine the external form of a salt diapir from seismic data. What is most clearly seen is the effect of the diapir on surrounding and overlying sediments. Study of external form by seismic means must be augmented by the use of profiles and maps derived from drilling data. The upper sketch in Fig. 7.1 is such a profile, although in this case does not represent a particular structure, but is a composite diagram consisting of various characteristics commonly seen in salt diapirs of the Gulf Coast area of the USA.

It has been learned from the study of thousands of diapirs in different parts of the world that their external shape is extremely variable. No two diapirs are ever alike, which is understandable when the many different factors governing their emplacement are taken into account.

Considered in plan view, salt diapirs vary from long salt walls through more restricted elongate, elliptical-shaped bodies to sub-circular structures. In seismic data, a salt wall diapir in vertical section across the short axis looks very much like a sub-circular diapir, and a sufficiently dense seismic grid is needed to define the shape and size of the structure.

Sub-circular diapirs are of common occurrence in the Gulf Coast area, and an example is shown in Fig. 7.6 from the coastal region of Louisiana. The upper sketch shows a profile across the diapir, and the lower sketch is a structure map contoured on an Upper Miocene horizon, and showing the line direction of the profile. The slight offset of the structure contours from the plan view of the diapir suggests that early salt movement produced the overburden doming seen, but that the salt later changed direction upwards. A great many examples of similar features occur in the literature of this area, and often they can be shown to rise from deeply buried salt ridges, like that in Fig. 7.5. The marked overhang of the salt seen on the NW side of this structure is very often present. This massive type of overhang appears to be due to a change in direction of the rising salt—perhaps related to overburden fracturing—at a relatively early stage; the fairly shallow, spreading overhang present at the summit of some diapirs is probably of different origin.

A section in the direction of the short axis of an elongate, elliptical diapir is seen in Fig. 7.7. The origin of this structure was discussed briefly in Chapter 5 (Fig. 5.10). In the direction of its long axis, it is about 12 km long, and shows considerable irregularity of development, which is thought to be due to the salt having injected the contiguous bedding in several places. Two locations at which clear splay of bedding in the 'country rocks' can be seen are marked S. The overburden loading at the time this lateral injection took place cannot have been large, and was probably less than that provided by about 600 m thickness of overburden.

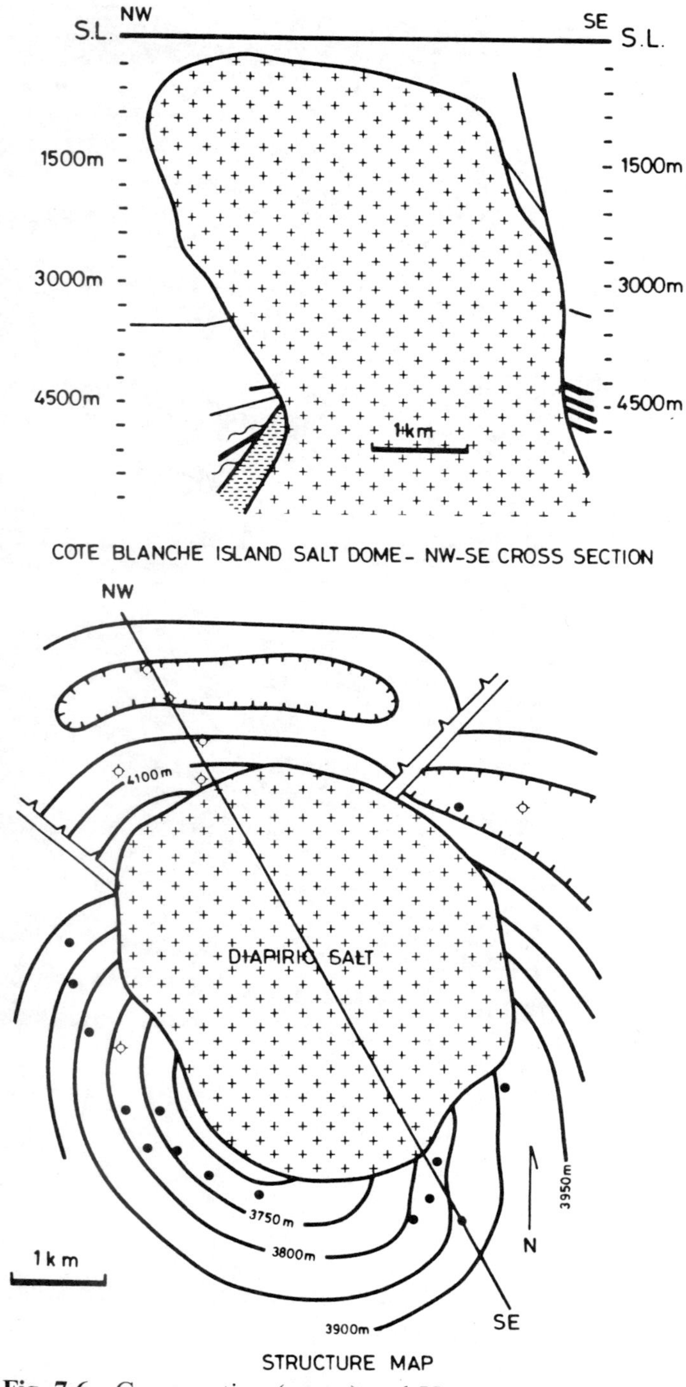

Fig. 7.6 Cross-section (upper) and Upper Miocene structure contours for surrounding sediments, Cote Blanche Island Salt Dome, Louisiana. (Redrawn, slightly simplified, from Woodbury *et al.* (1980), with acknowledgements to the CSPG.)

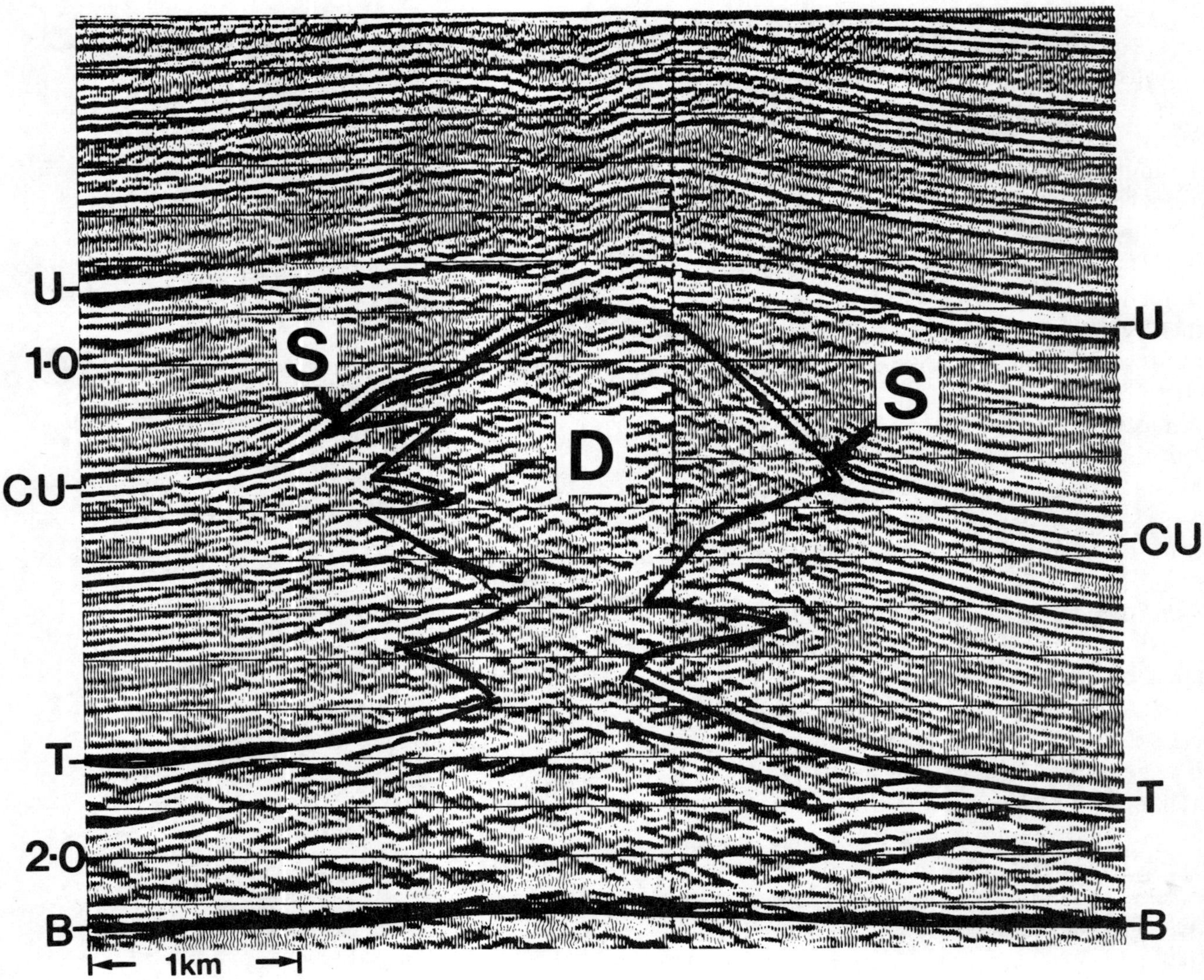

Fig. 7.7 Migrated section across the short axis of an elliptical diapir. Key: B–T, salt source interval; D, diapir; S, splay of bedding due to lateral injection of salt; CU, Late Cimmerian unconformity; U, Base Tertiary unconformity. (Courtesy of SS(E)L.)

In Fig. 7.8, five lines crossing a salt wall diapiric feature in the direction of the short axis are seen, together with a sketch map showing the relationship of the lines to the salt feature. Figure 7.9 shows a sixth line which crosses the long axis of the diapir, D, at a shallow angle. The estimated shape of the salt structure is outlined in each instance. This example is of particular interest since the diapir, half of which is seen at the end of line 5 at its full development, has been affected to a progressively increasing extent towards the ESE by dissolution and collapse into a large crestal graben which reaches its greatest development on line 1. The event marked C on lines 2 and 3 is believed to be the response to a cap rock development (see later in this chapter). In all sections, the top of the salt interval is T, and the base, B, the latter showing evidence beneath the diapir of strike-slip fault zones; this faulting was probably the initiating cause of the diapirism.

An interesting stratigraphic cross-section over the salt pillow at Mölme, in Northern Germany, is seen in Fig. 7.10, from Trusheim (1960). The salt in the pillow has formed a late diapiric feature, clearly due to the peripheral subsidence faulting produced by overburden subsidence into the withdrawal basin to the left of the pillow. This type of process, where a late piercement feature develops from a pillow due to faulting (of graben type in this instance), is seen in Fig. 5.26, and peripheral subsidence faulting in Figs. 5.30 and 5.31.

Another example, of unusual interest, from Trusheim (1960) is seen in Fig. 7.11. This shows a sec-

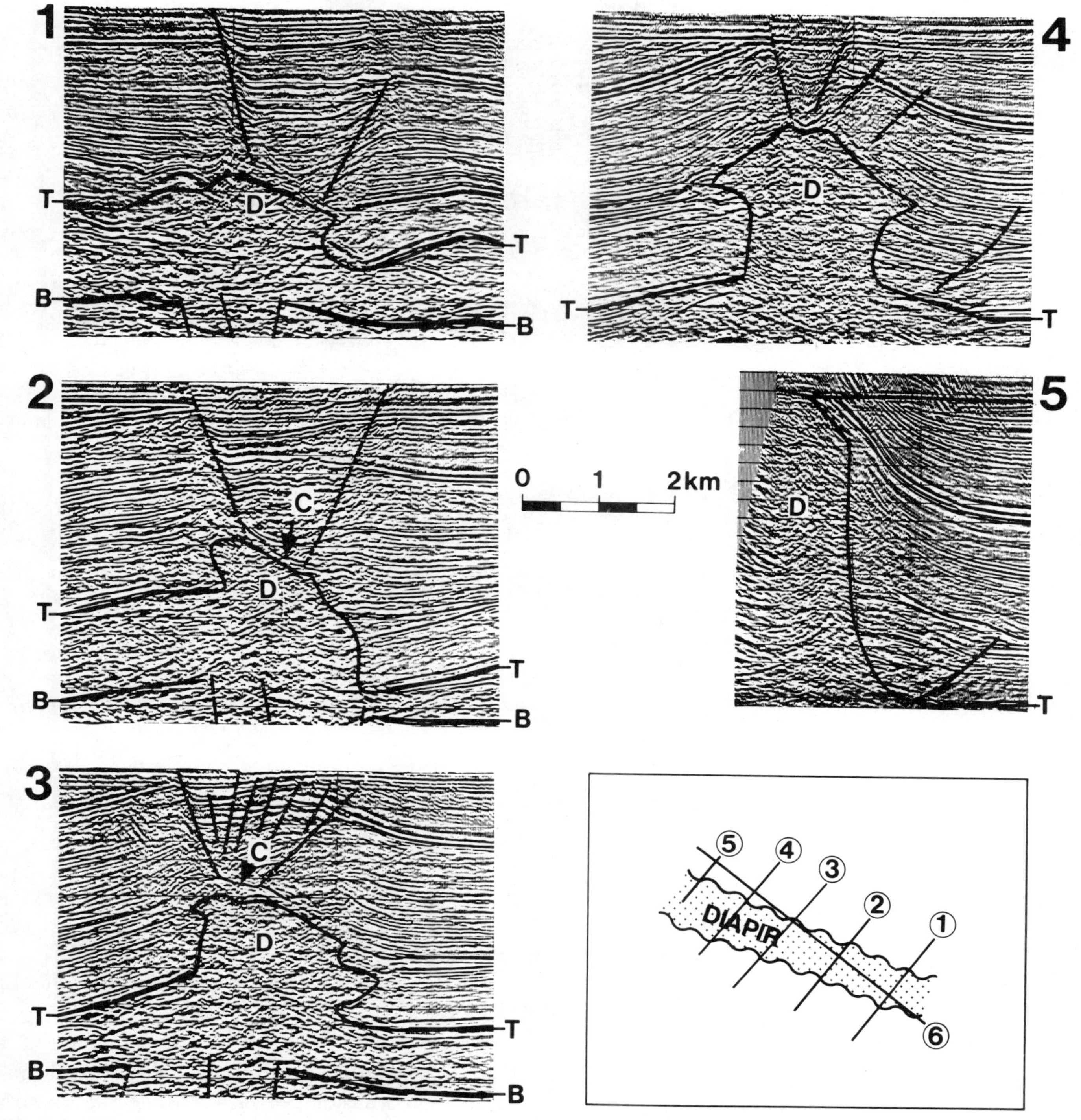

Fig. 7.8 Five lines across a salt wall diapir, D, which has been affected by dissolution and collapse to a progressively greater extent to the ESE (see key sketch map, lower right). A sixth line, running at a shallow angle across the long axis of the salt structure, is shown in Fig. 7.9. Key: B–T, salt source interval; C, reflection event believed to be the response to a cap rock development. (Courtesy of SS(E)L.)

tion across the Eilte salt structure, which has an oil accumulation in truncated beds below the SW overhang of the salt body. As interpreted, this salt mass has risen through the overburden, becoming at some stage detached from the source layer of salt. It is assumed that the salt has moved up the underlying fault zone, expanding to its present size in the low load-stress environment of the near-surface, beneath relatively thin overburden. This structure is, however, open to other interpretations, and doubts have been expressed regarding the process suggested above. The sub-salt faulting may actually have separated the

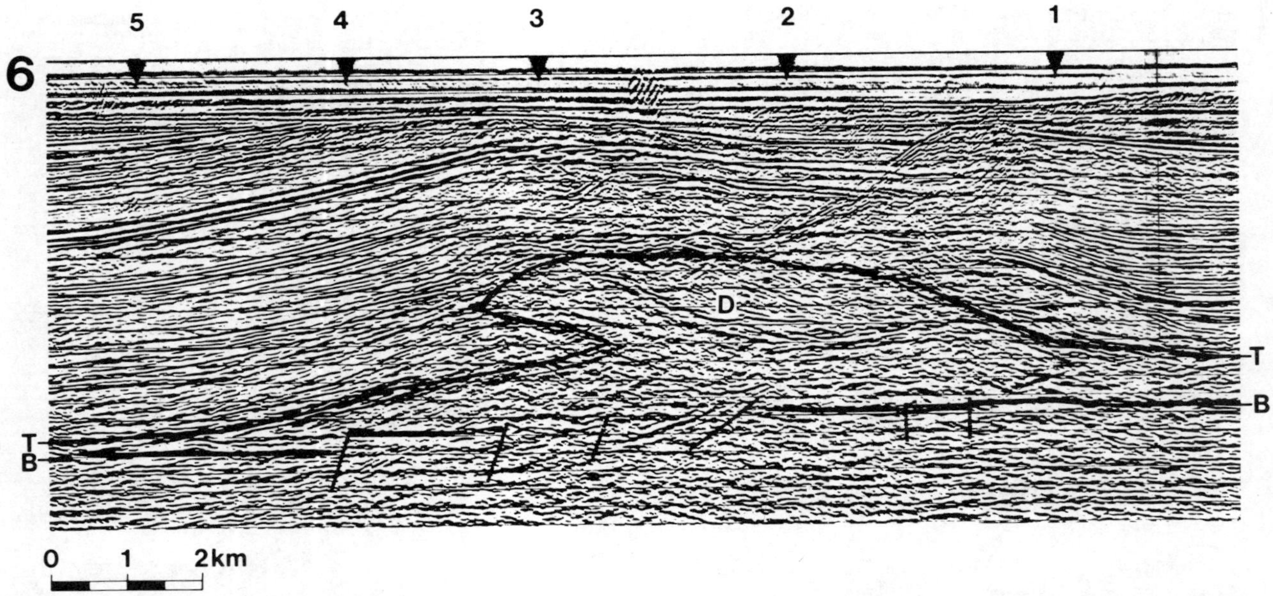

Fig. 7.9 Line 6 in the sketch map of Fig. 7.8. Key is the same, except that event C in Fig. 7.8 is not differentiated here. Intersections of lines in Fig. 7.8 are shown. (Courtesy of SS(E)L.)

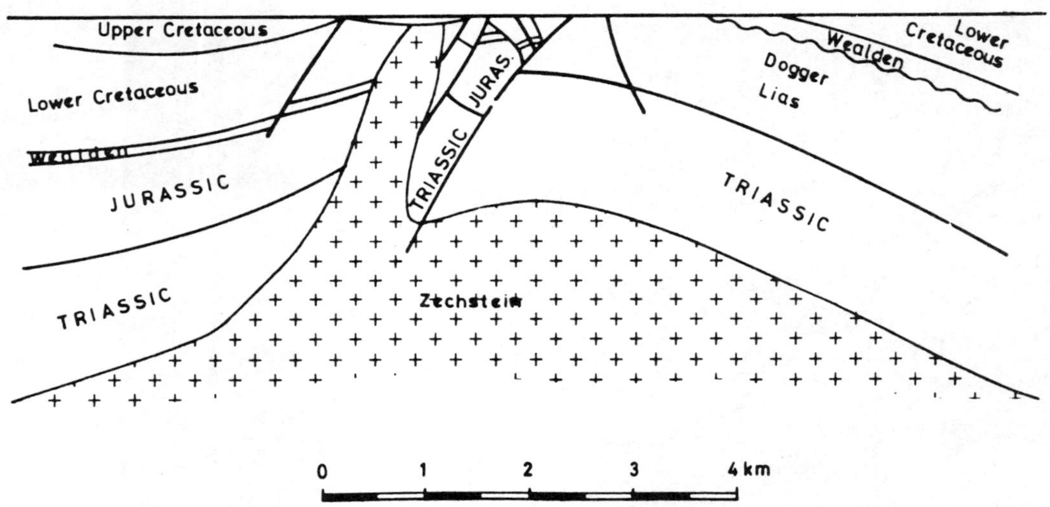

Fig. 7.10 Section through a salt pillow at Mölme, Northern Germany, which has developed a piercement structure at a late stage. (Redrawn after Trusheim (1960); courtesy of the AAPG.)

diapir from the underlying source layer. It would also be of interest to know the depth at which average sediment density equals that of salt.

The few examples illustrated here will serve to show the great variety of form exhibited by salt diapirs; numerous other examples will be seen in the rest of this work.

7.5.1 Diapir Flanks

As suggested by Figs. 7.6 and 7.7, the flanks of a salt diapir may vary from being relatively smooth, near-cylindrical surfaces, to irregular shapes with re-entrants and protruding segments. Model studies have suggested—and this is supported to some extent by field evidence—that the competency of the overburden materials through which the rising salt passes has a considerable bearing on this. Smooth-flanked, plug-like diapirs tend to be found emplaced in sequences of dense, compact lithology such as carbonates. Less dense and less competent clastic sequences, tending to occur in the shallower part of the stratigraphic section, appear to be subject to lateral pressure-injection of salt into bedding-planes (*vide* Fig. 7.7). The latter process can probably only take place under relatively

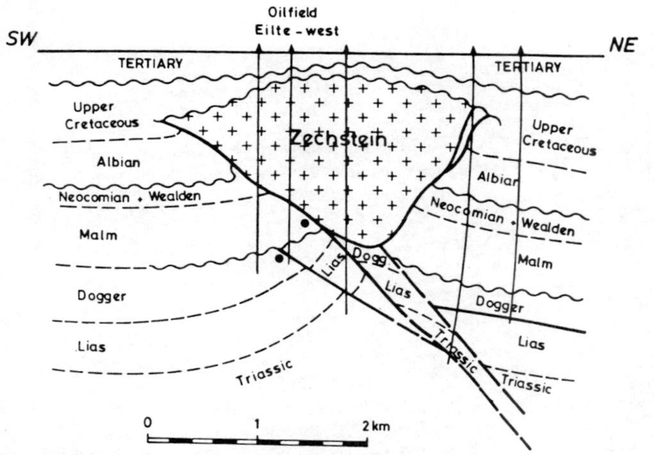

Fig. 7.11 Section through the Eilte salt structure, Northern Germany; an example of a salt diapir which has become separated from the source layer completely. (Redrawn after Trusheim (1960); courtesy of the AAPG.)

small overburden loading and, where it is observed, is a good indicator that the diapir was emplaced at quite shallow depth. The matter of lithological type becomes less important with increasing depth, since normal burial compaction will eventually result in a high degree of competence of most clastic rocks. With a typical clastic sequence, it is suggested that the critical depth below which bedding injection by the salt will not take place is between 800 and 1000 m.

7.5.2 The Crest of the Diapir

The crestal shape of diapirs is another very variable factor. In some areas, pronounced disruption of the crestal region is common, due to the development of spines, or to dissolution collapse following overburden fracturing and water circulation. Where the crestal region is intact, the diapir may have a flat, or dome-shaped top, often consisting of a cap rock development (see Section 7.5.3). In some circumstances, a spreading overhang—a mushroom-like lateral extension of the salt top—may develop. A sketch showing a German example is shown in Fig. 7.12. Trusheim (1960) noted that such features are quite common, and that North German examples seem to have been extruded under water, sometimes associated with outflow breccias formed by broken fragments of roof rocks. Gripp (1958) regarded the features as the remnants of 'salt glaciers' similar to the present day diapiric emergences in the Hormuz area of Iran (Kent, 1979, and others), and suggested that they may have been extruded below a layer of unconsolidated bottom sediments. The latter suggestion is of interest, since for a salt rock extrusion to survive, some unusual conditions must exist. If in the terrigenous situation, the climate would need to be extremely arid, as in present-day Iran, or a fine balance between erosion/dissolution and continuing

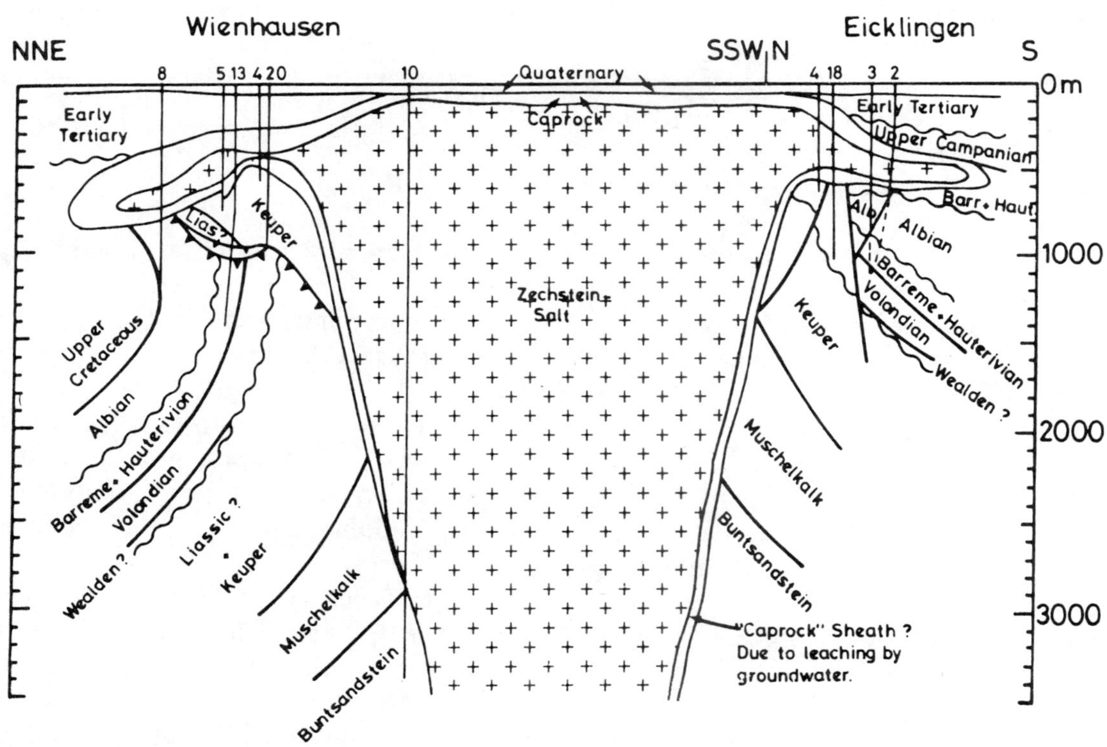

Fig. 7.12 Section through the Wienhausen salt diapir, North-west Germany, showing the development of a spreading overhang. (Redrawn from Gussow (1968), after Schott (1956); courtesy of the AAPG.)

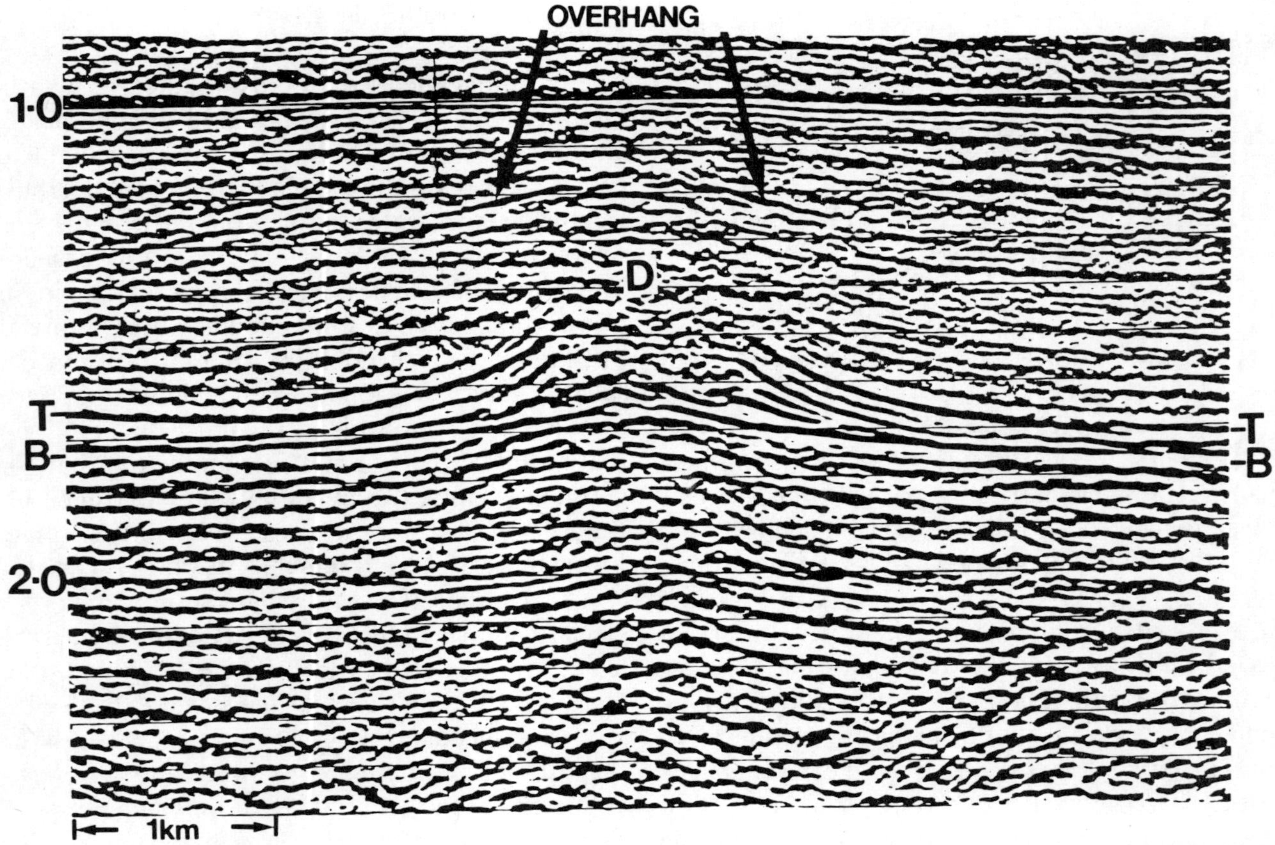

Fig. 7.13 Seismic section through a small diapir, D, in the Northern North Sea, showing a typical response to a spreading overhang as discussed in the text. Salt source interval is B–T. (Courtesy of SS(E)L.)

salt uplift would be required until net deposition covered the salt. In the marine situation, either the water would have to be saturated with respect to NaCl, or injection below a superficial layer as suggested by Gripp would be required, with the material overlying the salt being impermeable in order to prevent dissolution. This process is suggested to some extent in Fig. 6.20, in connection with later dissolution and collapse.

In Fig. 7.13, an example of what is believed to be the seismic response to a spreading overhang associated with a small diapir, D, is seen. Top and base of the salt layer are at T and B.

There is another possible explanation for overhang features of this type which derives from model studies. Certain experiments carried out by Parker and McDowell (1955), Ramberg (1981), and others suggested that if a rising body of mobile material encounters a particularly dense and competent layer, its upward movement is checked, and it tends to spread laterally beneath the competent layer. It is possible that the same effect may occur with a rising salt mass, particularly when it reaches the shallower part of the overburden. The energy of emplacement (either by injection or down-building) will be slackening off, and the density of contiguous sediments may be less than that of salt. In this case, lateral injection may occur.

7.5.3 Cap Rock

The term 'cap rock', which in general refers to the uppermost unit in a stratigraphic sequence, has a more specific meaning in the context of salt diapirism. Many salt diapirs have summits which consist of a flat, apparently layered structure sometimes several hundreds of metres thick, known as a cap rock. The layering normally includes some or all of the following materials in varying proportions: anhydrite, gypsum, calcite and sulphur, sometimes with minor occurrence of other minerals.

Cap rock is believed to be due to the accumulation of insoluble residues of the salt rock at its rising crest, where it is most likely to be subjected to dissolution by formation water in passing upwards through the overburden. The layered insolubles would be strongly compacted and indurated in all probability during this passage.

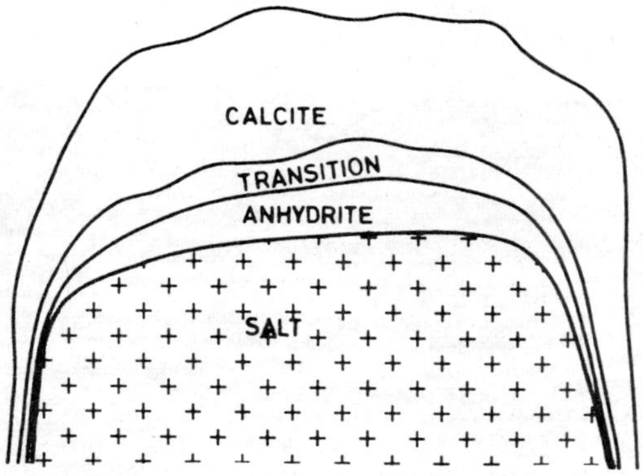

Fig. 7.14 Zonal configuration typical of many cap rock developments.

Figure 7.14 shows the separation of minerals frequently observed in the cap rock above a salt diapir. It could be more properly described as zoning, rather than layering, as the relationship between the zones can be irregular and complex. Taylor (1938) noted that in the cap rocks of Louisiana salt domes, a calcite zone is found above anhydrite in thicker, better developed cap rocks, and the two zones are separated by a transition which is generally a mixture of anhydrite, calcite, sulphur, gypsum and minor quantities of other minerals.

As regards the origin of the various minerals present, there is a consensus that the basic material present is accumulated anhydrite, and the other minerals are alteration products. Feely and Kulp (1957) suggested that, based on laboratory experiments, the sulphate ions of the anhydrite are reduced by bacterial action, releasing sulphur and H_2S. They also suggested oxidation of hydrocarbons as the origin of the carbon of calcite in cap rock, but this is rather more controversial.

Halbouty (1967) noted that, as may be expected from the mode of accumulation of the cap rock material, deep salt structures have less than shallow ones, but that it has been found on salt structures more than 10 000 ft deep; in the US Gulf Coast area, the average cap rock thickness is 300–400 ft, but in some examples, e.g. the Jefferson Island structure in Louisiana, over 1000 ft of cap rock has been penetrated.

If a cap rock is present intact, then normally it could be expected to provide a very strong acoustic impedance contrast with both overlying sediments and underlying salt because of the compact, high-density materials of which it is made up. Where the cap rock is sufficiently thick in relation to the dominant seismic wavelength at that level in the section, it should produce a strong reflection event at the diapir summit, if this has not been disrupted. Figure 7.15 shows an example of a small salt diapir in the Northern North Sea Zechstein Basin of the British sector, with a very strong reflection at C which is believed to be the response of a cap rock development and which stands out clearly in spite of severe interference from near-horizontal multiple reflections. The strong reflection marking the spreading overhang in Fig. 7.13 may also indicate the presence of a cap rock.

In many areas, such as the North Sea Southern Basin, cap rock developments appear to be absent. This could, however, be ascribed to the fracturing, dissolution and collapse of diapiric summits so typical in this area, and it has been suggested that cap rock structures are the norm in salt diapirs, and that, where absent, it is only because of the reasons just mentioned.

7.5.4 Overburden Displacement

The question of what happens to displaced overburden material during salt diapirism is similar to that posed by many large igneous intrusions and their displacement of the country rock. It is not easy to answer in many cases due to lack of evidence. Often, the overburden above a diapir has been so severely disrupted by fracturing, dissolution and collapse that no positive conclusions can be drawn.

Model work, by Nettleton (1934) and Parker and McDowell (1955) particularly, offers some proposals as to possible displacement processes. There is a suggestion that in some cases vertical displacement of a solid plug of overburden above the diapir may occur, while in other cases, roughly triangular flaps or segments of overburden may be pushed upwards and outwards from a central point above the diapir crest by the rising salt (see Chapter 5, Fig. 5.29).

Smith and Reeve (1970) studied 16 shallow salt diapirs, i.e. structures the crests of which were within 1200 m of the surface, in the Gulf Coast area of Texas. They discuss examples of salt diapirs showing both types of overburden displacement just mentioned, noting that large displacements in a vertical direction of overburden material above some diapirs have occurred. In one instance, in the White Castle Dome structure, Frio (Oligocene) material was observed which may be as much as 4000 m above its normal stratigraphic position related to the adjacent section, and Miocene cover is 1200 m high with respect to the flank section.

From the studies made, it appears possible that (i) the displacement of overburden in uplifted, triangular

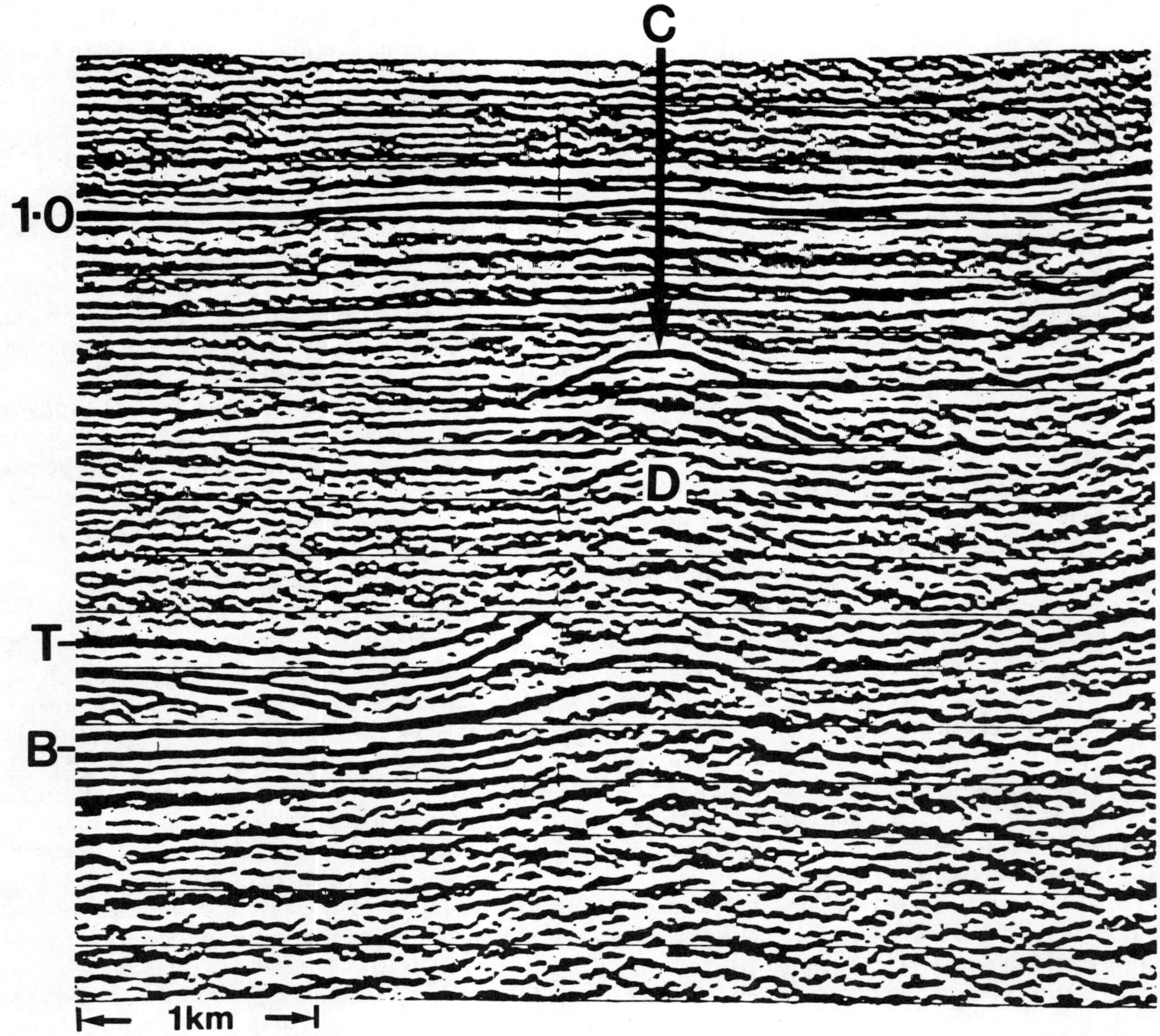

Fig. 7.15 The very strong convex-upwards reflection at C is interpreted as the response to a cap rock at the crest of diapir D in the Northern North Sea. Salt source interval at B–T. (Courtesy of SS(E)L.)

segments occurs with relatively thin overburden, or with less competent overburden materials, and is subject to marked erosional effects in many locations, and (ii) upthrust plugs of overburden (see Fig. 7.16) probably occur with thicker and/or more highly competent material above a plug-shaped diapir with a flat crest of large area, perhaps in association with a well-developed cap rock.

Following the analogy with igneous intrusions drawn earlier, the question must arise as to whether any process akin to magmatic stoping takes place during diapir emplacement, as a means of displacing overburden by engulfing it. From the writer's experience, the answer must be that although relatively minor stoping effects are seen in many seismic sections (see, for example, Chapter 2, Fig. 2.7), there is no good evidence that the process accounts for major material displacement. The two processes mentioned previously, associated with erosion and redistribution of the uplifted overburden material, probably account for most displacement effects.

7.6 DIAPIR EVOLUTION

First a note on terminology. Following Section 2.2 and the first paragraph of Section 7.2, it should be clear that in this book the term 'diapir' refers to a structure resulting from trans-formational invasion of overlying beds by the mobile substrate material involved—in this case, salt rock. Petroleum geologists—particularly in the USA—have come to use a term 'non-piercement diapir' to mean a structure which has not pierced potential hydrocarbon reservoir beds, although having pierced beds deeper in the stratigraphic section. Halbouty (1967) mentions the use of such terminology, and it appears in many papers (see, for example, Bishop, 1978; Woodbury *et al.*, 1980). While not arguing with this usage in the specialist sense, the writer feels that in a book which is not directed specifically at petroleum geologists, the use of a term which translates literally as 'non-piercement piercement' is not appropriate. The terms 'swell', 'pillow' and 'diapir' as defined in Chapter 2.2 will be employed throughout. The terms 'dome', 'stock' and 'boss' as alternatives for 'diapir' will not be used except in citations from the literature; although 'dome' has a long and honourable history as a term for salt diapir in the Gulf Coast region of the USA, elsewhere its usage is confined to the description of any structure showing quaquaversal dip, with no piercement connotations. Stock and boss, with their strong igneous associations, are not considered to be necessary for use in any special circumstances, and are therefore superfluous.

7.6.1 Stages of Diapir Evolution

Early physical model experiments by Nettleton (1934) using two immiscible liquids of differing densities as model mobile substrate and overburden showed that when a positive feature develops in the mobile substrate, it is surrounded by a depression in the surface of the mobile layer due to flowage into the feature (see Fig. 4.3). This depression was called the 'peripheral sink', and it has been established by drilling and other evidence that positive salt structures have such sink-like features associated with them.

Again on a matter of terminology, somewhat confusingly it has become usual to employ the terms 'peripheral sink' and 'rim syncline' (a depositional feature in the overburden related to the rise of a salt structure—to be discussed later) as though they are interchangeable. It is preferred here to use the term 'peripheral sink' in its original meaning, i.e. for the depression in the surface of the mobile layer, and 'rim syncline' for the later depositional features.

Trusheim (1960) in his study of Zechstein salt movement in Northern Germany showed that the

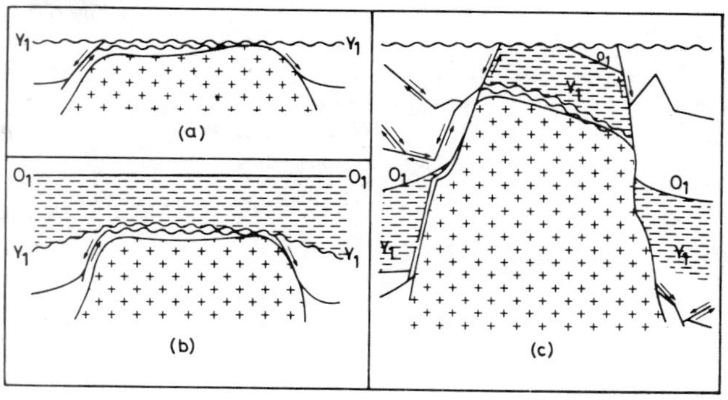

Fig. 7.16 Displacement of overburden by the upthrust of a massive plug of material above a salt diapir (cross symbol). This is the White Castle salt diapir of Louisiana, in which over 2000 ft of Miocene clastics above the diapir have been uplifted more than 4000 ft above their stratigraphic position in the flank section. Peripheral faulting in a concentric pattern surrounds the salt diapir. Three stages in the process, as interpreted by Smith and Reeve (1970), are shown. (Redrawn from Smith and Reeve (1970); courtesy of the AAPG.)

development of a diapir is a process in which ideally three stages can be recognised—those of pre-piercement, piercement and post-piercement.

The pre-piercement stage is represented diagrammatically in Fig. 7.17(a), and involves the development (for reasons not yet specified, but which might for instance involve base-salt faulting) of a thickened swell in the salt layer, with the concomitant evolution of a peripheral sink in the top of the salt surrounding the swell (if the latter is sub-circular) or parallel to it (if the swell is elongate). The volume of salt in the swell above the level of the adjacent undisturbed salt top will equal the volume of the peripheral sink below this level (although recent work suggests that this equality is only very approximate). It is important to notice that the peripheral sink is *not* a surface at which deposition contemporary with the salt movement is taking place—considerable thicknesses of sediment of post-salt age may have built up before the salt movement is initiated. At the depositional surface which is contemporary with the salt movement, and which may be many thousands of feet above the top salt surface, a subsidence depression known as the 'primary rim syncline' develops in which sediments are deposited. The general configuration of this depositional feature will follow that of the peripheral sink in relation to the shape of the salt swell.

The cross-sectional shape of the sediments deposited in the primary rim syncline, as seen in Fig. 7.17(a), should be noted, with the sediments thinning towards the salt swell from the trough of the syncline.

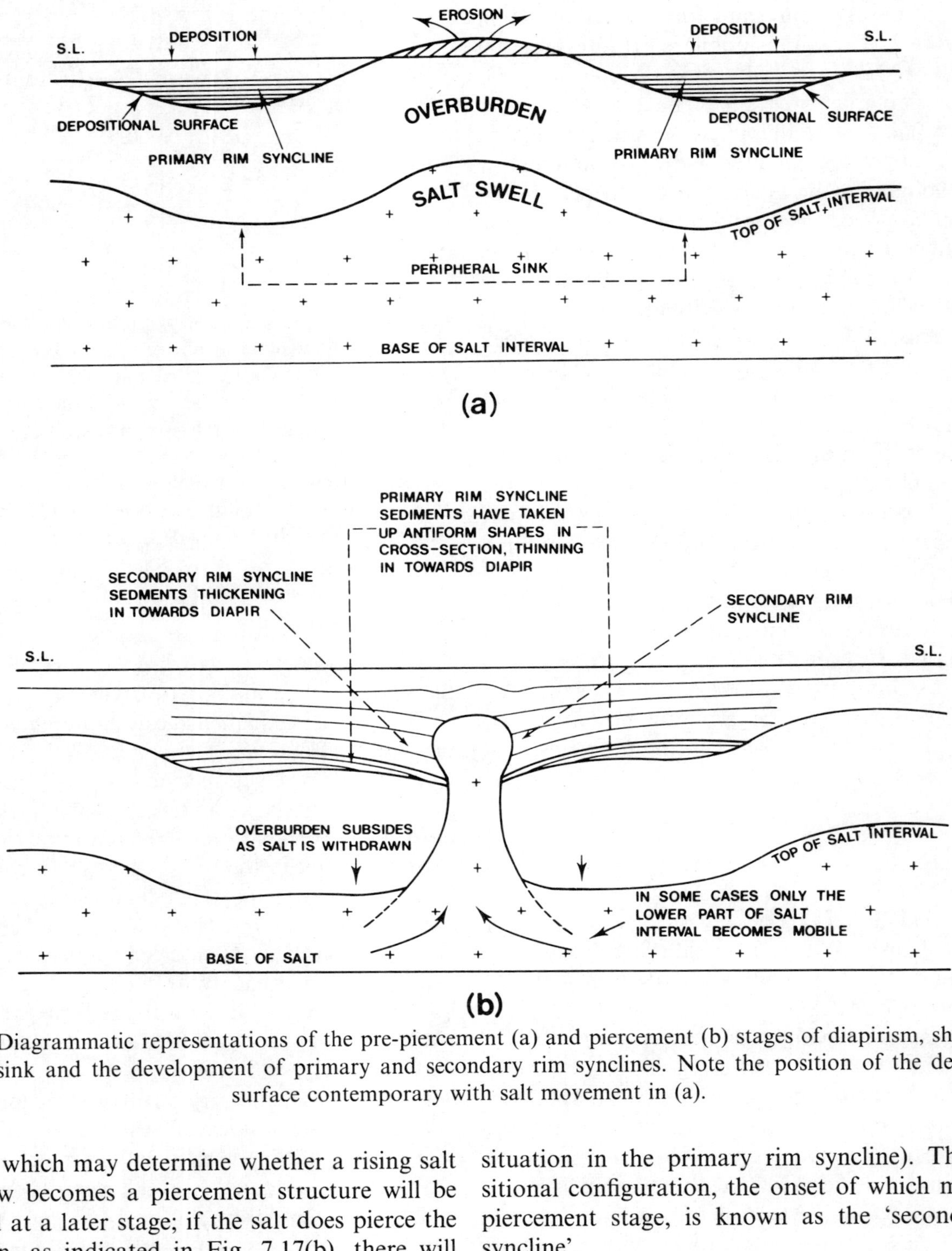

Fig. 7.17 Diagrammatic representations of the pre-piercement (a) and piercement (b) stages of diapirism, showing the peripheral sink and the development of primary and secondary rim synclines. Note the position of the depositional surface contemporary with salt movement in (a).

Factors which may determine whether a rising salt swell/pillow becomes a piercement structure will be considered at a later stage; if the salt does pierce the overburden, as indicated in Fig. 7.17(b), there will be a relatively rapid influx of salt from the zone immediately adjacent to the location of the piercement, causing much more rapid subsidence of the overburden immediately surrounding the position of the young diapir. At the depositional surface, this will result in rapid subsidence increasing towards the diapir, with deposition in the resultant developing depression of sediments which *thicken* towards the diapir because of the pattern and rate of the subsidence (unlike the situation in the primary rim syncline). This depositional configuration, the onset of which marks the piercement stage, is known as the 'secondary rim syncline'.

As suggested in the Fig. 7.17(b) diagram, the salt which flows into the diapir may not be drawn from the whole salt interval—there is evidence (examples will be shown later) that in some structures, only deeper parts of the salt interval have become mobile.

Another point illustrated by Fig. 7.17(a) is that, usually, the crest of the overburden high above a rising salt swell will be subjected to erosion, and normal winnowing and gravitational effects will lead to the

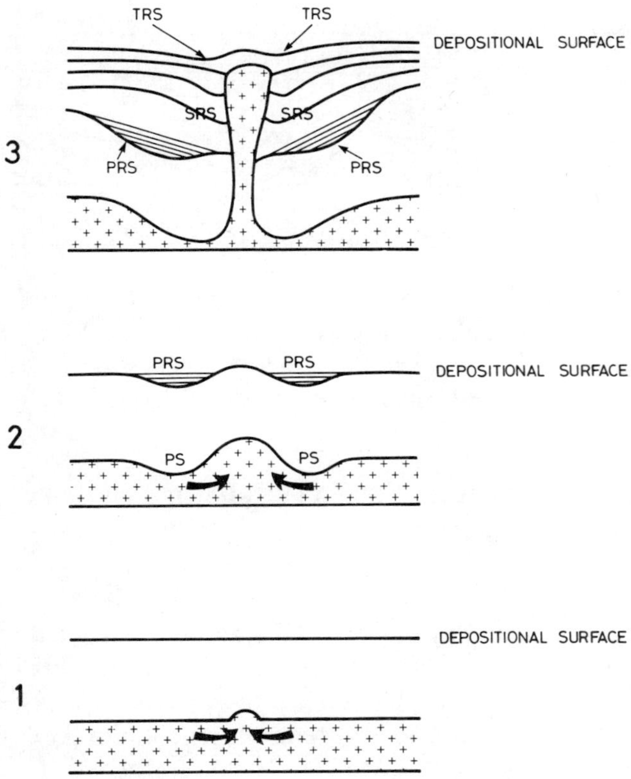

Fig. 7.18 Diagrammatic representation of all stages of evolution of a diapir, from (1) swell initiation, through (2) pre-piercement stage to (3) piercement and post-piercement stages. Key: Cross symbol, salt; PS, peripheral sink; PRS, primary rim syncline; SRS, secondary rim syncline; TRS, tertiary, or third-order rim syncline.

re-deposition of the erosional products in the adjacent primary rim syncline. This gradual decrease in overburden loading above the rising salt feature, and increased loading in the rim syncline, will have an accentuating effect on the flow of salt into the swell, accelerating the rate of rise of the structure.

Trusheim (1960) noted that the change from the swell/pillow to the piercement stage may be marked by a change in the facies of sediments deposited in the rim synclines from shallow-water to deep-water basin facies, and that this break, often marked by an emersive or transgressive surface, is often a good reflector in seismic data.

After the development of the piercement stage has largely concluded, some minor late stage rise of the salt occurs in some instances, often due to deposition of young sediments. The relatively minor salt uplift which results, and the usually superficial compensating subsidence in adjacent beds, can lead to the development of a 'tertiary' or third-order rim syncline. This is indicated in diagram (3) of Fig. 7.18 as TRS, with a simple recapitulation of the initiatory stage (diagram 1) and the pre-piercement stage (diagram 2), with the peripheral sink marked PS, the primary rim syncline PRS, and the secondary rim syncline SRS.

The tertiary rim syncline, if developed, is not often preserved due to the highly disruptive effects of crestal faulting, graben formation and dissolution collapse episodes. These latter phenomena are included as ancillary post-piercement effects, as are the development of spreading overhangs, and salt extrusion at the surface.

Two examples shown by Trusheim (1960) are of interest in that they show one example of a relatively symmetrical development of a diapir, and another example of a very asymmetric development. The first, seen in Fig. 7.19, is a line drawing of a seismic line through the Jaderberg diapir from north to south. Keuper and Lias sediments of the primary rim syncline are shown thinning on both sides towards the diapir, with the overlying Dogger/Malm sediments of the secondary rim syncline thickening symmetrically towards the diapir. What is probably a very minor tertiary rim syncline development is suggested at Base Tertiary level close to and just above the crest of the diapir.

The other example, seen in Fig. 7.20, is a similar line drawing of seismic data across the Gorleben diapir

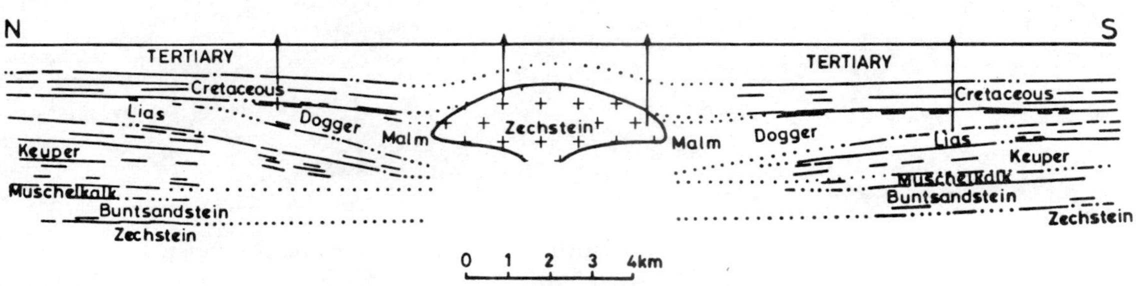

Fig. 7.19 Symmetrical development of the Jaderberg diapir, North Germany. Keuper and Lias primary rim syncline sediments thin towards the diapir on N and S sides, and Dogger/Malm sediments of secondary rim syncline thicken towards the diapir. What may be a very minor tertiary rim syncline is suggested at Base Tertiary level on each side of the diapir crest. (Redrawn from Trusheim (1960); courtesy of the AAPG.)

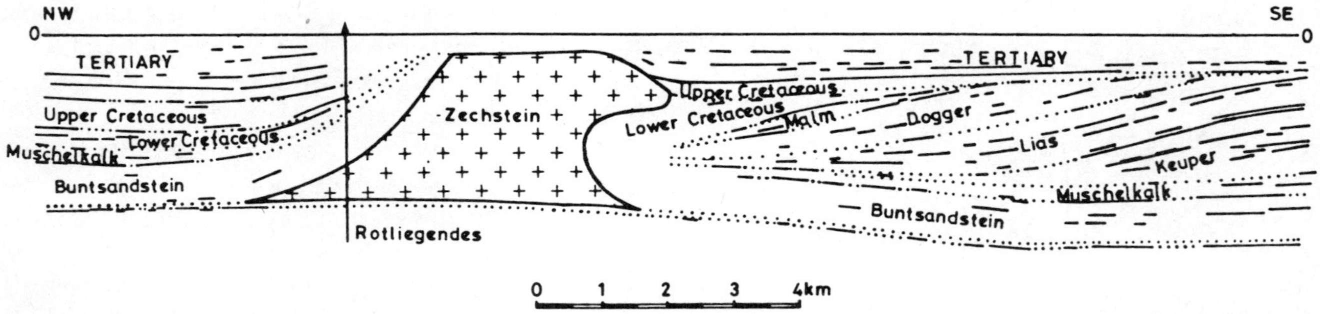

Fig. 7.20 Very asymmetrical deposition adjacent to the irregular Gorleben diapir of North Germany. To the SE of the diapir, there is a clearly-marked secondary rim syncline of Lower to Upper Cretaceous age, thickening towards the diapir. To the NW of the diapir, there is no equivalent development—only a simple salt withdrawal basin. (Redrawn from Trusheim (1960); courtesy of the AAPG.)

from SE to NW. In this example, a strong SRS (secondary rim syncline) is developed in Lower and Upper Cretaceous sediments to the SE of the structure, but not to the NW. The structure is of the type termed a 'duck's-head' or trapdoor diapir by Trusheim from the irregular shape of its upper region, which he ascribed to upward injection along inclined shear planes. This may explain the shape of some diapirs, but there are other factors which may influence asymmetric development, and this matter will be discussed further in Section 7.6.4.

Note that in the Gorleben example on the SE side, thinning units typical of PRS (primary rim syncline) deposition comprise the Keuper to Malm formations, while Lower and Upper Cretaceous units thicken towards the diapir in a SRS development. On the NW

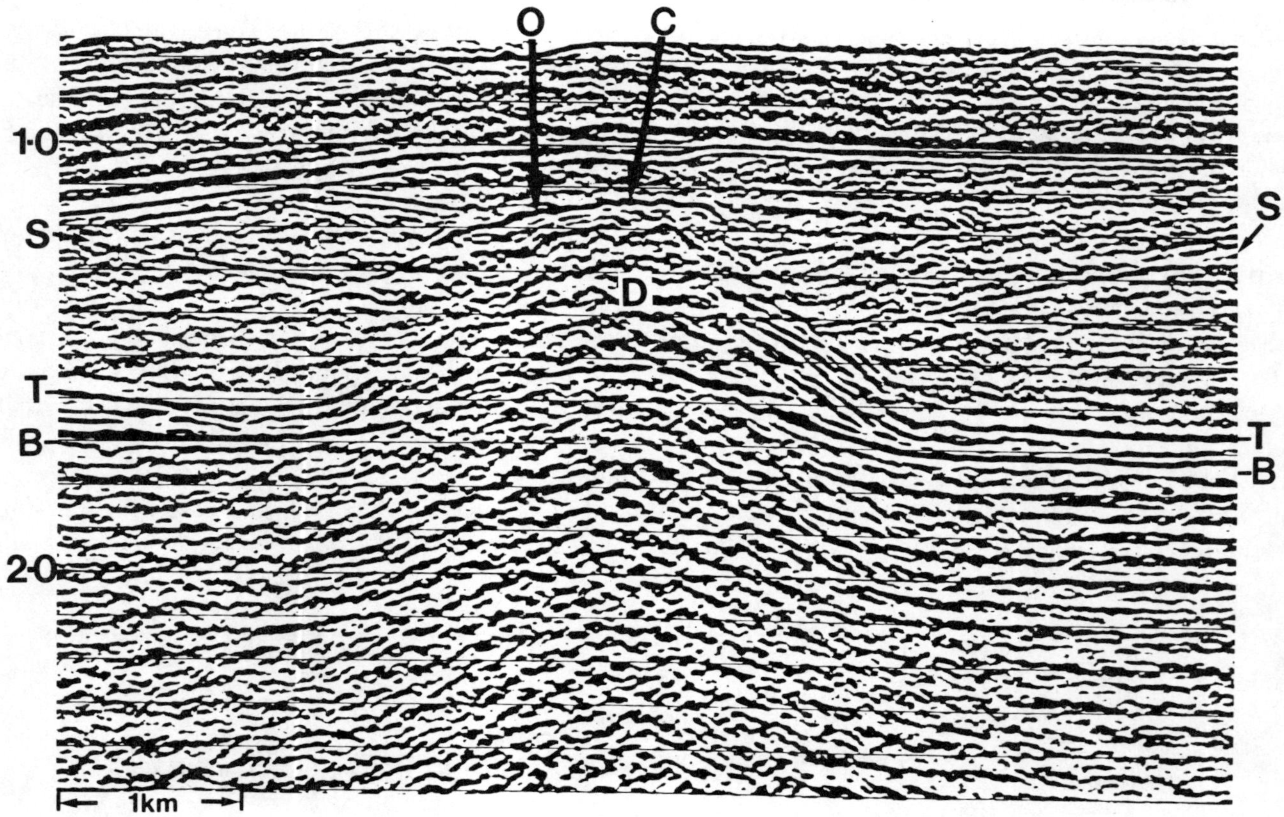

Fig. 7.21 Seismic section across a small diapir in the Northern North Sea; even on this noisy section, a strong reflection is seen at the base of the secondary rim syncline (S) on both sides of the diapir, D. A cap rock reflection, C, is present at the crest of the diapir, and the event extends to the left of the diapir, marking the presence of a spreading (O) overhang on this side. The overhang is disrupted on the right-hand side of the diapir, probably due to late subsidence into the secondary rim syncline, which is more strongly developed on this side, as can be seen by the relative positions of the event S to right and left of the diapir. (Courtesy of SS(E)L.)

side, however, no SRS development can be seen, but sequences from the Buntsandstein to Upper Cretaceous are all indicated as thinning towards the diapir, indicating very asymmetric development.

Trusheim's comment regarding the possibility of a strong reflection marking the break between primary and secondary stages in the diapiric development process receives much support in contemporary seismic data, even in problem areas. Figure 7.21 is an example of a small diapir from the Northern North Sea basin. In spite of very strong multiple reflection interference and other noise, the strong event marked S on both sides of the diapir D clearly indicates the base of the secondary rim syncline associated with this diapir. The salt interval is B–T. A cap rock event is present at C, and a spreading overhang is indicated on the left-hand side of the diapir at O, although this is not present on the right-hand side due to the strong subsidence effects of the asymmetrical secondary rim syncline here, which have disrupted the overhang on this side. A quite minor subsidence low is seen at the unconformity above C, probably due to some dissolution of the underlying cap rock or salt.

7.6.2 Variations on the Basic Theme—An Erosional Hiatus

It was suggested in the previous section that the simple three-stage development of a diapir is not always clearly apparent. Complications may occur to obscure the model which has been presented.

A major complication which is often present is an erosional hiatus within the period of development of the structure. In the British North Sea Southern Basin, an example is provided by the existence of the Late Cimmerian unconformity, which is the result of an erosional phase which cut down deeply into the Mesozoic sediments present over a large area covering the zone of Zechstein diapirism.

The sequence of events is represented in the sketch diagrams of Fig. 7.22, starting with sketch (1) in which the pre-deformation stage is shown, where a sequence of sedimentary units (1, 2, 3) is laid down on top of the salt layer (cross symbol). In sketch (2), a base-salt fault, f, is supposed to have initiated a salt swell, H, and deposition in the primary rim syncline, P, has occurred (unit 4). The salt rise is supposed to have slowed or stopped, and additional units (5, 6, 7) have been deposited. In sketch (3), a major erosional episode, U, has removed all the sediment above unit 3, including most or all of the primary rim syncline

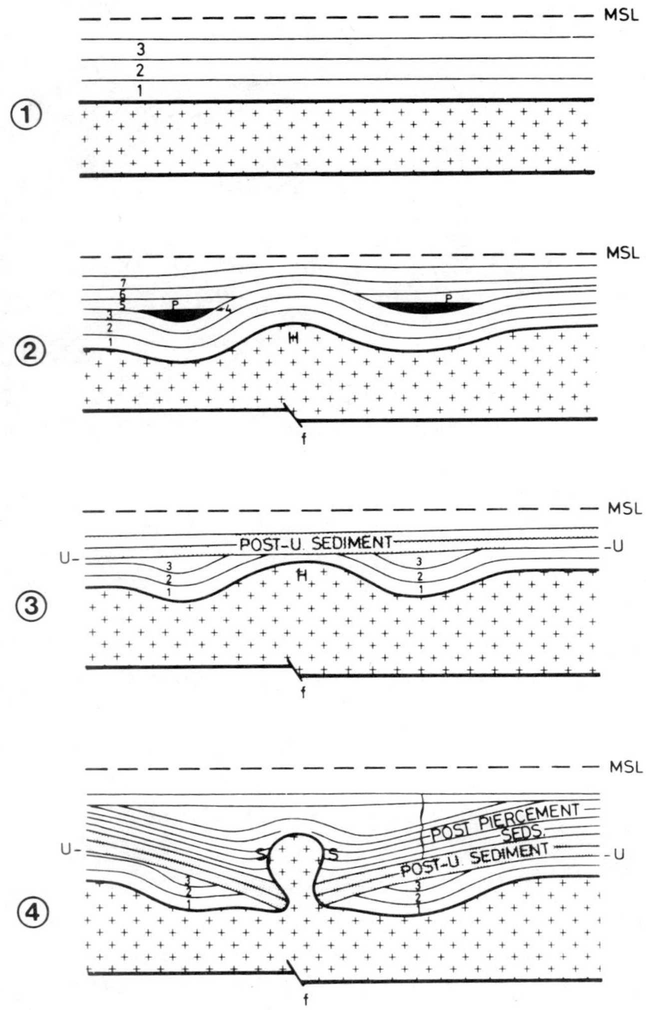

Fig. 7.22 Diagrams showing the complications of the basic multi-stage diapiric process which are introduced by the presence of a major erosional hiatus.

deposits, and a new series of post-U sediments has been laid down. At some time during or after this period, the salt recommences its uplift. In sketch (4), piercement has taken place, strong subsidence has occurred in the vicinity of the diapir due to the accelerated flow of salt into the piercement, and secondary rim syncline deposits S have been laid down, thickening towards the diapir.

If any primary rim syncline deposits remain, not having all been removed by erosion as is the case in the sketches, the strong subsidence associated with the piercement stage will have caused them to take up a semi-antiform shape with thinning in towards the diapir (see Figs. 7.17 and 7.18), unlike the secondary rim syncline deposits which thicken towards the diapir, and the pre-movement sediments (units 1 to 3), which are of uniform thickness.

A real example of this process is shown in the seismic section of Fig. 7.23. In this example, some

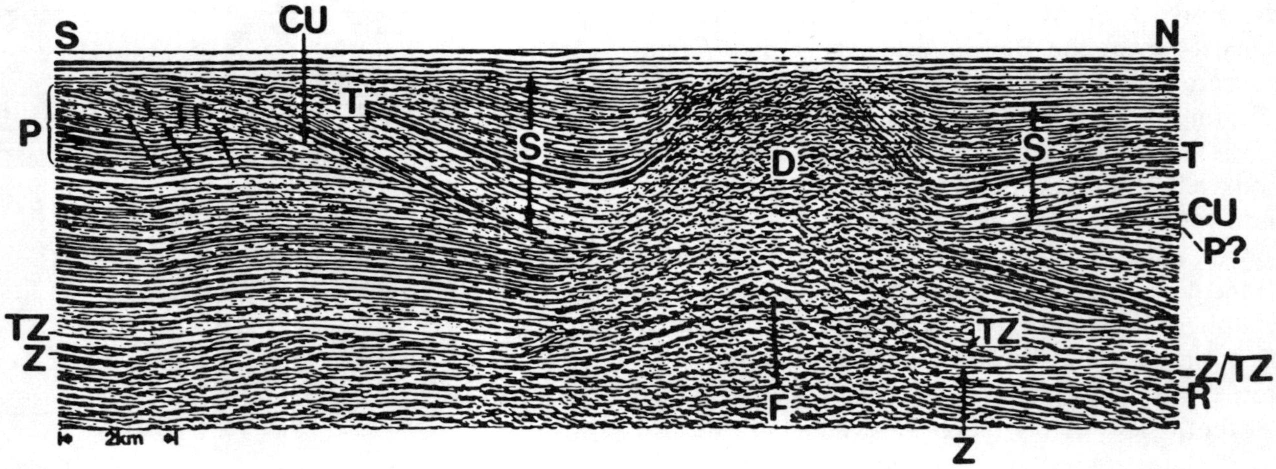

Fig. 7.23 Seismic section showing a real-earth example of the situation sketched in Fig. 7.22. See text for key and discussion. (Courtesy of SS(E)L.)

primary rim syncline sediments have survived (lettered P), and are particularly clear on the south side of the diapir, where they can be seen as a sequence thinning towards the diapir, D. The strong subsidence of the piercement stage has caused them to take up a semi-antiform shape, and the deformation accompanying this has resulted in gravitational normal faulting affecting the sequence. The diapir in this example is a salt wall seen in cross-section, and the periods of salt pillow formation, piercement and secondary rim syncline sedimentation can be dated with fair accuracy, depending on the detailed stratigraphic correlations which can be made with the seismic horizons. The salt interval is between Z and TZ; to the north of the diapir, all the salt has flowed into the diapir, and the base and top of the salt interval cannot now be seen, and only a single strong event marked Z/TZ remains, associated with Zechstein basinal carbonates normally at the base of the salt interval and possibly some thin anhydrite/carbonate units from the top of the salt interval. The separation between top and base of the salt is again visible a little way in from the right-hand end of the section. Event R is the base of the Rotliegendes (Lower Permian), which is only thinly developed here. F is the estimated location of a strike-slip fault at base-salt level which probably initiated the salt feature here, and from the change in level of the Z/TZ horizons across the section, it seems that the apparent 'downthrow' associated with this fault was to the north of the diapir. The Late Cimmerian unconformity is marked CU, the secondary rim syncline depositional interval is arrowed at S, and the Base Tertiary is at T. Just above, and to the right of the crest of the diapir D, a slight dip of the Late Tertiary/Quaternary sediments overlying the main Tertiary interval unconformably may indicate the development of a minor 'tertiary' or third-order rim syncline.

The strong truncation of the events between TZ and P at the CU unconformity is significant. If the unconformity can be seen in the mind's eye as restored to its original horizontal position, the strong dip of the truncated, uniform thickness sequence of pre-movement age of deposition indicates the uplifted pre-piercement swell/pillow development. The baselap shown by the events immediately above the CU surface indicates the prograding deposition which took place towards the diapir in the deepening secondary rim syncline basin.

The dating of the sequence of events is thus: swell/pillow formation began at the base of the P sequence—in this case, at some time during the Late Trias–Early Jurassic (a good well correlation nearby would give a more precise dating). The piercement stage came during the Late Cimmerian erosional phase (Late Jurassic/Early Cretaceous), and continued through to Tertiary times, but later on the south side of the diapir than on the north side. The evidence of the latter observation is based on the later-continuing subsidence effects in the secondary rim syncline on the south side compared to that on the north side. The asymmetry of development indicated by this discrepancy in the period of subsidence, and the fact that all the salt has gone from the source interval on the north side while some still remains on the south side, is a matter which will be pursued further in the next section.

In Fig. 7.24, another example of a diapir from the same general area is shown. This time, the diapir D is a sub-circular 'bullseye' structure. The relatively

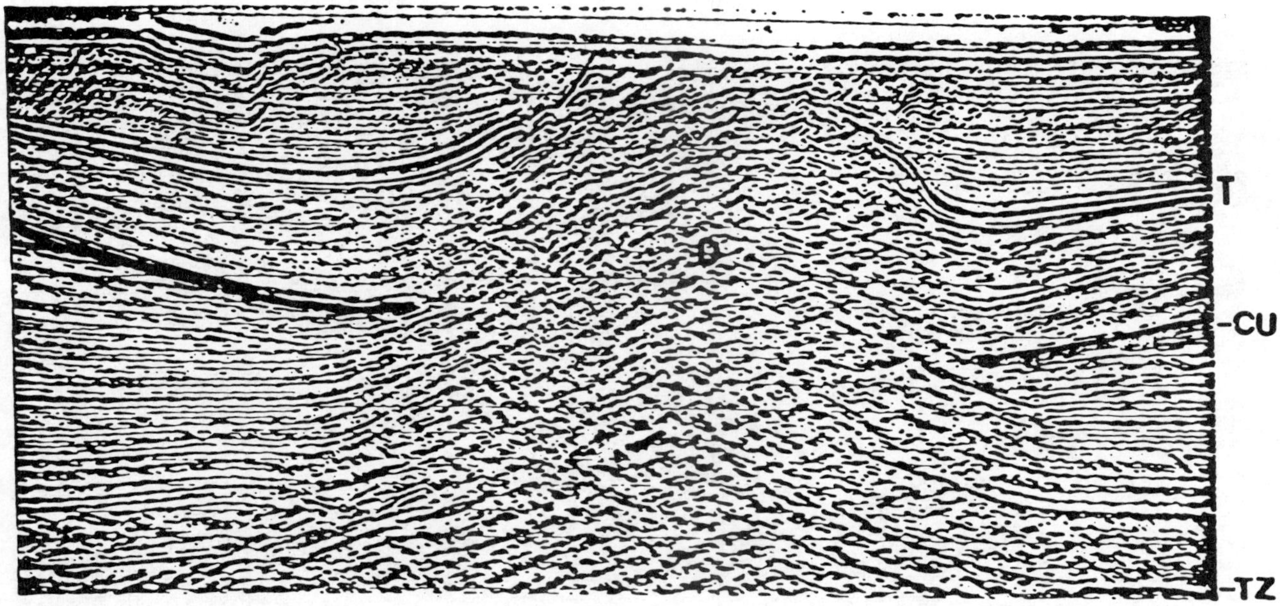

Fig. 7.24 Seismic section across a 'bullseye' (sub-circular) diapir from the Southern North Sea. (Courtesy of SS(E)L.)

shallow angle of truncation of the events below the unconformity CU indicates that the swell developed here in the pre-piercement stage was of rather low amplitude when piercement occurred, compared to that in Fig. 7.23. There are no signs of primary rim syncline events, as here the CU erosion has removed these deposits, but the secondary rim syncline events thickening towards the diapir above CU are clear. The top of the salt interval is marked TZ.

7.6.3 'Turtle' Structures

Where a group of two or more salt diapirs evolve more or less coevally, during the period of subsidence of the secondary rim syncline the semi-antiform configuration of the primary rim syncline sediments associated with a single diapir may become a full antiform structure between two or more diapirs. From its shape in cross-section, this has been termed a 'turtle' structure. Depending upon the geometry and relative positions of the diapirs involved in the group, this high structure may become a closure, in which case it could provide excellent hydrocarbon trapping potential.

7.6.4 Asymmetry of Diapir Development

The frequently asymmetric development of diapirs has been referred to on several occasions, and has been noted by various workers. Boldreel (1985) recently drew attention to a group of diapirs in the salt province of NW Jutland, Denmark, all of which are positioned over, or in the immediate neighbourhood of, base salt faults. The diapirs which have associated secondary rim synclines all show asymmetry of development. Most of the diapirs present in the Southern Zechstein Basin of the British North Sea show some degree of developmental asymmetry, and are similarly associated with base salt faults.

Trusheim (1960) mentioned asymmetry in connection with a number of structures in Northern Germany (see Fig. 7.20), ascribing it to the injection by the salt of inclined shear planes in the overburden, as discussed in Chapter 5, and referring to the resulting diapirs as 'duck's-head' or trapdoor structures, from the final shapes developed by the salt. This would certainly result in a very asymmetric shape of the emplaced salt, but does not necessarily explain the effects seen in many cases of strongly unequal subsidence of the overburden on each side of the diapir. Although the passage of salt upwards through the overburden is inclined to the vertical, if the process were under gravitational control, and with uniform overburden loading, salt should still flow symmetrically into the diapir from both sides (where an elongate structure is involved).

Figure 7.25 shows a slightly enlarged view of the central part of the seismic section of Fig. 7.23 across a large salt wall which has been emplaced directly upwards with no inclination away from the vertical. The diapir and the effects in the overburden indicate considerable asymmetry of development.

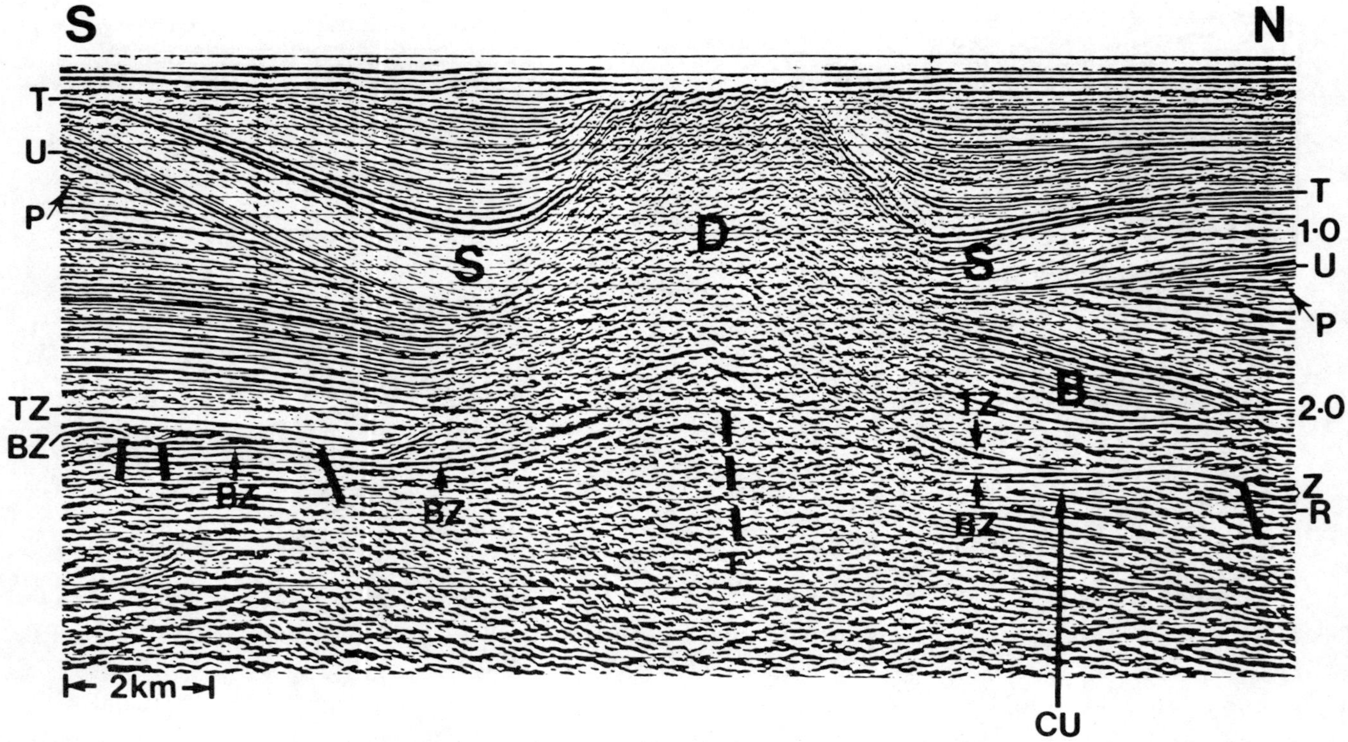

Fig. 7.25 An enlarged portion of the example of Fig. 7.23, showing the asymmetry of development of this diapir. (Courtesy of SS(E)L.)

The reflection configuration in the U–T interval, and above the unconformity at T, provide evidence of the sequence of events. In the U–T interval, reflections are downlapping onto the unconformity surface and prograding inwards towards the diapir on both sides, indicating sedimentation in the deepening secondary rim syncline during the piercement stage.

Above the T unconformity to the right of the diapir, however, reflections adjacent to the salt in the deepest part of the secondary rim syncline indicate *onlap infill* of this deep trough. This proves that subsidence on the right-hand side of the diapir had ceased before the onset of Tertiary deposition (above T, the Base Tertiary). The reason for this seems likely to be the exhaustion of the salt supply on this side, as seen by the situation at the Zechstein interval below.

On the left of the diapir, above T, there is no sign of onlap infill; uniform sedimentation in a steadily deepening trough is indicated.

The evidence thus points to most of the early salt flow into the diapir having come from the right-hand (N) side of the feature. This flow ceased in Late Cretaceous/Early Tertiary times, and the continued rise of the diapir through the Tertiary period resulting in the steep updrag of the Tertiary reflection events against the salt was fed entirely by salt flow from the Zechstein section (BZ–TZ) to the left of the diapir. There is still some salt left within the Zechstein on this side, and dissolution removal of salt at the diapir crest may be balancing any remnant salt flow into the diapir.

The difference in level of the basal Zechstein across the section suggests that a large fault (estimated at F) affected this level beneath the present location of the diapir, and that the effective downthrow ('effective' because it is probably a strike-slip fault) is to the right. In these circumstances, the differential loading mechanism discussed in Chapter 5 may be adequate to explain why salt flowed initially from the right to form an uplift which later became a piercement. The odd lens-like segment of overburden marked B to the right of the diapir may represent a small withdrawal basin associated with this early salt flow; note that there is no equivalent to B on the left of the diapir. If so, the deposits in B are probably of Mid-Triassic age.

The main pillow uplift took place in Late Triassic to Early Jurassic times, the evidence for this being the stratigraphic position of the primary rim syncline sediments at P (seen better in Fig. 7.23), with salt now flowing into the feature from both sides. The early salt flow from the right, plus later flow into the main pillow, seems to have resulted in the early exhaustion of the salt in the Zechstein on the right. The surface levelling which took place during the Late Cimmerian

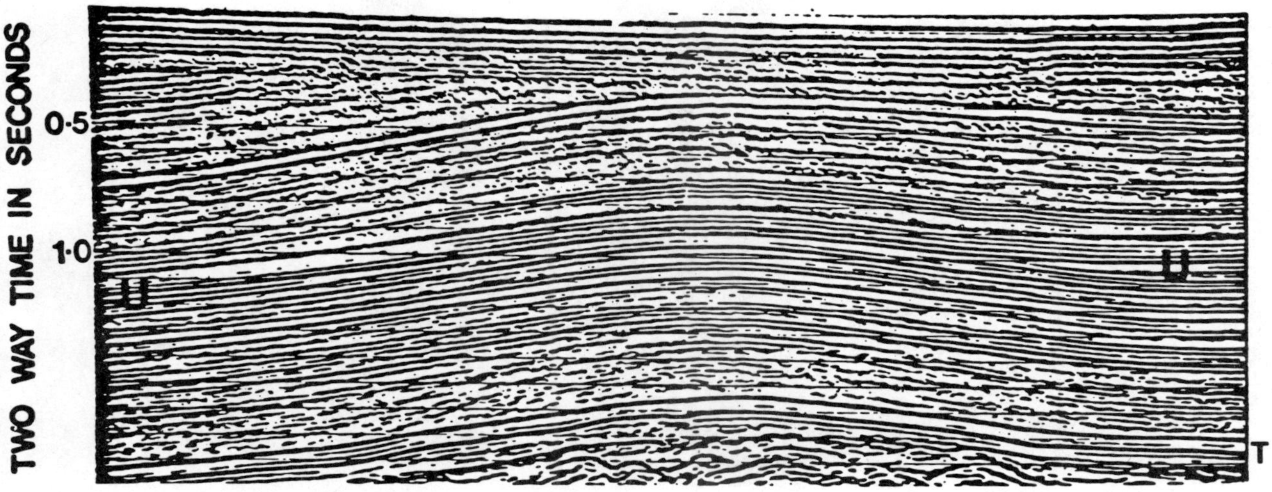

Fig. 7.26 Seismic section showing two periods of salt movement—pre- and post-unconformity U. Top salt at T. (Courtesy of SS(E)L.)

erosional phase could have produced the major differential loading in the salt/fault system which led to the piercement stage of the diapir. The reason for the development of the earlier load differential which caused the initial salt flow from the right cannot be determined, but presumably stemmed from surface levelling either by erosion or sedimentation during the Early to Middle Triassic.

At present, the process suggested is purely conjectural; it is attractive in that it offers a relatively simple explanation for a phenomenon seen in the development of many diapirs in different parts of the world, particularly those of elongate form which are associated with base salt faulting. It is also possible that there is more than one reason for the development of asymmetry.

7.7 THE AGE DETERMINATION OF SALT MOVEMENT

Some indications have already been given as to how the period of different stages in salt movement may be determined by the depositional patterns present in the overburden. In particular, reasonably precise determinations can be made in seismic data of the various stages of development of a salt diapir, provided that good stratigraphic correlations with seismic events can be made, and that all stages can be seen in the data. There may be difficulty as regards the latter point, as we have seen, where for instance a major erosional episode has removed the primary rim syncline deposits in their entirety. The situation becomes complex when there are several major unconformities in the section, and there has been a series of salt movements with varying rates of rise—perhaps sometimes with complete cessation of movement for a time.

The other problem to be addressed is that of the determination of movement period in structures which have not reached the piercement stage—swells and pillows in the salt interval.

The two most useful indicators of salt movement period in the seismic section are (i) truncation at unconformities, and (ii) thinning over the crest of a rising salt feature.

An example of the first effect of quite subtle nature is seen in Fig. 7.26, where a small salt swell (salt top at T) has resulted in slight toplap truncation of overburden units beneath the unconformity U above the flanks of the swell, although this is not apparent over the crestal region of the high. Restoration of the unconformity to its original horizontal position would show clearly that a swell-like thickening had taken place in the salt before the erosional phase. The shape of the unconformity, and the sequences overlying it, show that a second phase of salt uplift took place in post-unconformity time. In the shallow part of the section, another unconformity truncates the overburden units, and this time there is no evidence of renewed salt movement above the unconformity.

In schematic form, Fig. 7.27 shows a situation in which two salt pillows have developed in adjacent locations A and B, but not coevally. The thinning of overlying unit 3 above pillow A and that of unit 4 above pillow B show that the rise of pillow A preceded that of pillow B. The salt movements can be quite accurately dated provided that good stratigraphic correlations can be established for units 3 and 4.

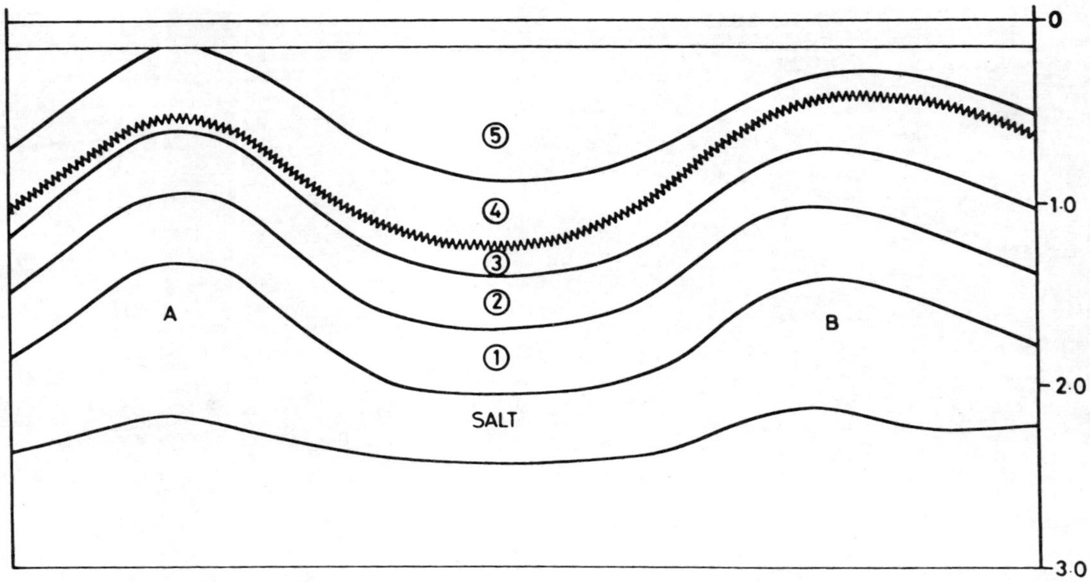

Fig. 7.27 Diagram illustrating the principle underlying the determination of movement period of salt in two pillows, A and B.

The seismic section in Fig. 7.28 shows a salt pillow P formed within the salt interval BZ–TZ in the Northern North Sea. The onlapping event C, and the configuration of events A and B on the right of the structure, strongly suggest supratenuous drape over the pillow. This evidence of syndepositional rise of the salt indicates that the crest of the salt pillow cannot have been very far below the contemporary depositional surface at some time before the erosional phase took place which is represented by the unconformity U_1. Effects at and above the shallower unconformity U_2 show that salt rise continued later, with

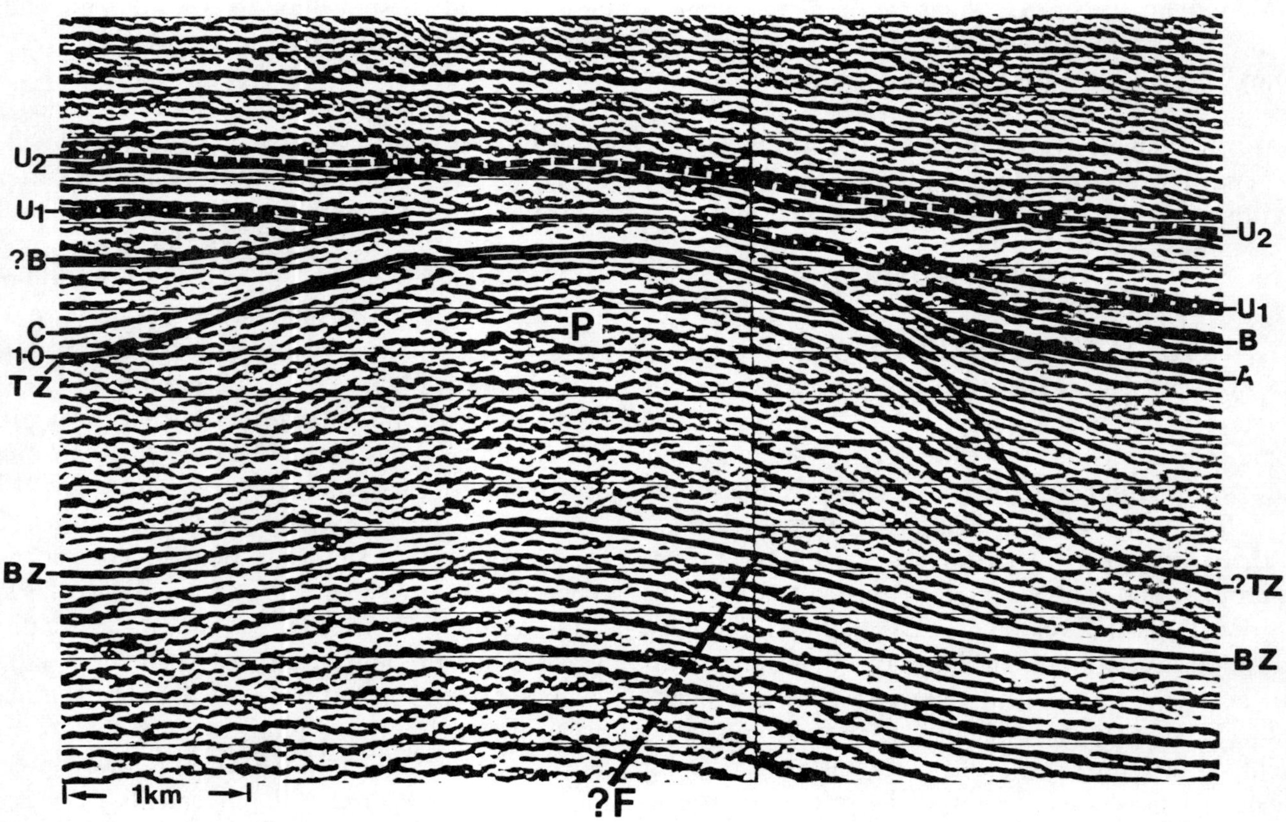

Fig. 7.28 Movement periods of a structure, P, which is partly a pillow and partly a piercement structure. See text for key and discussion. (Courtesy of SS(E)L.)

perhaps a temporary cessation and some dissolution between U_1 and U_2. Although termed a pillow, the structure has in fact pierced the overlying sequences on the right, though not on the left. Again, with good stratigraphic information, the salt movement periods can be determined with some accuracy here.

The age determination of many salt structures can be made by the application of the simple ideas just illustrated. Further comments on this matter, including problems presented by more complex features, will be made later.

7.8 THE INTERNAL STRUCTURE OF SALT DIAPIRS

Where diapiric salt masses have been mined, or investigated by boreholes, complex internal structures have been revealed. Flow-banding and folding of the salt sequences are normal, and highly deformed but distinct stratification is present. The existence of this stratification was cited as a non-fluid-like characteristic in the discussion after Nettleton's (1934) paper on the application of fluid dynamics to salt diapirism; were mobile salt to behave as a fluid, then intimate mixing could be expected to take place, with resultant modifying effects at interstratal boundaries not normally seen in salt bodies.

Kupfer (1968) and others have remarked on the tendency of diapiric salt bodies in some areas, e.g. the Gulf Coast, to contain relatively pure salt in contrast to the source interval from which they flowed, which may contain thin detrital sediment bands and other competent lithologies. The suggestion was made that the more mobile salt layers have flowed out selectively into the diapir, leaving behind the less mobile units in the source interval.

In many cases, bands of contrasting lithology are certainly present within diapirs, and their presence may be of importance in regard to the integrity of the salt body, and hence its suitability as a storage feature for, e.g., radioactive waste. This matter will be discussed further in Chapter 9.

A frequently observed phenomenon which accentuates the appearance of stratification within the salt is the alternation of bands of relatively pure halite with others of anhydritic halite, the latter being of darker colour. A sketch from part of a photograph taken in the Grand Saline salt mine, Texas, USA, is shown in Fig. 7.29, where this effect picks out very clearly the formation of similar-type flow-folding in the salt. The camera was pointed vertically upwards at the roof of

Fig. 7.29 Sketch traced from a photograph of flow folding in the roof of a salt mine cavern. (After Muehlberger and Clabaugh (1968); courtesy of the AAPG.)

the cavern, which was more than 100 ft above. The axes of these folds are vertical, or near-vertical, and the development of such folding was suggested in the model experiments of Escher and Kuenen (1929) discussed in Chapter 4; this can be seen in the cross-sections A–A′ and B–B′ in the lower sketch of Fig. 4.1, with the fold complexity increasing from the periphery to the centre of the model. The vertical section through the model shown in this same figure can be compared to Fig. 7.30, which is a vertical section through the potash mine of Ronnenberg, Germany. Borehole data have given a detailed picture of the complex deformation of the Zechstein Z2, Z3 and Z4 cycles of salt, anhydrite and potash intervals.

The flow-folding within salt may lead to the formation of isoclinal-type folds, and the repeated re-folding of these with pinching-out of some limbs so that an extremely complex series of structures evolves. There is also some evidence of vein-like features filled with larger-than-average halite crystals which have been interpreted as fractures within the salt body which have developed at certain stages, to be sealed later by the recrystallised halite. It may be that such fracturing can develop at only moderate or shallow burial depths, with the re-sealing taking place at greater burial depths.

For several papers with interesting discussions of internal features of salt diapirs, the reader is referred to AAPG Memoir No. 8 (1968).

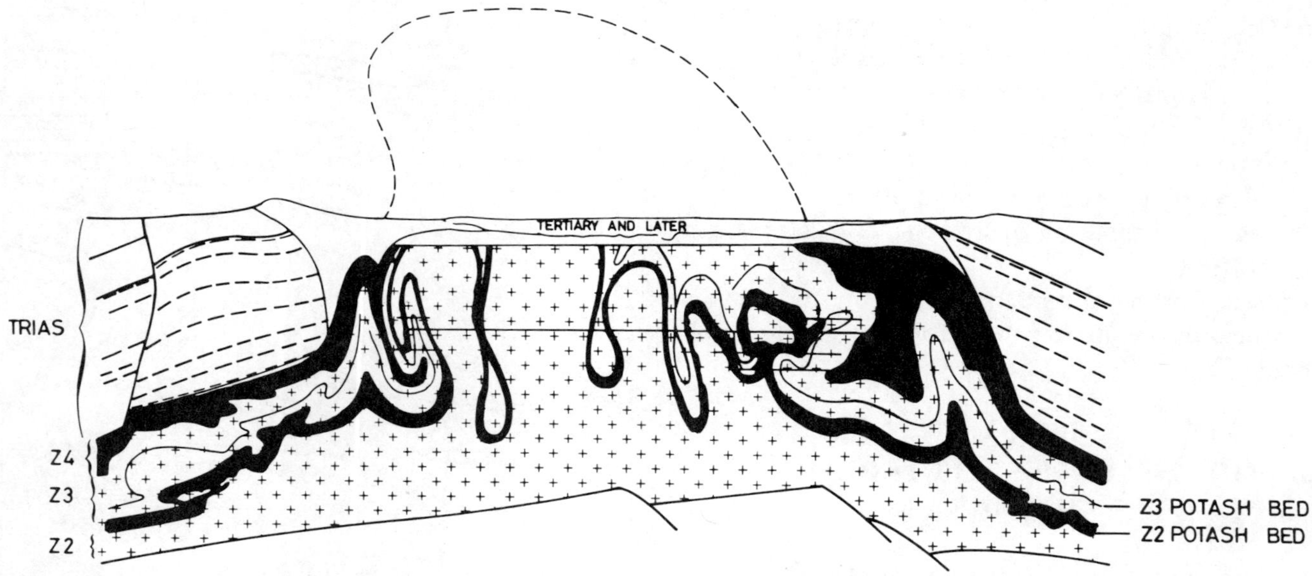

Fig. 7.30 Internal deformation within a salt diapir: the Ronnenberg potash mine, Germany. The internal structure of this diapir is known in detail from mining and borehole information, and the complex flow folding is picked out well by anhydrite and potash marker beds. (Redrawn from Baar (1977) after Lotze (1957), with acknowledgements to the Elsevier Scientific Publishing Company.)

7.9 OVERBURDEN LOADING AND GRAVITATIONAL INSTABILITY

7.9.1 Introduction

With regard to salt movement and uplift by halokinetic processes, it is now appropriate to examine in more detail some of the basic ideas touched on in Chapters 2 and 4. It is felt to be necessary to question some of the views currently held among geologists regarding the reasons for the initiation and development of at least some salt structures. This will be done in an attempt to highlight, if not resolve, inconsistencies which exist in the observed facts in various salt areas if these current views are correct.

When considering the behaviour of a salt/overburden system under the control of gravity, two quite distinct questions must be addressed. These may be summarised as follows:

(i) What factors govern the change of state of salt from the elastic to the plastic?
(ii) When salt is in the plastic state and thus *potentially* mobile, what are the factors which govern the initiation and continuation of its movement both locally and regionally?

The principal factors involved in question (i)—temperature, differential pressure and the presence of free water—were discussed in Chapter 3. So far, as regards the pressure parameter, it has been tacitly assumed that the mass of the overburden is the sole factor involved. However, as will emerge later, there are reasons for thinking that lateral pressure gradients *within* a salt interval, and related lateral salt movement, must also be considered in this context.

The factors involved in question (ii) above will now be discussed.

7.9.2 The Question of Overburden Loading

As previously remarked, halite which is the mineral forming the bulk of salt rock has a density which is effectively invariant with depth of burial, which means that it is not subject to compaction. Unlike halite, the clastic sediments forming the overburden associated with salt intervals do undergo compaction to varying degrees with increasing burial depth. Figure 7.31 shows a curve of increasing shale densities as burial depth increases, sketched from figures produced by Dickinson (1953) for the Gulf Coast region of Louisiana, together with a dashed line showing the constant density of salt over the same depth range. The shale and salt densities are equal at a depth of some 4000 ft.

If a salt interval is present in this area at a greater depth than 4000 ft, then a density inversion will exist at the top of the salt layer, resulting in a gravitationally unstable situation with denser sediments resting on relatively buoyant salt.

Note that in any adjacent area, where the over-

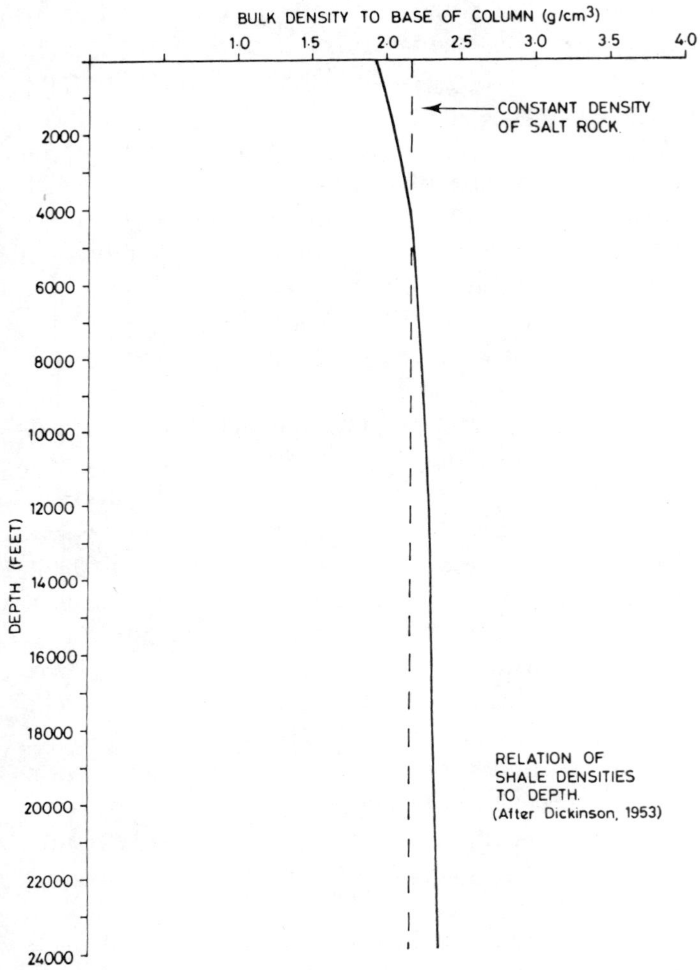

Fig. 7.31 Curve of shale densities plotted against burial depth. Constant density of salt shown as broken line. (Information derived from figures produced by Dickinson (1953) for shales in the Louisiana Gulf Coast area.)

burden lithofacies could be different, the situation outlined may occur at a different geological time, or not at all, depending on the depth of the salt layer and the bulk density of the overlying sedimentary column.

Where gravitational instability occurs as a result of the situation outlined, a measure of it will be afforded by the product of the density differential and the overburden thickness. The density differential is the difference between the average overburden density (the bulk density to the base of the sedimentary column overlying the salt) and the salt density. Thus, for the same thickness of overburden, the degree of instability will become larger with increasing density differential.

The condition of gravitational instability cannot, of itself, generate any salt movement; any of the salt movements described in detail require the development of a horizontal pressure gradient by mechanisms discussed in this book. The rate of salt movement will depend on the magnitude of this pressure gradient, and the strength of the material containing the salt. The scale of salt movement depends (apart from the obvious factor of the amount of salt available) on the length of time that the pressure differential operates, and the distance that the salt can move before encountering resistance which is sufficiently strong to prevent further progress.

7.9.3 The Question of Gravitational Instability

Thus, if the question is put: 'Where a density inversion occurs at the top of a salt layer, is the resulting state of gravitational instability sufficient to initiate movement in the salt?', then the answer must be, 'No—if it is the sole factor'. The evidence for this is strong and widespread. Kent (1979) in discussing the emergent salt diapirs of Hormuz in Iran notes that while the salt source layer is of Cambrian and infra-Cambrian age, with an overburden of 10 000 to 20 000 ft of nearly continuous marine limestones, evidence points to the salt movement and diapirism having taken place largely no earlier than the Cretaceous, and in some cases the Tertiary.

With such an overburden, consisting mainly of dense carbonate rocks, it would be incredible to suggest that a density inversion at the salt top did not exist long before the Cretaceous period in this area. A considerable number of other salt areas exist, such as the Michigan and Williston basins in North America, and parts of the North German Zechstein Basin, where salt layers of substantial thickness lie under thick overburden sequences, but show no signs of plastic deformation. In the case of the latter area, Sanneman (1968) noted that a 1000 m thick salt layer in one location shows no sign of movement although overlain by 2000–3000 m of sediments, and geologically continuous with an adjacent area in which salt pillows and diapirs have developed. In a nearby area, a similarly thick salt layer has flowed into pillows, but can be shown to have remained immobile for a geologically long period beneath an overburden not less than 2000 m thick. The cause of eventual mobility of this salt is believed by Sanneman to have been a tectonic pulse.

Halbouty (1967) presents information in two tables which list, for the Eastern and Western Hemispheres, salt basins and provinces in which salt structures produced by plastic deformation (a) do, and (b) do not, occur. For the Eastern Hemisphere, 33 salt basins/provinces appear, of which only 16 contain halokinetic-type salt features (three of which are categorised as 'probable'). In the Western Hemisphere, the corresponding figures are 28 salt areas, only 10 of which have structures of this type, one of which is a 'probable'. The lists are interesting in that there seems to be a rather general relationship between the areas in which plastic deformation structures have occurred, and continental margins or tectonically active zones. Also, it is of interest to note that of all the salt areas listed, nearly 60% contain non-mobile salt.

The demarcation between the two different types of basin is very strong in the continental USA including the Gulf of Mexico. Of the 15 salt areas listed by Halbouty, only 4 have mobile salt structures. To this list could now be added the Carolina Trough, offshore east coast USA—again a continental margin basin—in which salt diapirs have been identified since Halbouty's classic work was written. Of the 4 basins listed as having mobile salt structures, the Paradox Basin of Utah and Colorado is probably a special case in that the several major salt anticlines present appear to be of halotectonic origin.

In many—probably most—of the non-mobile salt areas which exist, density inversion at the salt layer has certainly occurred during some part of the burial history. How can this be squared with current views on overburden loading as a prerequisite for the initiation of salt mobility? Is the salt layer thickness also important, or is some differential in the applied stress required? These factors will be examined in the following sections.

7.9.4 Salt Thickness

The current view appears to be that (as in the case of the overburden) there is some minimum or critical thickness of salt required before movement will take place. Although there is no general agreement as to what this thickness is, the figure of 300 m of salt is often encountered in the literature. In the European area recently, Boldreel (1985) mentions in the case of a diapir province of Jutland, Denmark that movement of Zechstein salt deposits is small where the thickness is 300–600 m, and no movement is seen where the thickness is less than 300 m, remarking that this may indicate that the Zechstein sequence does not consist of pure halite. On the other hand, Taylor (1984) states that in the Zechstein salt deposits of the British North Sea, salt structures are common where the salt thickness is at least 130 m.

To the writer, there seems little or no justification for trying to establish some critical minimum thickness of salt. The same physical modes of deformation apply to halite which is 30 m thick as to halite which is 300 m thick. The only difference of any significance which arises is in the size of feature resulting from a thick or thin source layer. In the limiting case of a very thin salt layer, the latter will probably act purely as a lubricant between floor and roof sequences in the case of mobility being initiated. As the salt source layer increases in thickness, any features resulting from salt mobility will increase in size. There seems little point in drawing an arbitrary line through some stage in this process. The suggestion by Boldreel just quoted, that the degree of purity of the halite may affect the ability of the salt to flow, may be relevant, however. The effect of disseminated impurities such as anhydrite, or substantial interbedding of carbonates or non-mobile evaporites with halite may have important inhibitory effects on salt movement; as far as is known, such effects have not been studied in detail. The presence or absence of free water in the form of fluid inclusions—perhaps to a great extent depending upon whether the halite is primary or re-precipitated—may also have great importance as regards the yield strength of the mineral, as noted in Chapter 3. However, it is not believed that the thickness of a salt layer *per se* has any significant influence on the ability of the salt layer to become mobile under the right conditions.

7.9.5 Initiation of Salt Movement in General, and of Diapirism

Attempts made in the past to establish a paradigm for the initiation of salt movement, and of piercement structures, have led to many inconsistencies between various concepts. It is probable that there is no single, simple set of circumstances which result in the development of piercement structures. As regards more general, lateral movement of salt, certain simple mechanisms depending on the basinal configuration and overburden depositional patterns were suggested earlier; there still remains the question as to whether some even more fundamental cause underlies these basinal movements.

In some areas, isostatic (halokinetic) forces seem to be responsible for salt structures, while in others, compressive tectonic stress has undoubtedly initiated mobility and structures in salt. Elsewhere, both types of forces seem to have acted in concert.

The idea that *differential* stress must be present is

well established. In order to initiate mobility in salt, this may include variations in overburden loading either by lateral variations in thickness or density, geothermal gradient (since the viscosity of halite decreases with increasing temperature), or a gravitational gradient brought about, for example, by tectonic tilting of an area. There also seems to be the possibility that salt rock may, while under applied stress which exceeds the theoretical yield stress of the material, remain in either a supercritical or metastable state without becoming plastic until 'triggered' into movement by tectonism in the form of (for instance) faulting or lithospheric plate movement.

Referring again to the belief currently held that gravitational instability produced by a density inversion at the salt layer is a necessary development for the initiation of salt mobility, consider the situation in an area of the offshore Gulf Coast of the USA as described by Humphris (1979). From the density logs of offshore wells it has been determined that an overburden thickness of 1200–1500 m of the Tertiary and younger sediments present would be required to produce a density inversion at the salt layer present. However, on the lower continental slope, minor salt swells have been identified by sparker surveys with an overlying sediment thickness differential of only 550 m between salt highs and lows, with only an additional thickness of 350 m of sediment above the structural highs.

Humphris argues that the structural growth of these swells must be due to lateral salt flow resulting from stress exerted by greater sediment loading updip on the slope (see sketches of Fig. 7.32), and there seems no good reason to doubt this view.

The interesting implication here is that the stress imposed by lateral salt flow from upslope is quite adequate to replace the effect of the vertical pressure of a thick and denser overburden in initiating salt movement. Thus, not even the much greater compressive stresses resulting from tectonism are required to cause the salt in this case to deform plastically. Another point of significance is that without knowledge of the bulk density of the overburden, it would be extremely difficult in this instance to differentiate between the effects due to lateral stress application in the salt layer, and those due to purely gravitational, halokinetic processes. *This could be the case elsewhere.*

It seems, therefore, that movement by plastic deformation in salt may be intiated by either vertical or lateral stress differentials resulting from a variety of causes. More specifically, upward and diapiric movement in salt, based on field observations, model studies and seismic evidence, may be due to the following causes:

(1) Lateral variations in the bulk density of the overburden, which may or may not be linked with thickness variations.
(2) Irregularities at the floor or roof of the salt interval. In practice, this usually means faulting, and the differential load stress developed across such faults.
(3) Post-salt faulting, which (unlike the differential stress flow mechanism suggested in (2)) provides an immediate upward stress relief route for already mobile salt.
(4) Compressive tectonic stress, either of a slowly developing kind, or as a 'pulse' of geologically short duration. Such stress may be related to lithospheric plate movement, and 'pulsation tectonics'.
(5) Laterally applied stress within the salt interval.

Laterally-moving salt meeting some form of obstruction—the 'buttress' or damming effect—may lead to upward movement. The damming effect may be provided by, for example, a facies change as in the basin-edge diapirism to be discussed in Chapter 8, by thinning in the salt interval, or by increasing viscosity in the salt layer as salt moves from a high stress to a lower stress zone. Stress relief considerations in cases like these will lead to upward movement of the salt. The laterally applied stress within the salt interval may be produced in various ways: by regional or basinal movement of salt, by salt movement induced by a gravitational gradient, such as during tectonic tilting of an area, or by geothermal gradient effects on salt viscosity combined with other factors, for instance.

It is pointed out that although the above-mentioned situations have been categorised as possible causes of the initiation of piercement structures, in fact they are really only possible causes of upward movement, as opposed to lateral movement, of salt rock which is mobile. As noted in Chapter 4 (see Fig. 4.15), diapiric structures very frequently do *not* follow the formation of a swell/pillow structure in the salt. Sometimes, as in the illustration just quoted, major salt pillows are formed with crestal heights 1000 m or more above the salt source layer, but with no evidence of piercement of the overburden being seen. On the other hand, many diapirs have passed through the piercement stage when the amplitude of the precursor pillow is much less than this. Where large pillows are formed, a linear 'stretch' factor of several percent over distances

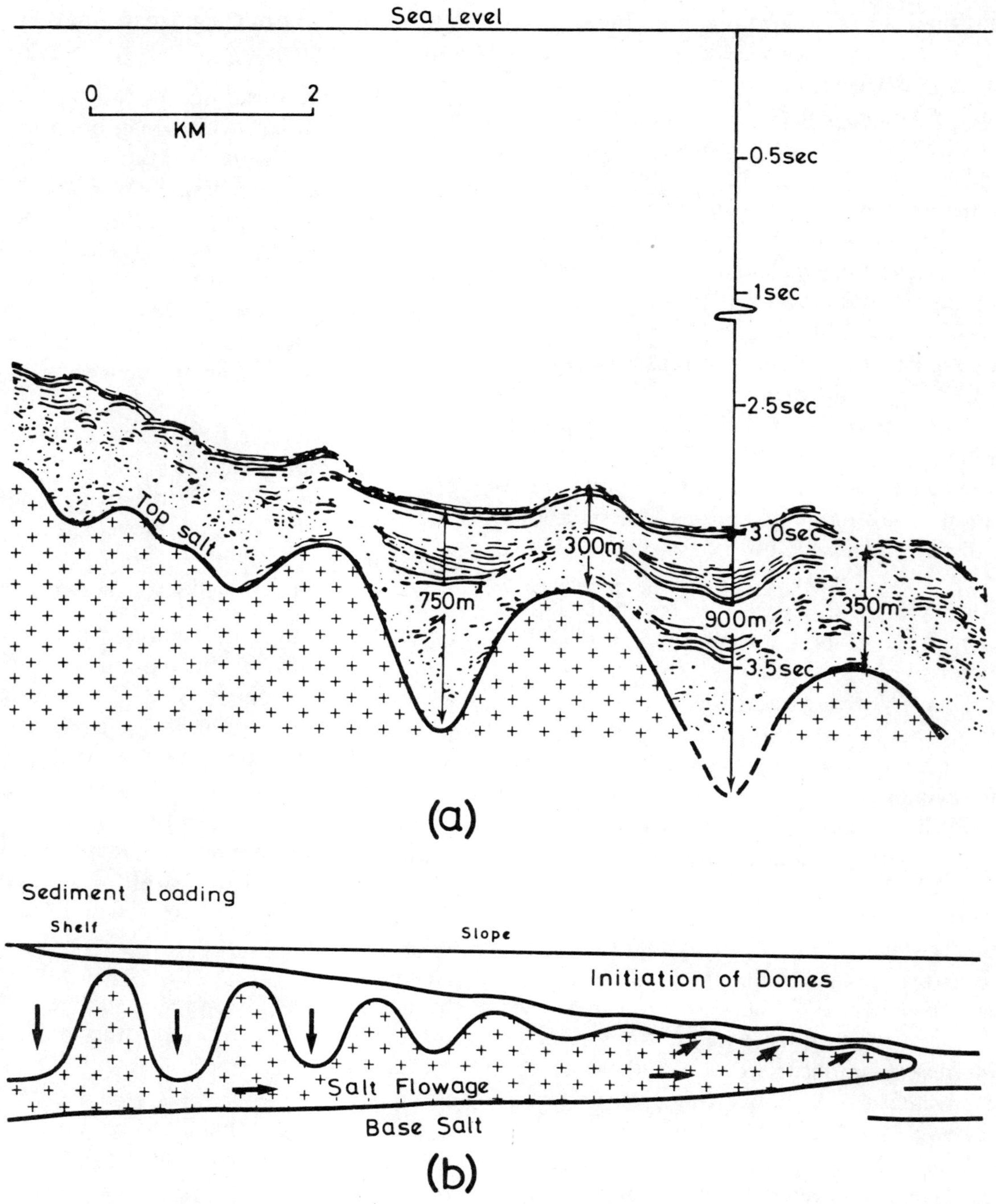

Fig. 7.32 Part sketch, part tracing, of a sparker section (a) near the Sigsbee Escarpment, offshore Gulf Coast of the USA, showing salt swells developed beneath an overburden less dense than the salt (cross symbol). Sketch (b) shows the interpretation given by Humphris for the mode of origin of these salt swells. (Redrawn from Humphris (1979); courtesy of the AAPG.)

of 15–20 km can often be measured in the salt roof sequence as related to the salt floor, with no piercement effects.

Clearly, differential load stress is not the sole factor determining the development of a piercement. Other factors which influence this must be identified.

The salt supply is of importance; the salt source layer may be thick enough to allow pillow development, but too thin to sustain piercement structures. Sometimes the geometry of a developing salt structure may itself eventually isolate the structure from the main salt supply. (An example of this, where overburden subsidence effects around a developing diapir have isolated the feature from the surrounding source layer, will be seen in Section 8.3.)

Another important element is the competency of

the salt roof sequence—where very large amplitude pillows have formed without piercement, and the salt source is still thick, this must be an overriding factor. This factor is equivalent to the 'plunging strength' of overburden material in the centrifuged model studies of Ramberg (1981) discussed in Chapter 4.

The rate and pattern of deposition above the salt is a further matter which must have a bearing on the type of salt structure which evolves. Salt structures frequently show considerable intermittency of development which is controlled largely by rates of basinal subsidence and rates of sedimentation, it is believed. It may be that the evolution of diapiric structures from pillows is sometimes inhibited by a period of rapid deposition. Model studies by Parker and McDowell (1955) and others have shown that the growth of a model diapir can be slowed, stopped and restarted by manipulating the amount of model overburden.

In general, it appears therefore that plastic deformation of salt rock, and both lateral and vertical movement of salt, can be brought about by a variety of causes, all of which involve the application of stress differentials. The initiation of salt movement in general, and of upward salt movement, may be brought about either halokinetically or halotectonically; once initiated, the movement may be controlled largely by gravitational forces, but in some instances movement may be controlled by laterally applied stress if the overburden bulk density is less than that of the salt.

7.10 RATE OF GROWTH OF SALT STRUCTURES

Determining the rate of growth of a salt structure would seem to be an easy matter, provided its geometry was reasonably well defined, and good stratigraphic determinations of the start and finish of movement could be made. A simple calculation should then give, for example, an average rate of rise of the feature expressed in millimetres per year, or metres per million years (m/Ma).

In practice, such a calculation would have no value, since it makes the tacit assumptions of a constant rate of overburden deposition, and of the salt body as a closed system. Neither of these assumptions is reasonable: deposition rates vary enormously, while the current size of a salt feature may be the result of extraneous effects such as dissolution or surface extrusion, and may bear no relationship to the total amount of salt which has flowed.

It is usual to relate the growth rate of a salt structure to the patterns of deposition in primary and later rim synclines on the basis that deposition in such salt withdrawal basins can be equated volumetrically with salt flow into a related structure. As is normal practice in structural geology, such volumes can be expressed as areas on correctly scaled cross-sections of either stratigraphic or seismic-derived origin. A material balance equation can be developed relating excess sediment thicknesses in the withdrawal basins to mobile salt volume, applying corrections (with geological input from adjacent areas if necessary and available) for erosional hiatuses and compaction in the sediments, and salt losses by dissolution and surface extrusion, etc.

Where determinations have been made, growth rates of salt swells, pillows and diapirs are found to vary quite widely within a normal range from 10 to several hundreds of metres per Ma. Variations occur related to various stages of development of a single structure, to different areas, and to different parts of the same area, often depending on the location of the structures within the basinal setting, and in turn on regional subsidence rates and rates of deposition in the post-salt section.

Ewing and Ewing (1962) studied small elevations in the sea floor detected by precision depth recorder above salt diapirs of the Sigsbee Knoll group. These structures are present in the area of the abyssal plain south of the Sigsbee Escarpment in the Gulf of Mexico. From ^{14}C dating of material in cores, and rates of deposition in the mainly pelagic sediments forming the sea floor, it was concluded that deformation of the latter had taken place at a rate of about 1 mm per year (1000 m/Ma) over the past 11 000 years, and that this probably represented the minimum rate of rise of the underlying salt features. The authors noted that it was not possible to differentiate between absolute upward salt movement and any effects of sediment compaction. From seismic profile data, the crests of the underlying salt structures were determined to be at depths between 3520 and 3640 m below MSL.

Ramberg (1981) made an interesting study of diapir growth rates involving the model ratio of time. Model ratios were calculated for a centrifuged model consisting of a silicone putty diapir rising through alternating layers of putty and modelling clay as overburden. The model was scaled so as to be as nearly as possible equivalent to the natural situation for a diapir rising from a 500 m thick salt source layer beneath a 2000 m thick overburden. Briefly, the results suggested a time of 1 Ma for the salt in a real diapir of the scaled dimensions to rise 2000 m through the overburden.

In Ramberg's opinion, this rise rate of 2000 m/Ma is not unreasonable. However, it must be pointed out that this represents an *average* rate of growth. It has been established that diapiric growth shows considerable variation in the different stages, being rapid at some periods (usually the early stages after piercement), and slow at others. Thus an *average* rate of growth of 2000 m/Ma implies substantially greater rise rates than this at some stages; in view of rise rates of some real-earth diapirs as calculated in other ways, this may be somewhat high. If so, the error is in a sense which would imply (as one possible factor) an underestimation of the competency of the overburden—the approximate equivalent of Ramberg's *plunging strength* in a model.

Seni and Jackson (1983, 1984) have made extremely thorough and detailed studies of rates of salt movement and diapir growth in the East Texas diapir province. By measuring syndepositional variations in the thickness of overburden sediments, they were able to relate these to salt movement periods and patterns, and to measured volumes and rates of salt flow, the methodology including volumetric measurements by planimetry. Some basic assumptions, variously proved, unproved, and of a simplifying nature were adopted.

The study distinguished between net rates of pillow growth (measured by equating rate of pillow crest uplift with either (a) rate of sediment thinning over pillow crest, or (b) maximum deposition rate in PRS—primary rim syncline), net rates of diapir growth measured by equating the rise of the diapir crest with maximum rates of sediment accumulation in SRS and TRS (secondary and tertiary rim synclines), and gross rates of diapir growth measured by the volume of salt passing through the maximum cross-sectional area of the diapir as calculated from gravity-derived diapir shape models.

Figures for net pillow growth in the East Texas Basin vary from 40 to 100 m/Ma when calculated by supracrestal sediment thinning, and 100 to 130 m/Ma when derived from sedimentation rates in the PRS. Both methods show fastest growth rate was in the early stage of development. Net rates of diapir growth varied between 150 and 230 m/Ma in the early stages (coinciding with high regional sedimentation rates) and a mean rate of 30 m/Ma in later stages. Gross growth rates for some diapirs in the area fall within the range of 42–530 m/Ma in the early stages, decreasing to 10–20 m/Ma at later periods. To illustrate the practical difference between net and gross growth rates for a diapir, the authors make the comparison between salt moving upwards in a diapir (at the gross rate of growth) while the diapiric crest remains stationary due to dissolution (zero net rate of growth).

An important point to note regarding one of the assumptions mentioned above, made by Seni and Jackson, stems from the discussion of Humphris' (1979) paper earlier in this chapter. Growth rates of salt structures related to overburden depositional patterns can only be validly used where it is reasonable to link the salt rise with gravitational effects of sediment deposition. Where the rise of a salt feature is demonstrably due to some other effect, such as lateral stress applied within the salt interval, i.e. where the overburden bulk density may be less than that of salt, as in the case quoted by Humphris, this approach is invalid, since the salt uplift is not related to the depositional rates of overlying sediments. Therefore, before any calculations based on sediment volumes in rim synclines is undertaken, the condition that a density inversion existed at the top of the salt layer when movement commenced must be satisfied, and the assumption made of equivalence of sediment volume in the withdrawal basin to volume of salt which has flowed into the diapir during basinal deposition.

In the Southern North Sea Basin, the writer recently made approximate calculations of the rise rates of three Zechstein salt pillows, using an approach similar to the net growth rate calculation, method (b), for pillows as derived by Seni and Jackson. Figures of 10, 46 and 50 m/Ma were arrived at for the three structures. These are approximate only since, due to a lack of available well data in the immediate vicinity, depths had to be calculated from seismic processing velocities; the figures must therefore be regarded as a rough indication.

In the North German Zechstein area, Trusheim (1960) gives an average figure of 300 m/Ma for salt movement in the diapir stage, and Sanneman (1960) quotes the same figure (0·3 mm/year), together with the somewhat surprising conclusion that this mean rate of flow is constant both in the stage of lateral flow into a pillow and vertical flow during the later diapiric stage.

Sanneman's (1960) paper is of great interest, particularly for the chart (reproduced also as Fig. 23 in Trusheim's 1960 paper) depicting the time periods of development of pillow and diapir stages of eight North German salt structures (one of these, Wildes-Moor, having only developed to the pillow stage). Following from the conclusion mentioned above, this indicates that the pillow and diapir stages in each structure took up approximately equal periods of time.

Recent work by Sørensen (1986) has suggested that the basic assumption, regarding a simple relationship between sediment thickening in withdrawal basins and the volumetric growth of a salt feature, is inadequate. To arrive at a more rigorous model of the processes involved, the interrelation of a number of different volume terms must be taken into account. Some subsurface volumes undergo a change of role during salt movement with sediment being first displaced by salt and then salt by sediment, or acting first as a sediment source, then as a sediment sink. What Sørensen calls the *excess volume* (i.e. the additional volume of sediment in the salt withdrawal basin, as compared to a 'norm' sediment volume over the same period) he relates to the volume swept out by the deforming sediment surface and the mobile salt/sediment interface.

The complex and interesting study of growth rate calculation provides many insights into the behaviour of the salt/overburden system during diapir development. For a full account of the subject, the works cited earlier by Sanneman, Trusheim, Seni and Jackson (1983 and 1984), and Sørensen (1986) should be referred to.

7.11 SUMMING UP

Several of the views expressed in this chapter are at variance with concepts of salt movement held in the past and still current to a greater or lesser extent.

One of these concepts is that of some universal 'critical thickness' of overburden which will initiate movement in a buried salt layer, and another is that of some critical minimum thickness of salt which must be present before mobility can occur. Reasons have been given why the writer does not regard either of these concepts as being valid.

Another idea which is regarded with suspicion is that of the spontaneous initiation of salt features—pillows and diapirs—when the mass of the overburden above a salt layer exceeds some critical value. The view taken in this book is that before salt will commence to flow—even when it is in a plastic and potentially mobile state—there must be a lateral pressure gradient within the salt layer which can only be generated by differential stress. Various situations in which such differential stress may develop have been described. This idea is supported by several examples known to exist of salt intervals buried under large overburden loads which have at some time developed density inversions at the salt layer, but where the salt has not become mobile since the required stress differential has not evolved. One exception to this is where an overburden fault affects the top of the salt interval in a gravitationally unstable situation, in which case salt already in a plastic state will inject the fault zone and develop into a diapir. This stress-relief process will result in a diapiric feature which shows no evidence of the pre-piercement (pillow) stage in the overburden. Examples of this type of diapir are known or inferred in the North Sea area, and provide good evidence that salt can be in a plastic state without flowing, until the intervention of some other event such as the faulting mentioned.

Where a lateral pressure gradient is developed in the salt layer, and lateral salt flow begins, then some additional mechanism—several of which have been listed earlier—is required to cause a local salt thickening/uplift to take place which may or may not develop eventually into a diapir.

Such lateral pressure gradients in the salt may lead to structural developments even beneath an overburden which has an average density which is less than that of the salt. Unless the average overburden density is known, it will be difficult or impossible to distinguish between such structures and those which are due to lateral density/thickness variations in a denser overburden.

The importance must be stressed of being able to observe (normally by subsurface geophysical methods) the source salt interval, and the depositional structures in the overburden resulting from any salt movement which has occurred. In areas where the source salt layer and some if not all of the movement effects in the overburden are too deep for observation, there may be serious problems in determining the origin and development details of any salt structures present. Assumptions may be made which are not valid, regarding for instance the presence or absence of a pre-piercement stage or of an unseen major unconformity at depth. Erroneous conclusions may be drawn regarding the mode of development of individual features, and the causes of salt movement on a regional scale.

If the views expressed in this chapter are correct then, taken in conjunction with the strong correlation noted between mobile salt basin locations and continental margins or tectonically active areas, they must lead inevitably to a reassessment of present ideas. In particular, current views on the causes of initiation of salt movement and of diapirism on a regional—and perhaps even global—scale need to be re-examined. A further comment will be made on this matter in the final chapter.

CHAPTER 8

Diapirism—2

8.1 INTRODUCTION

In this chapter, some unusual forms of diapiric salt structures will be considered with a view to ascertaining whether, in spite of (or perhaps because of) their unusual nature, these structures shed any light on salt movement and piercement processes of a more frequently encountered kind.

Subjects included in the discussion will be a form of diapirism associated with the periphery of a basin, a hypothesis concerning the possibility of patterns or families of salt structures developing from a primary feature, differential and preferential movement in salts including the effects of applied stress on the more deformable of the evaporites, which are precipitated later than halite, and the effects to be expected where stress is applied to two potentially mobile salt intervals separated by competent lithological units.

8.2 BASIN-EDGE DIAPIRISM

In their pioneer paper on the Permian evaporite geology of the southern British North Sea basin, Brunstrom and Walmsley (1969) drew attention to a region in the northern part of the Zechstein sub-basin between the main area of salt structures and the Mid North Sea High which forms the northerly margin of the sub-basin. In this area, marked A in Brunstrom and Walmsley's map which is shown in Fig. 8.1, the authors understandably expressed surprise that although the Zechstein salt has developed some broad, elongate pillows, there are virtually no diapiric structures. This is an odd and possibly significant fact when it is considered that this area coincides with the maximum primary development of Zechstein salt (over 1000 m in thickness) in this general area, the salt being overlain by more than 2500 m of overburden clastics at present, this overburden having been, even as early as the end of the Triassic period, of the order of 1200 m thick.

In the northerly part of this area (B in Fig. 8.1), there are virtually no salt structures seen at all on the map, a situation which is confirmed by more recent seismic data. However, lateral salt movement within the Zechstein interval is indicated by deformation of the competent Plattendolomit (cf. Chapter 2, Figs. 2.4 and 2.5).

In the more southerly area, where the main diapirism

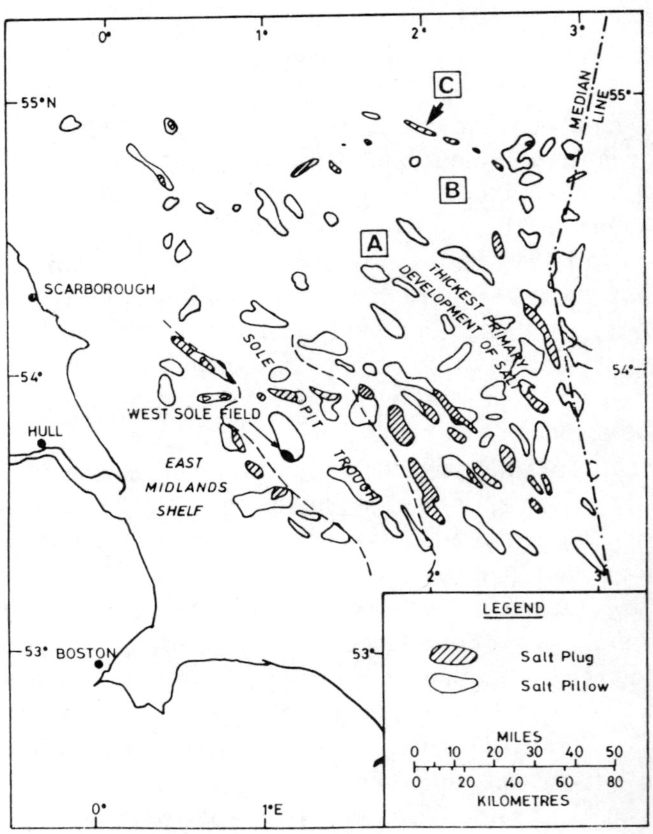

Fig. 8.1 Sketch map of the area of diapiric Zechstein salt in the Southern Basin of the British North Sea. Areas A, B and C are referred to in the text. (From Brunstrom and Walmsley (1969); courtesy of the AAPG.)

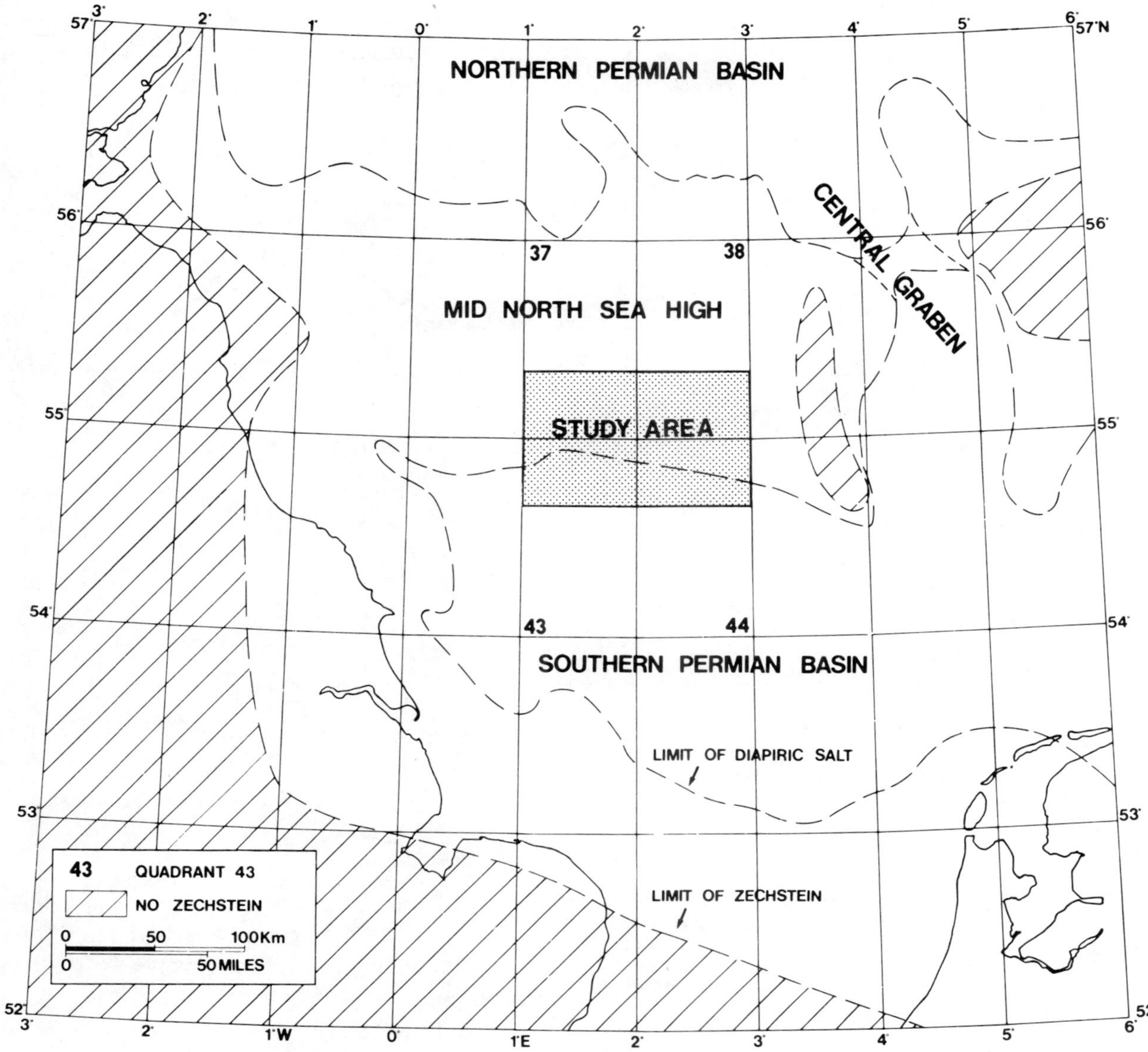

Fig. 8.2 Sketch map of the Southern Permian Basin, British North Sea, with the study area of basin edge diapirism dot-shaded, and major regional features labelled.

and pillow formation has taken place, the structures trend parallel to pre-Permian features and faulting rather than to the basin margins. Movement of these deeper faults is believed to have initiated (and controlled) the vertical salt movements here, the main movement period being during the Jurassic, and perhaps more specifically, during the Late Cimmerian erosional phase of Late Jurassic/Early Cretaceous times. With the sparse information available at that time, the authors of the paper cited expressed the opinion that outward movement of salt from the central part of the basin towards the margins had had only minor effects within the general context of salt movement in the basin.

More recent information suggests that centrifugal salt movement due to the overburden load differential in the basin does seem to have taken place, at least in the northerly direction towards the Mid North Sea High. This movement is believed to be responsible for the line of elongate features shown at C in Brunstrom and Walmsley's map (Fig. 8.1).

These features have been studied using recent seismic data (Jenyon, 1985a), the study area being outlined across the southern margin of the Mid North Sea High in Fig. 8.2. This sketch map also shows the major regional structures related to the Permian in this area. The schematic of Fig. 8.3 outlines various features associated with the zone marked C in Brunstrom and Walmsley's map; these features will be referred to later, but the three zones present should be noted. Zone 1 in the south comprises thick basinal facies Zechstein salt, Zone 3 in the north has only

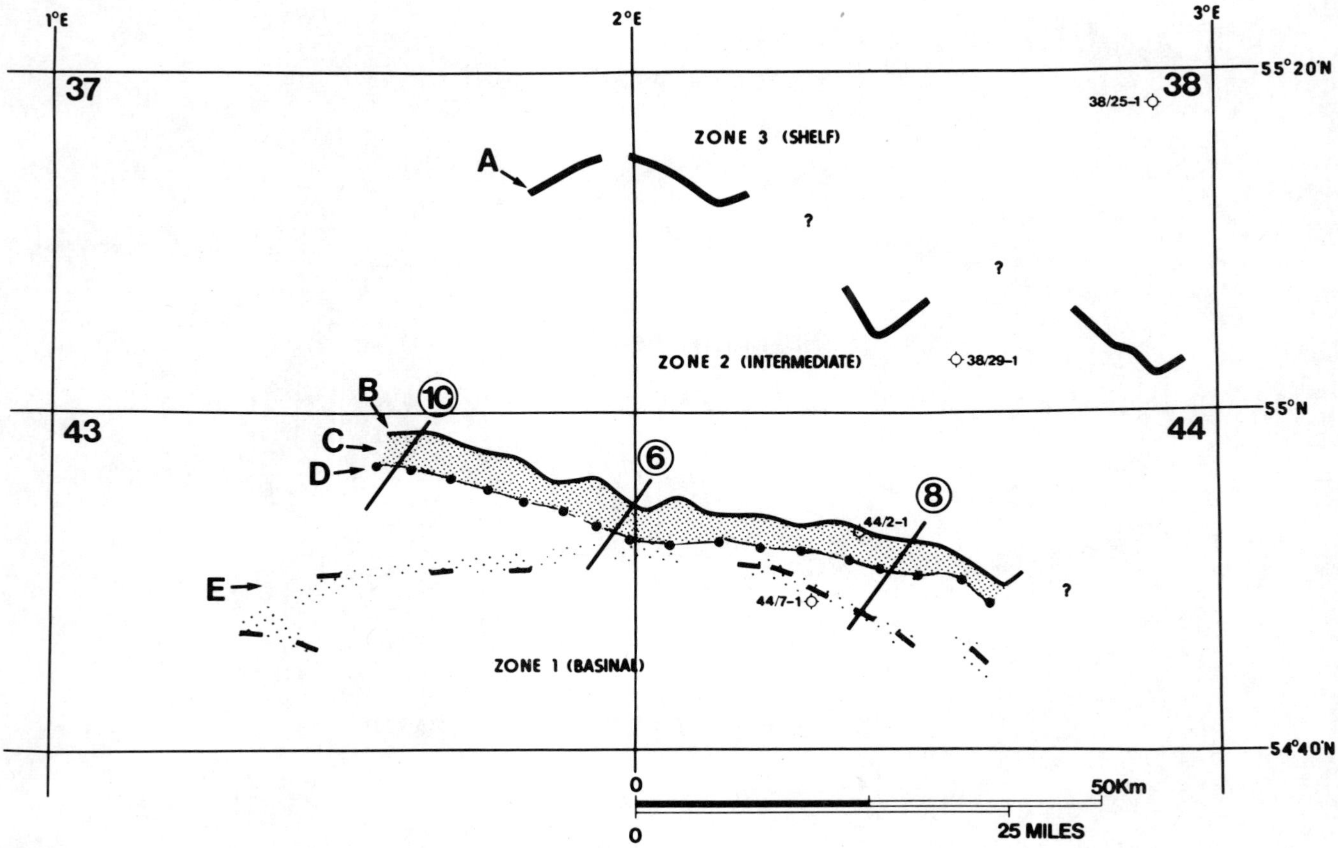

Fig. 8.3 Sketch of linear zone of edge diapirism discussed in the text. This is the same as the zone marked C in Fig. 8.1

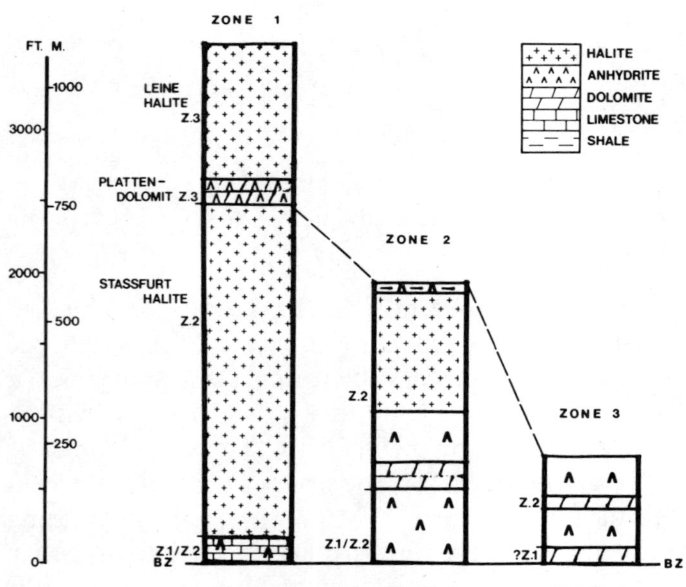

Fig. 8.4 Schematic Zechstein well sections from the three zones (basin, intermediate, shelf) seen in Fig. 8.3.

shelf facies Zechstein rocks, principally dolomites and anhydrite, while Zone 2 is intermediate, consisting mainly of shelf facies carbonate and anhydrite rocks, with some shelf-type halite which is much thinner than the basinal halite of Zone 1, and in general does not show features indicating any marked mobility. The edge of the intermediate shelf zone at B is fault-controlled; that of the shelf proper at A may be fault-controlled, but the evidence is less positive.

The lithological differences in the Zechstein of the three zones is indicated by the well schematics of Fig. 8.4. These show thicknesses referred to the base of the Zechstein as datum for the three wells which are shown in Fig. 8.3.

The sequence of events which is believed to have resulted in the zone of elongate structures parallel to shelf-edge B is shown diagrammatically in Fig. 8.5 (which should be looked at in conjunction with Fig. 8.3). As noted previously, the basinal salt here (zone B in Brunstrom and Walmsley's map, Fig. 8.1) shows virtually no evidence of vertical salt movement in a band some 40 km wide before reaching the zone of elongate features. However, deformation of the Plattendolomit competent band within the total salt interval indicates that some lateral salt movement up the basinal dip from the direction of the region of thickest salt (and overburden) towards the basin edge (fault-controlled shelf-edge B in Figs. 8.3 and 8.5) was

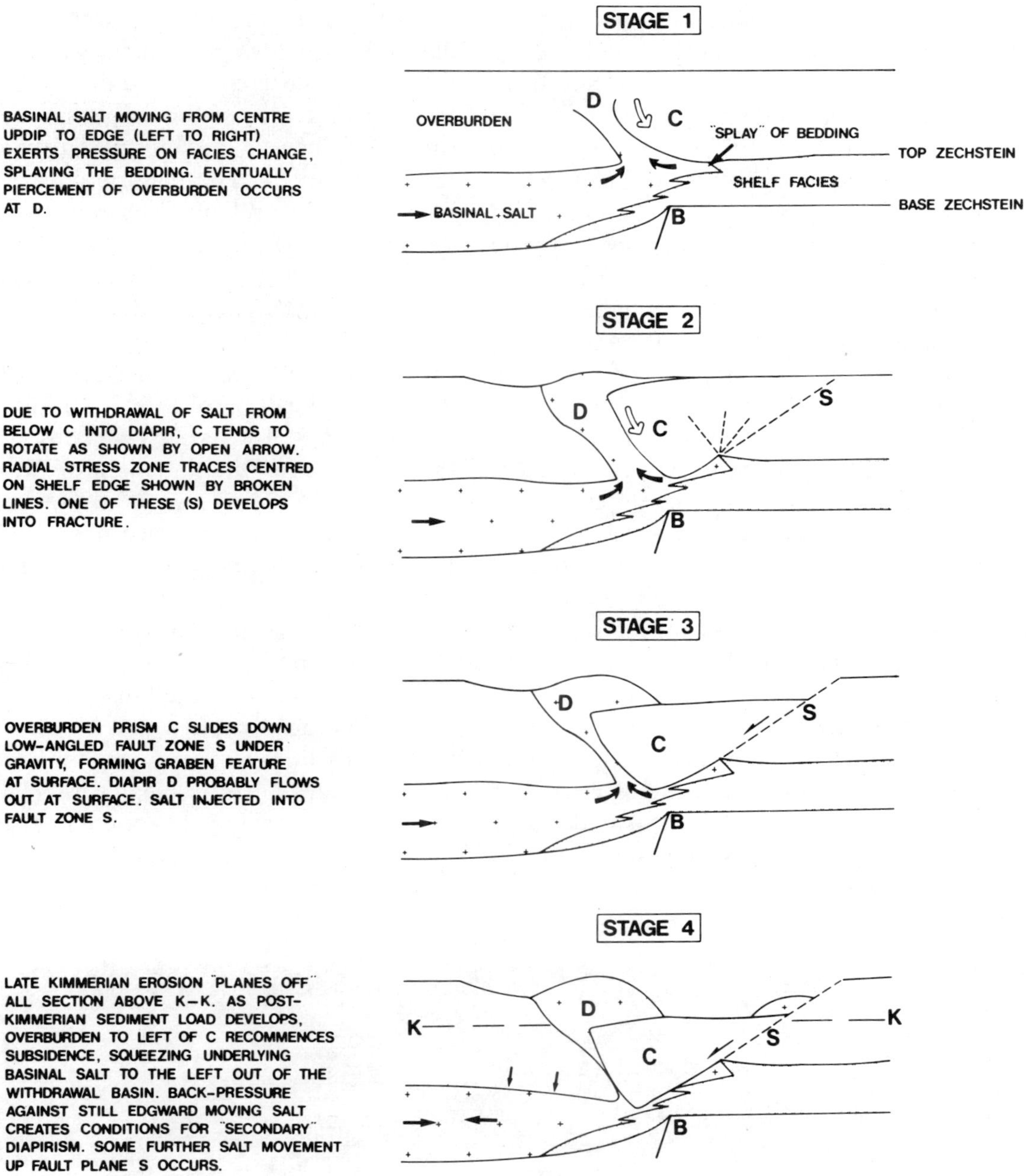

BASIN-EDGE DIAPIRISM - FOUR STAGES IN THE SUGGESTED PROCESS.

Fig. 8.5 Diagrammatic representation of four stages in the process of basin edge diapirism based on the interpretation discussed in the text.

taking place. At this shelf edge, a facies change is present, with the basinal salt passing into the shelf facies Zechstein by interdigitation of the salt with the shelf facies units. The increasing pressure on this narrow zone of facies change is suggested in the seismic data by effects of 'splay', or forcing-apart of the shelf facies bedding. The seismic data also indicate that the shelf facies units do not stop at the basin edge B, but step down into the basin in an abruptly thinning wedge. Apart from any facies change, the rather rapid constriction of the total Zechstein interval in passing from the basin to the shelf is also likely to be an important physical factor in producing resistance to edgewards movement of the basinal salt. Another factor involved is probably the passage from basinal salt in a mobile state to much more viscous, or non-mobile, shelf halite.

As the back-pressure built up in the salt, a linear salt swell parallel with the basin edge is supposed to have formed which became a piercement in Late

Triassic or Jurassic times, parallel with, and some 5 to 6 km basinward of the facies change, and breaking through to the contemporary surface. The 5 to 6 km of overburden between the diapir and the shelf edge is important, because as salt withdrew from beneath it into the diapir, the gradual withdrawal of support must have caused this segment of overburden (C in Stage 2 of Fig. 8.5) to develop a tendency to rotate counterclockwise (viewed in the direction of Fig. 8.5) about the supporting shelf edge which would act as a pivot. This tendency would set up tensional stresses in the overburden in zones above, and parallel with the shelf edge in plan view. In cross-section, the traces of these zones would radiate from the shelf edge as centre, as indicated in Stage 2 of Fig. 8.5. One of these zones, S, in the latter figure, later developed into a fracture, and then a low-angled gravitational normal fault (Stage 3, Fig. 8.5) and a massive wedge-shaped prism of overburden (C in Fig. 8.5, Stage 3) between the fault and the primary diapir D moved down the fault zone, coming to rest on the basal part of the Zechstein eventually.

The subsidence of the large overburden prism C must have resulted in a major graben feature at the contemporary earth surface, with extrusion of considerable volumes of salt. Subsequently, the Late Cimmerian (alternative spelling Kimmerian is used in Fig. 8.5) erosional phase removed at least 600 m of sediment, including the bulk of the salt in the primary diapir D, which would explain why, with such major subsidence effects in the withdrawal basin, only relatively small traces of the diapir can be seen adjacent to the basinward face of the subsided prism C in the seismic data. The diapiric effects seen probably represent only a relatively narrow 'feeder' zone of the diapir, as suggested in Stages 3 and 4 of Fig. 8.5.

The reason why the low-angled fracture developed into the fault S is believed to be a combination of gravitational effect and the geometry of the step-down of the facies change into the basin, which prevented the development into a fault of the more steeply-dipping zones of weakness, which is also suggested in the diagrams of Stages 3 and 4 of Fig. 8.5.

During the subsidence of the prism, and also at later stages, salt was injected not only into the diapir D, but also into the fault zone S. Evidence for this lies in the pattern of acoustic impedance contrasts in zone S, and also in marked uplift of strata above the fault zone in post-K times which can best be explained by salt movement up the fault well into the Tertiary. Post-unconformity movement of salt is also indicated by upward bowing above the subsided prism. Such late edge-ward salt movement, with related subsidence/drape effects, continued to cause late-stage features in the overburden until nearly the Middle Tertiary.

The subsided prism C shows rotation clockwise (in

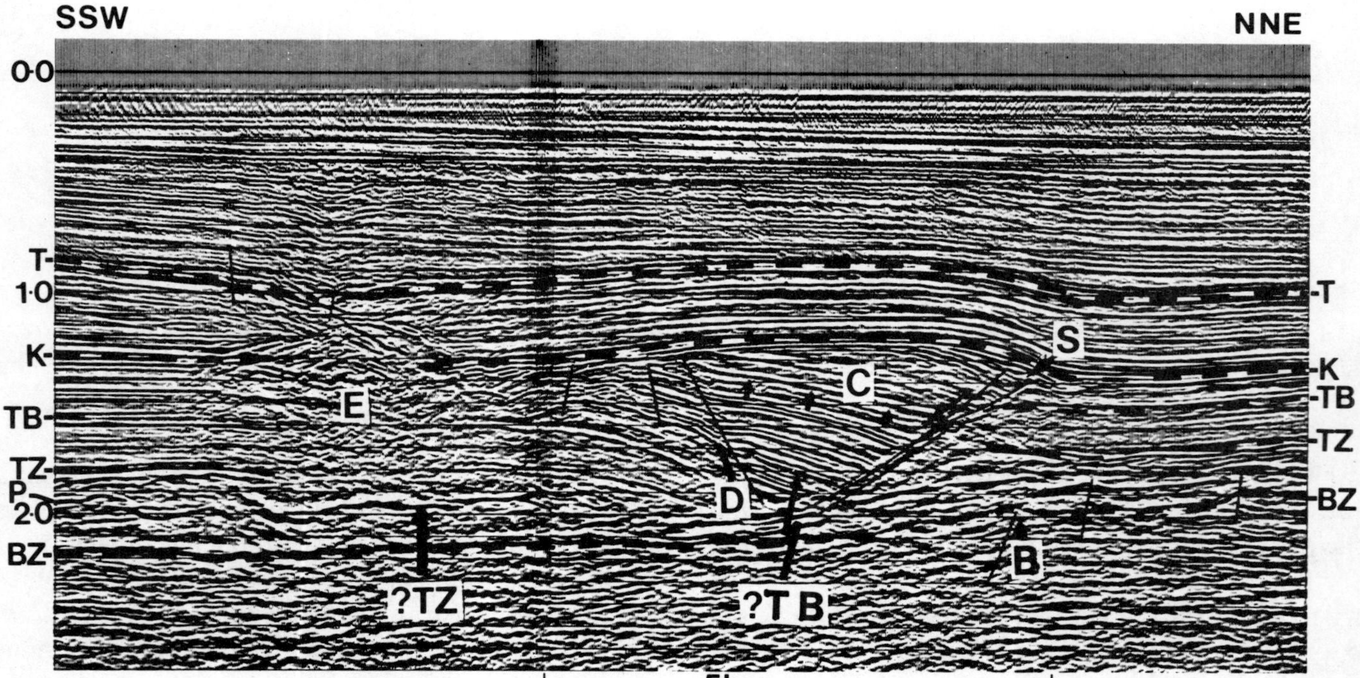

VERTICAL SCALE EXAGGERATION = 2.5:1 ON ALL SEISMIC EXAMPLES

Fig. 8.6 Migrated seismic section, circled 6 in Fig. 8.3. (Courtesy of SS(E)L.)

the direction as seen in Fig. 8.5) in some locations, but not in others. This may be related to the torque applied to the prism by salt flowing from beneath it up into the primary diapir D; it is likely that the volume of the salt flow varied in different locations along the complete linear structure.

Seismic evidence for the complex series of events suggested is seen in Fig. 8.6, for the seismic line shown as a circled 6 in Fig. 8.3. At the SSW end of the section, the Zechstein interval BZ–TZ consists mainly of basinal halite, containing the deformed competent band of Plattendolomit, P. At the NNE end of the section, the shelf facies Zechstein is seen, and the fault-controlled shelf edge B. Above and to the left of B, the shelf facies units can be seen in a wedge stepping down into the basin, and splay of the shelf facies bedding at TZ level above B indicates the damming effect of the facies change on movement of the basinal salt towards the margin. The subsided overburden prism C is very clear, with the low-angled fault zone down which it slid indicated at S and showing good fault plane reflection events. The arrowed zone marked D is believed to be the location of the feeder zone to the primary edge diapir, the latter being no longer in evidence due to the erosional phase represented by unconformity K. Immediately to the left of D, the strong subsidence effects in the overburden resulting from the salt flow into the primary diapir are obvious. Note that in this example, the subsided prism C has completely closed off the feeder zone to the diapir, as suggested in Stage 4 of Fig. 8.5.

The small arrows within the subsided prism C indicate the estimated position of the basal part of a Mesozoic sequence which is mainly Upper Triassic, but possibly with a thin Jurassic interval at the top, that has been removed by erosion from the adjacent areas, but which is present here due to early subsidence of the prism. The estimated position of the top of the Bacton (Lower Triassic) is shown as ?TB within the subsided wedge. The amount of material removed by erosion in adjacent areas is shown by comparing the situation in Well 44/2-1 (Fig. 8.7), the only well to be drilled into the subsided overburden here, with that in Well 44/7-1, which lies only 10 km to the SW of the subsidence. In the latter well, the unconformity has cut down into the Bunter Shales, implying the removal of at least 600 m (2000 ft) of material during the erosional phase.

The secondary diapir shown at E in Fig. 8.6 is of much later development than the primary diapirism, having occurred during the Cretaceous (K–T interval). It is suggested that after the K erosional phase,

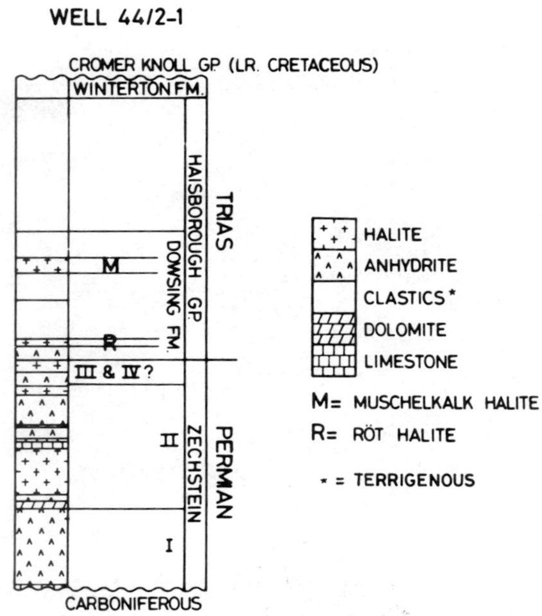

Fig. 8.7 Schematic Permo-Triassic well section for the only well to have been drilled through the subsided overburden prism.

developing sedimentary loading above the primary withdrawal basin caused further subsidence, and salt movement to the NNE. This produced the later effects in the vicinity of the subsided prism such as uplift of the latter as seen by the antiform shape in the K–T interval above it, and probably further salt flow up the fault zone S. Movement of salt out of the primary withdrawal basin towards the SSW met salt still flowing edgeward, and this resulted in the formation of elongate pillows, and subsequently diapirs, parallel to the primary diapiric features. The immediate cause of the piercement in the arcuate zone of these secondary diapirs which developed (see Fig. 8.3) was probably overburden fracturing due to the subsidence into the withdrawal basin, as discussed in Chapter 5, but in certain locations they are also associated with Base Zechstein faults. A very interesting effect of these two modes of piercement initiation is seen in Fig. 8.8 (line circled 8 in Fig. 8.3) which is a line about 25 km to the east of that in Fig. 8.6. Here the secondary diapir E is located immediately basinward of a strong salt pillow, which is a cross-section of the secondary elongate structure mentioned previously. Beneath E, a fault is present in the Base Zechstein, and it is suggested that flow of salt in a SSW direction out of the primary salt withdrawal basin met edgeward-moving salt and produced this pillow during the K–T interval (note truncations beneath unconformity T, the Base Tertiary). During this same period, a fault either occurred, or was reactivated, at Base Zechstein

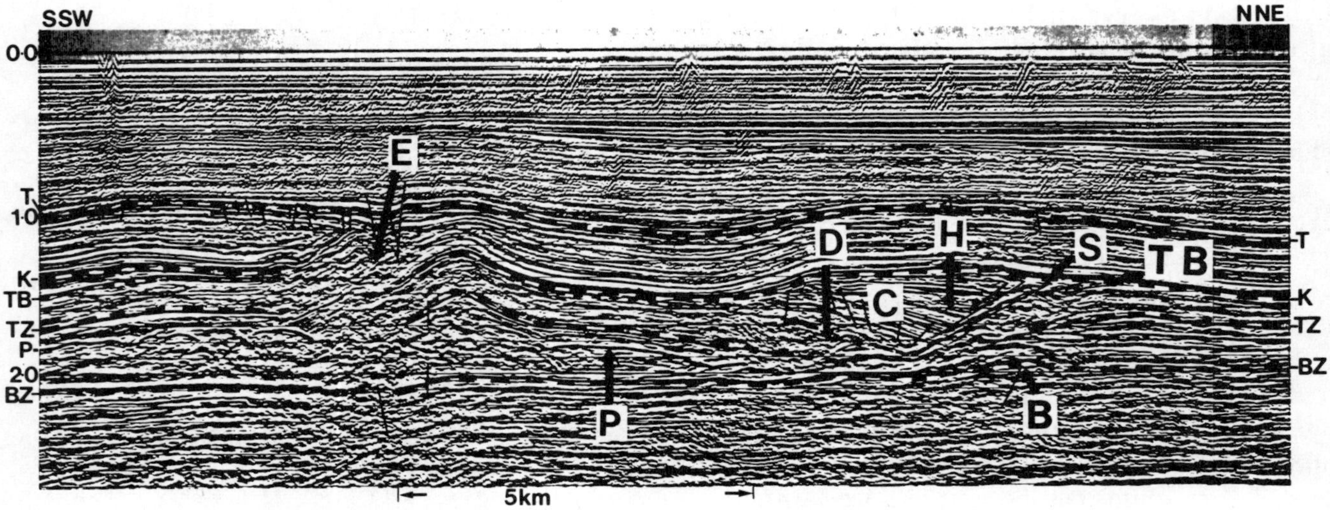

Fig. 8.8 Migrated seismic section, circled 8 in Fig. 8.3. (Courtesy of SS(E)L.)

level immediately to the SSW of the pillow which became an 'instant' focus of upward salt movement, robbing the pillow of the salt supply at a crucial time, leaving it as a failed secondary diapir.

The structures associated with the subsided prism C in Fig. 8.8 are also of interest, and an enlarged portion of the section is shown in Fig. 8.9. Referring to both figures, the subsided prism C is shown to be affected by a series of faults radiating apparently from approximately the same point in cross-section. It is believed that these fault traces are the real-earth equivalents of the fracture traces shown by the broken lines in Stage 2 of Fig. 8.5, and before subsidence of the prism, radiated outwards from a point at the top of the TZ at the shelf edge. The main fault, S, has a very strong fault-plane reflection, and this together

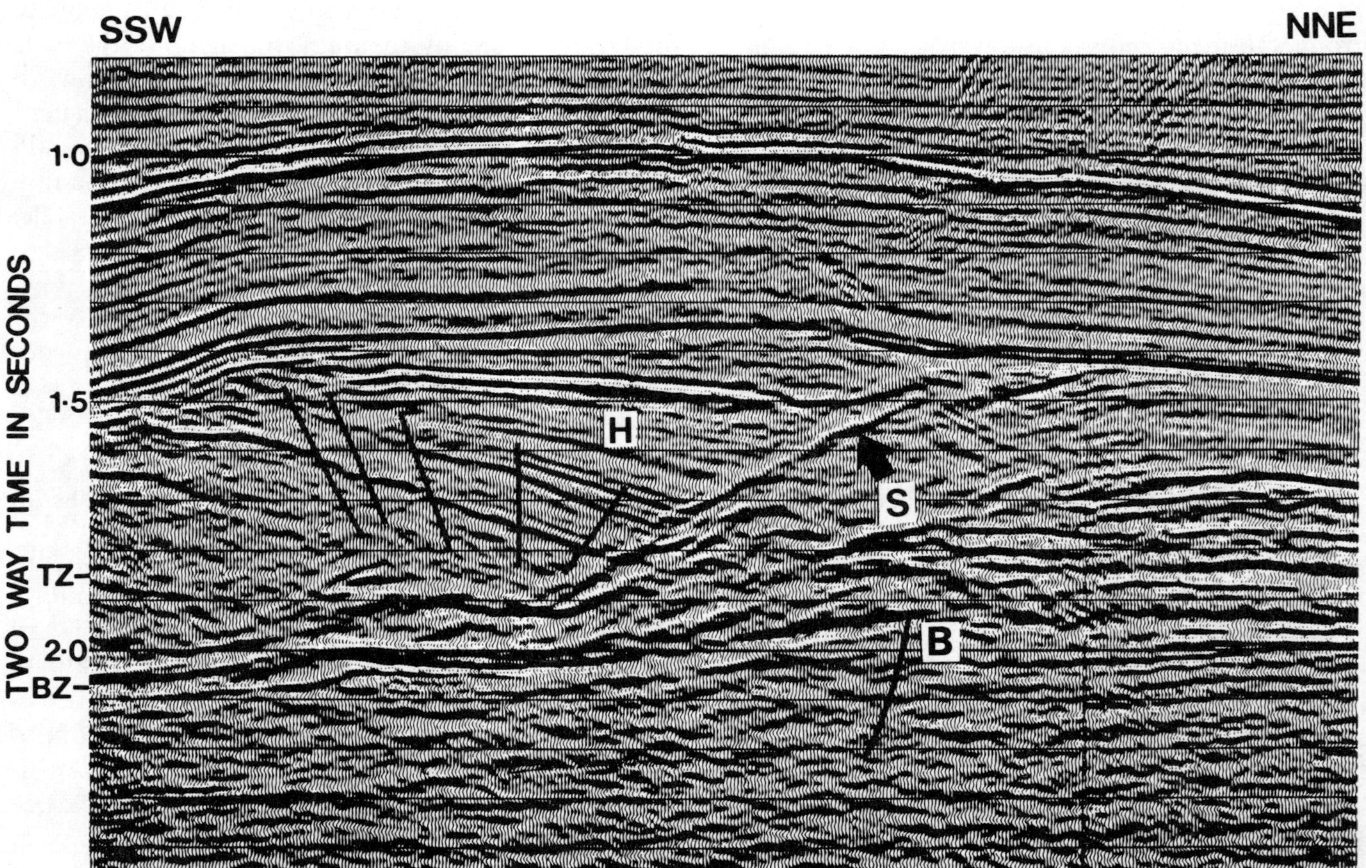

Fig. 8.9 Enlarged portion of the section shown in Fig. 8.8.

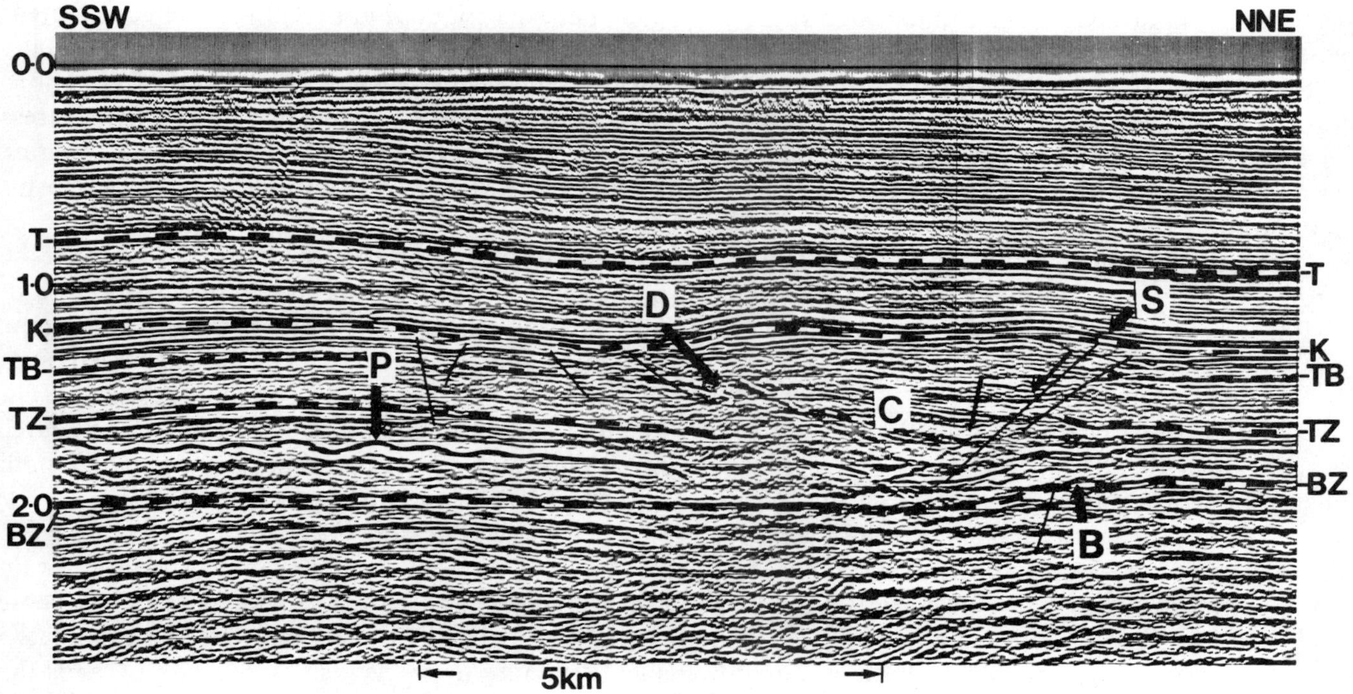

Fig. 8.10 Migrated seismic section, circled 10 in Fig. 8.3. (Courtesy of SS(E)L. Figures 8.2 to 8.10 after Jenyon (1985a) with acknowledgements to the AAPG.)

with considerations of acoustic impedance contrasts, or the lack thereof, seems to indicate that the fault zone was strongly injected by salt. Injection of this salt from the fault zone along the bedding planes may have caused the divergence observed at H.

A further seismic example, from the western extremity of the elongate primary feature, is shown in Fig. 8.10 (line circled as 10 in Fig. 8.3). The principal familiar features are all present, except that a secondary edge diapir is not present within the section, although such a feature is present off to the SSW of the example. Of special interest is the deformation of the Plattendolomit P to the SSW of the primary features, which with the lack of deformation of the Top Zechstein (TZ) units indicates lateral movement within the salt interval, the splay effect seen at and below TZ level abutting the NNE face of the subsided prism C, and the step-down of the shelf facies units above and to the left of B, into the basin.

From the seismic examples shown, and the sketch diagram of Fig. 8.3, it is clear that the subsided prism of overburden can be traced for a very long distance in essentially similar form. It is now known to extend westwards out of the study area and can be traced from east to west for some 90 km. Over this length its shape changes slightly, but the basic wedge-shape lying in a V-shaped cradle of salt is everywhere recognisable. In a way, it could be regarded as an unusual type of linear rim syncline, produced as it was by overburden subsidence following salt withdrawal from the deeper section. The analogy cannot be carried far, however; this feature is controlled by a basin edge facies change acting as a dam, or buttress and, more importantly, the period of deposition of the sediments in the subsided prism is not related to the period of salt movement and diapiric development.

There are several points about this group of features of complex origin which are relevant to salt tectonics in general. The diapirism involved is plainly of halotectonic rather than halokinetic origin, although gravitational effects controlled the development of the features. The initial effects seem certain to have had their origin in the lateral updip movement of salt within the Zechstein interval, rather than in any differential overburden loading locally, although the cause of the general salt movement from the direction of the basin centre towards the margin was probably the result of a lateral pressure gradient produced by overburden subsidence under gravity ('probably' is used since other effects may be responsible for lateral movement of salt on a regional scale).

The development of both primary and secondary diapirs in this context is of interest in relation to the mechanisms of fault-induced upward salt movement as discussed in Chapter 5, including those processes associated with base salt faults and with fault effects related to the margins of salt withdrawal basins. The examples shown also indicate the reality of the injection of a fault zone by salt under high pressure. The subsidence effect associated with the salt movement in this case has led to the preservation of a Mesozoic sequence no longer present in adjacent areas due to erosion, and this could be of significance in hydrocarbon exploration, depending upon the periods of maturation and migration of any hydrocarbons present in such an area. In passing, the difficulty is noted of reconciling edge-diapirism with the 'downbuilding' hypothesis (Chapter 4).

8.3 PATTERNED DIAPIRISM

The question of whether secondary salt features can evolve in an ordered way from the interaction of source layer salt movement and overburden subsidence related to the formation of a primary salt structure still remains an open one for many geologists. The possibility of such processes has been mooted by several authorities, nevertheless, and deserves attention. The concept does not seem to have gained great credence, perhaps because it is too 'simple'; many believe that the complexities engendered by variations in overburden density or thickness, and external stress factors due to tectonism and other causes, militate against the undistorted operation of the mechanisms proposed.

An early suggestion that outward flow towards the margin of a peripheral sink may occur in addition to inward flow towards a central, primary diapir, came in Nettleton's 1934 paper, when he calculated that, theoretically, pressure in the trough of the sink might be greater than in the adjacent part of the salt source layer, depending on the overburden material strength and the depth of the sink trough.

The effect was demonstrated in model studies later, by Parker and McDowell (1955), as mentioned in Chapter 4 (see Figs. 4.10, 4.11 and related text). These authors speculated at the time that the Spindletop Field structure in Texas (associated with perhaps the earliest major oilfield to be found) may be an example

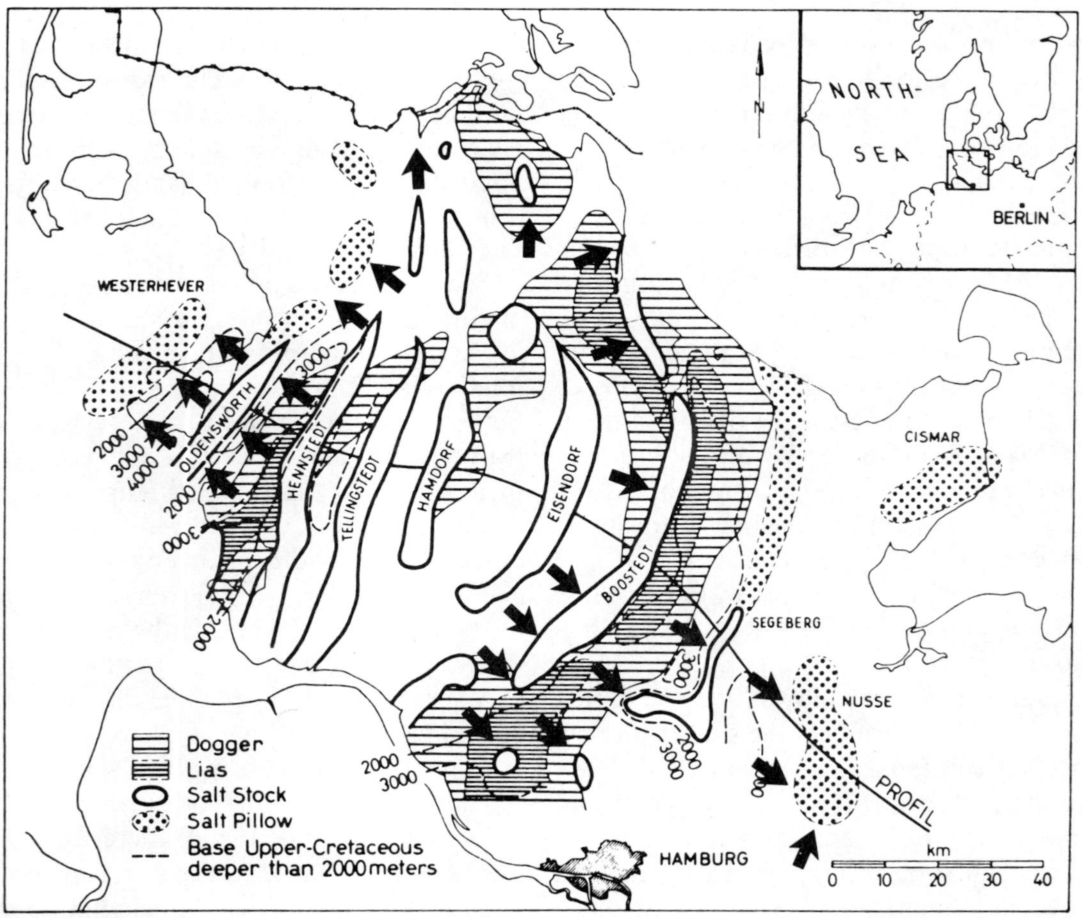

Fig. 8.11 Sketch map of a group of salt diapirs and pillows NW of Hamburg, Germany, shown as if with Cretaceous and Tertiary removed. Arrows show Sanneman's interpretation of the directions of salt migration in the group, based on the overlap of rim synclines. Cf. Fig. 8.12. (From Sanneman (1968), Figs. 7 and 9, courtesy of the AAPG.)

of the 'master' or primary diapir in a patterned group, which had led to the evolution of a ring of salt features around it, all lying approximately on the circumference of a circle with Spindletop Dome as centre. These marginal secondary structures were supposed to have formed by centrifugal flow of salt from the trough of the peripheral sink surrounding the central diapir. Not everyone agrees with this interpretation, although Halbouty (1967) gave it qualified approval ('qualified' since in the case of a similar group of structures at Calcasieu Lake, Louisiana, the central feature was shown to be of later development than the ring of structures around it).

Sanneman's papers of 1960 and 1968, supported by Trusheim's work, suggest that a well-known group of diapirs and salt pillows in Northern Germany (Fig. 8.11) is the result of the formation of a patterned 'family' of salt structures, the oldest—of salt wall type—being in the centre of the group, and the other elongate diapirs and salt pillows being progressively younger outwards. The Fig. 8.11 illustration conforms in general to Trusheim's analysis of salt movement patterns in this basinal area as producing large vertical salt walls in the basin centre, surrounded by diapirs, pillows and swells successively outwards towards the basin margin, as sketched in Fig. 4.9.

Mathematical treatments of salt/sediment instability under gravitational control, following earlier work by De Golyer (1926) and Nettleton (1934), were effected by Daneš (1964), Biot and Odé (1965), Selig (1965) and Selig and Wermund (1966). The latter paper is of particular interest since it leads to predictions that a family of structures would develop from an initial disturbance in a plane interface between salt and sediment, and would spread radially outwards in a regular pattern from the location of the initial perturbation; also that the spacing of these secondary features is dependent on the thickness of the source layer and the ratio of viscosities of overburden sediments to the salt (see Fig. 8.12).

This paper also suggests that the concept of a central primary feature with an annular peripheral sink, and

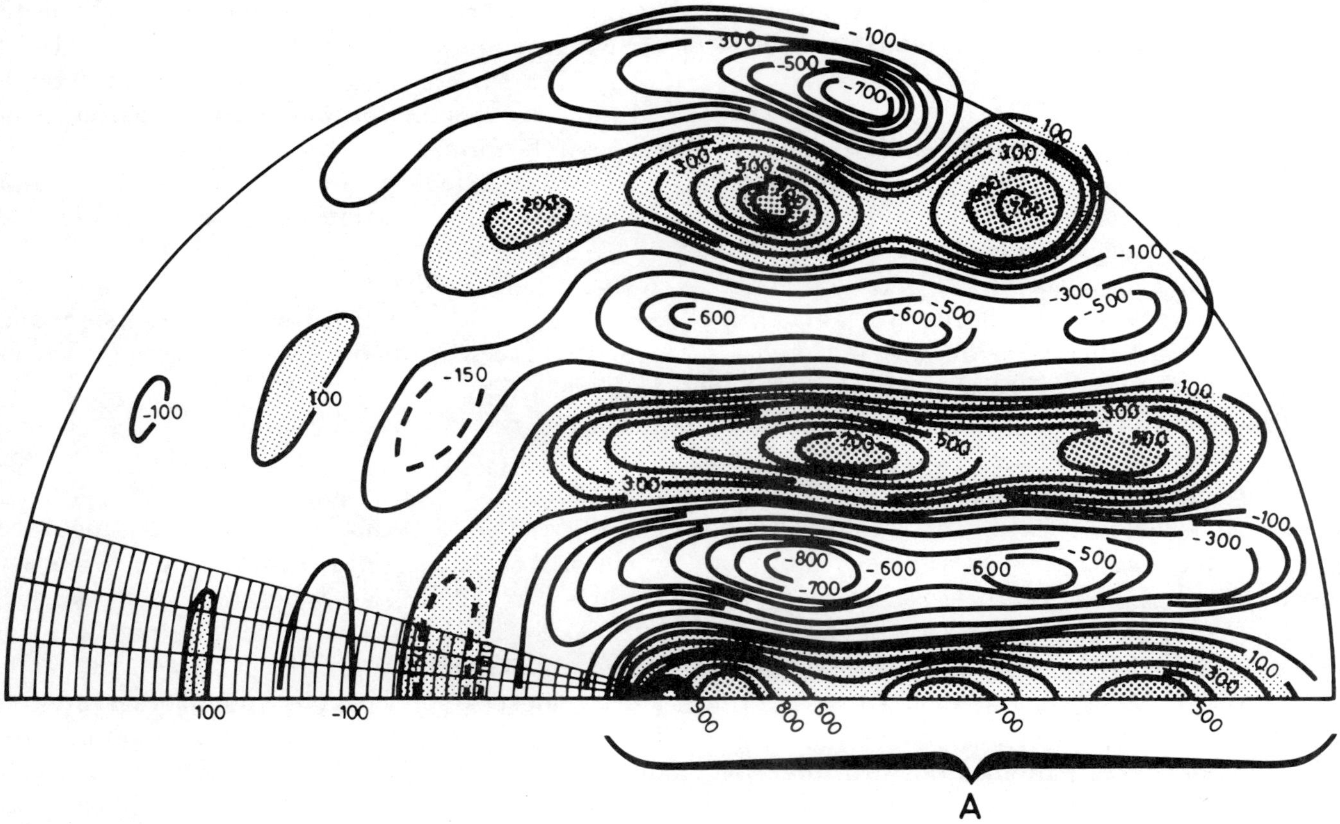

Fig. 8.12 A map of one of the mathematical model situations developed from Rayleigh-Taylor instability theory by Selig and Wermund (1966). From an initial ridge-like disturbance (A) scaled to represent 40 miles, a series of high (dot-shaded) and low elongate features is developed. Heights and depths of contours scaled in feet related to 'salt surface' value of 0. Original illustration plotted on a semi-circular grid, part of which is shown. Note the striking similarity to the right-hand half of the group in Fig. 8.11, if the Tellingstedt feature is regarded as the 'initial disturbance'. (After Selig and Wermund (1966); courtesy of the Society of Exploration Geophysicists.)

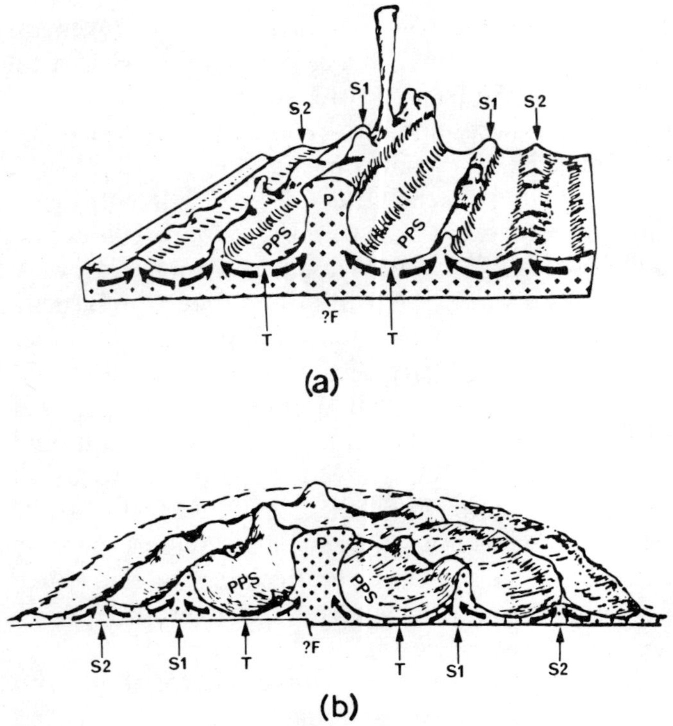

Fig. 8.13 Diagrams illustrating the hypothetical development of patterns of secondary salt features from (a) a primary linear salt wall, and (b) a primary sub-circular diapir.

secondary features developing around the outer margin of the latter, could be regarded as the limiting case of a more general situation involving an initial disturbance of linear, elongate form. Maps constructed of their mathematical models show a series of elongate secondary disturbances paralleling the initial ridge-like model perturbation. The authors noted that in the real earth, development of such patterns would be dependent on factors other than gravitational control, such as the thickness of the salt source layer (and hence the supply of salt), and the absence of any 'superposed instability' (presumably including tectonism). Also, the theoretical treatment says nothing of whether the secondary disturbances—or the primary for that matter—would remain in the swell/pillow stage dealt with by the model, or whether diapirs would develop, this being also subject to extraneous factors.

If the hypothesis of patterned salt structure evolution has any factual basis, regular development of the resulting secondary structures (whether they are formed at the outer margin of an annular or linear peripheral sink) cannot be expected. This was found by Parker and McDowell in the model studies referred to, where the secondary uplifts were discontinuous. As regards linear peripheral sink developments, it seems appropriate to consider the real earth situation of the Bay Marchand salt ridge, offshore Louisiana, as described by Frey and Grimes (1970). They suggested that the individual diapirs along the ridge formed by 'cannibalising' the adjacent salt. Initial irregularities control the formation of structures by the operation of differential overburden pressure at each point along the ridge, with any upward-leading locations along the ridge gaining salt by flow from adjacent locations under greater overburden load. This seems to be a likely analogy to effects at the crest of a marginal peripheral sink uplift, whether of annular or elongate form.

These ideas are embodied in the sketches of Fig. 8.13, which suggest the hypothetical development of patterns of secondary features from (a) an elongate salt ridge and (b) a central sub-circular diapir. The two sketches are block diagrams (the lower sketch showing only half the group) of the salt interval as though with the overburden removed. The salt cross-section is marked by the cross symbols, with the primary structure at P, the trough point of the PPS (peripheral sink) indicated at T, and the development of two generations of secondary features marked at S1 and S2. Arrows mark the supposed directions of salt flow inwards and outwards from the troughs of successively formed peripheral sinks, and a fault is shown at ?F to indicate a possible initiating cause of the primary feature. As noted, the primary salt feature in (a) and the secondary features in both (a) and (b) are indicated as having developed instability in upward growth due to differential overburden pressure, and actual or incipient diapiric spines are suggested as evolving.

What is possibly a real example of the process is shown in Fig. 8.14, in a seismic section from the Northern North Sea. Although this section is noisy and the resolution mediocre, the symmetry about the small primary diapir, P, as shown by the secondary features SA and SB, is very striking (compare this with the S1–P–S1 part of the sketch in Fig. 8.13(b)). The source salt interval is present between B and T, and it is most likely—from observation of this interval particularly to the left of SA and to the right of SB—that complete salt depletion has occurred. The salt within the original Zechstein interval (B–T) can only have been quite thin, and the strong reflection events now seen within the remnant interval are typical of the response of carbonate and anhydrite units of the total Zechstein interval.

The salt depletion is probably the reason why the primary diapir P has not risen further upward through

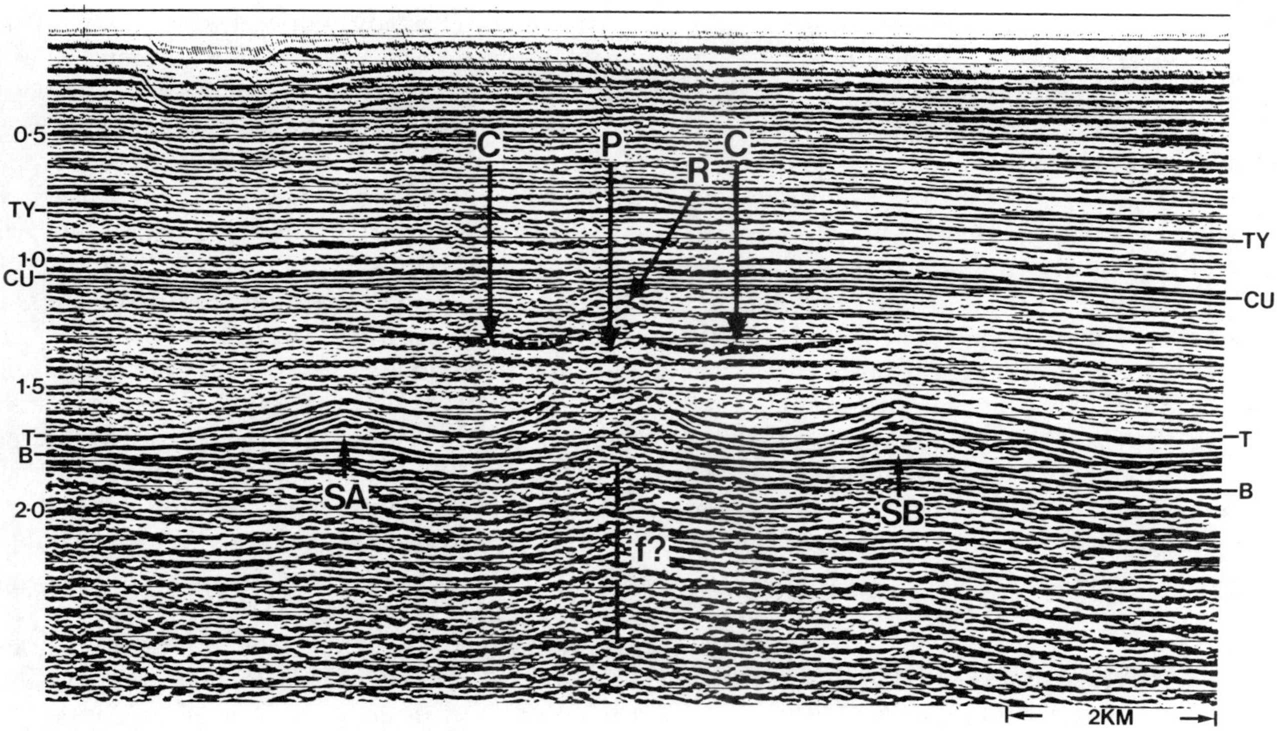

Fig. 8.14 Migrated seismic section showing a possible example of a patterned group which has reached the stage of 'first generation' secondary features SA and SB. (Courtesy of SS(E)L.)

the section, and also why the secondary features SA and SB have remained at the pillow stage without developing into piercements. The salt supply has been the overriding control factor in this group. A faint, shallow secondary rim syncline development can just be made out at C–C, appropriate in its dimensions to the quite minor subsidence resulting from the small primary piercement, and there is an event at R interpreted as the response to a cap rock. The Late Cimmerian unconformity is present at CU, with the Base Tertiary at TY.

Due to the low density of seismic lines in this area, it is not possible to establish this group unequivocally as an example of the processes under discussion; nor is it possible, assuming that the group is indeed an example of such processes, to determine whether it represents a concentric or a linear configuration. However, in the writer's opinion, the striking two-dimensional symmetry about P shown by the secondary features, and the reasonable assumption based on the visual evidence (and the width of the peripheral sink with respect to the primary diapir base width) that SA and SB are the result of salt flow related closely to the formation of diapir P and its peripheral sink, make it more than likely that the pattern process is responsible for the group. Here it has advanced only as far as an incipient 'first generation' of secondary features due to the limitations of salt supply.

The implications of this mechanism, if and when it occurs, may be of importance in hydrocarbon exploration, and a correct understanding of it is required in order to decide where the best trap locations and reservoir/seal properties may be found in the group. Due to the progressive outward development of such a group over an appreciable time span, a 'spectrum' of rim syncline sediments of varying burial depth histories and lithological type will be present. Also, as each successive generation of secondary features develops in a large group, the troughs of the peripheral sinks will subside in turn, and may come to rest on the base of the salt interval, effectively sealing off each part of the group in turn from further salt flow; this will have the effect of varying depositional rates in the overlying rim synclines, if it occurs. If these groups exist in reality, and are produced by the processes suggested, there is another important aspect to their development. The mode of evolution of such a group does not appear to be very robust—any disturbing factor (as suggested by Selig and Wermund) would seem likely to disrupt the train of events necessary, depending upon the stage reached, to the extent that the resulting features may not be recognisable as a product of the pattern process. They would thus provide good evidence, if intact, of an absence of any major tectonism in an area during the period of development of the structures.

Much more evidence of a detailed nature is required before this hypothetical phenomenon can be established as a real process. Conjecturally, the writer believes that the process suggested does indeed occur, and suspects strongly that a number of pattern groups of this kind may be present in the Northern North Sea Zechstein sub-basin. They do not appear to exist now in the Southern Basin, which was probably subjected to severe tectonic disturbance during the main period of Zechstein salt movement. If this is the case, a reason for the different forms of diapirism in the two sub-basins must be sought: again conjecturally, it may be that the Mid North Sea High, and the Pennine and Grampian Highs offered some shielding effect to the Zechstein salt of the Northern sub-basin, allowing the undisturbed halokinetic development of the secondary features in groups of this type.

8.4 DIFFERENTIAL AND PREFERENTIAL SALT MOVEMENT

The two types of deformation to be considered involve (i) relative movement between two parts of the same salt interval, and (ii) relative movement between different types of evaporite salts.

If the proposition is accepted that salt rock deforms plastically when subjected to sufficient vertical load stress combined with temperature effects, then logically it must be expected that the lower part of a horizontal salt interval should become mobile before the upper part. This happens because of the additional 'self-loading' factor provided by the bulk of the salt itself, and the probability that the lower part of the interval is at a higher temperature than the upper, tending to reduce the viscosity.

When the lower part of a salt interval begins to flow laterally and, for reasons discussed previously, to thicken into a swell structure, it will uplift the non-mobile upper part of the salt towards the surface more or less intact, depending upon the stress pattern which develops. Such effects of differential salt movement can be of some importance in certain types of mining.

This mechanism has been described by Coelewij *et al.* (1978) in connection with magnesium-salt exploration in the NE Netherlands Zechstein Basin, undertaken with a view to subsequent solution mining of the bittern salt. A Zechstein salt pillow at Veendam, near to Groningen, was found to have undergone differential salt movement in which the lower part of the Zechstein salt interval had become mobile, uplifting the non-mobile salt above without any extensive disruption or serious faulting taking place in the salt roof. This effect was combined with preferential flow of bittern salts, including the 'target' mineral, the magnesium salt bischofite. The authors use the term preferential flow to indicate the movement of evaporite salts with differing rheological properties relatively to each other. Most of the late-stage bittern salts, which are found in the uppermost part of an evaporite cycle, are more amenable to deformation under stress than is halite. They are usually of appreciably lower density and higher solubility than the latter mineral. As might be expected, the lower the density of the mineral concerned, the lower is its yield stress. In the paper just cited it is mentioned that determinations of the yield stress of several evaporite minerals under simulated subsurface conditions (2400 m depth, and 100°C) were made at the Shell Research laboratory in Rijswijk, Netherlands. The value for halite was found to be 40 kg/cm^2, while those for carnallite (hydrated potassium magnesium chloride) and bischofite (hydrated magnesium chloride) were 4 kg/cm^2 and 1 kg/cm^2 respectively. Thus, bittern salts of this type will deform much more than halite under any given stress, and will also deform plastically under a value of stress which leaves halite still an elastic solid.

A combination of the two types of salt movement has the effect, in the case mentioned, of lifting the Mg-salt-rich upper part of the salt interval towards the surface in a pillow structure, and also causes the bittern salts (which, occurring as they do in the latest part of an evaporite cycle, would therefore, if present, be in the top part of the salt interval) to flow preferentially into the pillow crest.

Provided the crest of the structure remains intact, there may result a 'pod' of bittern salts which, with the uplift towards the surface resulting from differential movement, will facilitate solution mining. The integrity of the pillow crest is vital, since due to the high solubility of the bitterns, any appreciable disruption of the crest by faulting or piercement will allow rapid dissolution and loss of these salts.

An effect of differential salt movement is shown in Fig. 8.15, where in the Zechstein interval (BZ–TZ) NE of diapir D, the Zechstein Z2 salt (interval BZ–P) has flowed into the feature, but the overlying Z3 salt (within interval P–TZ) has not. Strictly speaking these two salts are not in the same interval, since they are separated by the competent Plattendolomit Equivalent band P, but nevertheless the effect of movement of a deeper salt relative to a shallower salt within a total salt interval is demonstrated.

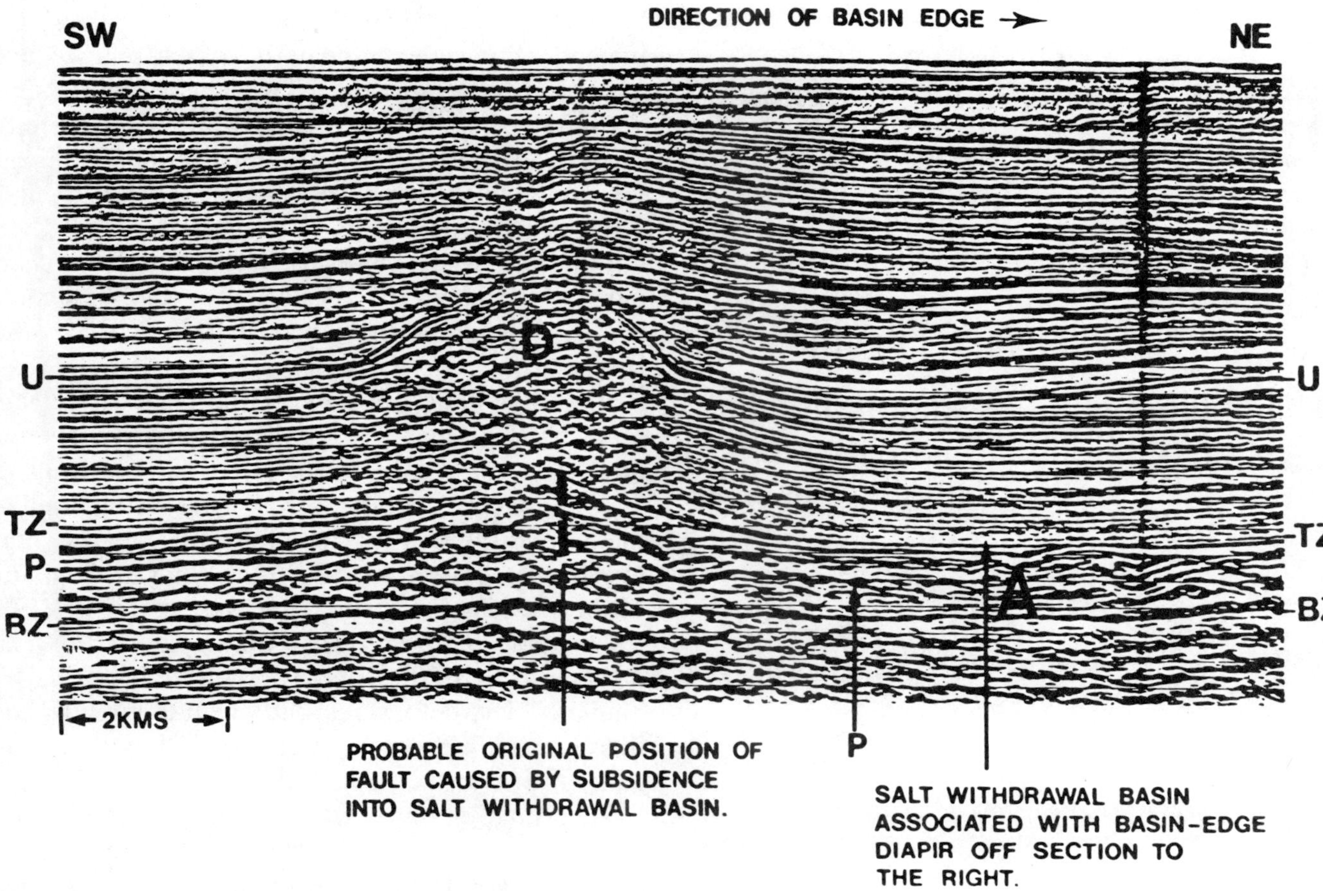

Fig. 8.15 Migrated seismic section showing an example of apparent differential salt movement. U, unconformity. Remainder of key in text. (Courtesy of SS(E)L.)

This same example illustrates an anomalous situation, however, in that neither the Z2 nor the Z3 salt to the SW of the diapir seems to have moved at all. The reason for this cannot be established with certainty, but the following suggestion is offered. The Z2 salt on the NE side of the diapir was mobilised partly by lateral stress related to an edge-diapir off section to the NE, with salt flowing out of the salt withdrawal basin towards D, and partly by overburden load stress. The combined stress has been adequate to cause flow in the deeper salt on the NE side of the diapir, but the gradient in the lateral stress component must have reduced below a value sufficient to maintain flow in the shallower salt (P–TZ interval) to the SW of location A. Continuing reduction in this lateral stress gradient has also meant that the combined stress was insufficient to cause mobility in any part of the BZ–TZ salt intervals on the SW side of the diapir. The diapir itself was probably initiated when subsidence into the salt withdrawal basin caused an overburden fault to affect the top of the salt interval in the location indicated, and the Z2 salt from the NE side then injected the fault zone. The undisturbed nature of the event P, and the lack of overburden subsidence effects on the SW side of the diapir, indicate the absence of any salt movement into the diapir from this direction. This example suggests that differential salt movement may be related to lateral stress gradients within a salt interval as well as to vertical overburden load stress, although the relationship may be complex. It also illustrates the important fact that lateral stress may be the crucial element in diapir development in some areas; had this been a diapir of halokinetic development, the salt flow under purely gravitational control would have been symmetrically from both sides into the diapir.

As regards preferential salt movement, some of the characteristics of the bittern salts have already been mentioned. Due to their high solubility they are the last minerals to be precipitated when a basin evaporates to dryness. It has been said that the most remarkable thing about them is the fact that they ever survive at all—many of them become deliquescent even in damp air—and when they are found, they have normally been re-dissolved and re-precipitated many times. In many basins they are not found, due

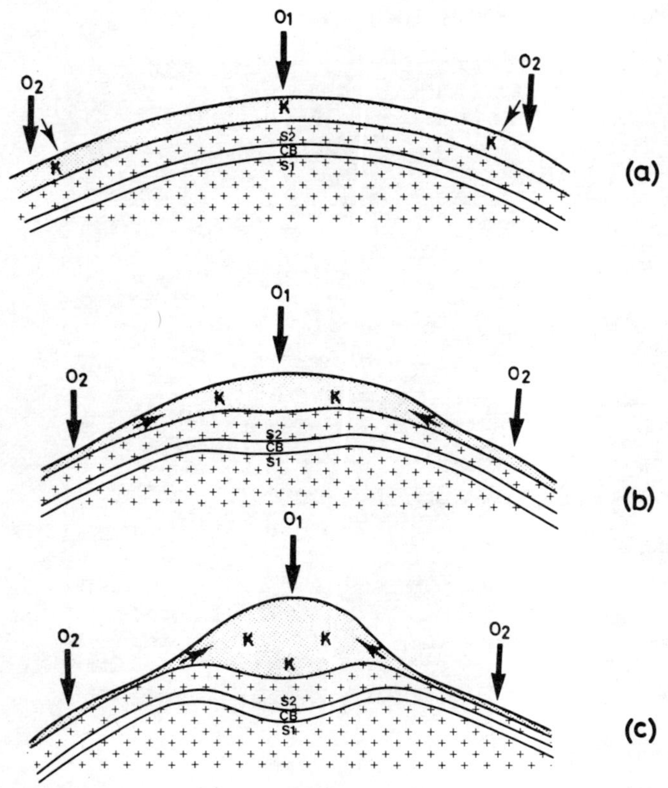

Fig. 8.16 Diagrams representing progressive stages in 'flow concentration' of potash salts in the crest of a salt swell.

to the original brine never having reached such a late stage of concentration (and being removed by reflux), or due to dissolution and dissemination. When they are found, most are not primary precipitates, but are alteration products of other less stable minerals. Their position in an evaporite cycle has been mentioned, and is indicated in the schematic of Fig. 2.2. They tend to be found towards the centre of a basin, although frequently their distribution is rather capricious. They may exist in bands or lenses of relatively pure nature, or as intergrowths with other minerals—particularly halite.

Even as intergrowths, they are the 'weak link in the chain' as far as deformation is concerned, and will flow preferentially within the intergrowth.

The schematic sketches of Fig. 8.16 show what may happen when a potash-rich interval occurs in an evaporite unit which is undergoing pillow formation.

Two halite intervals, S1 and S2, are shown (a), separated by a competent band CB of carbonate or anhydrite, and halite S2 is shown as having a potash-rich upper part (K). (A note on terminology: the word 'potash' is frequently used as a portmanteau term for any bittern salts, whether they contain potassium or no!)

As deformation of the main pillow continues, overburden stress differentials (01, 02), with stress components acting perpendicularly to the pillow flanks and inwards towards the axial plane, cause the more deformable potash salts to flow from the higher to the lower load stress regions, thinning on the flanks and thickening in the crest (b).

This crestal thickening is accommodated partly by uplift (c) of the crest into a tight antiform feature, and partly by downward displacement of the halites, causing the competent band CB to flatten, and then to assume a synform shape.

The 'flow concentration' of the more deformable salts discussed here is analogous to the suggestion quoted in the previous chapter to explain why the halite in some diapirs is very pure compared to the source interval of the salt, which may contain interbedded detrital material not found in the diapir. These bands of 'impurities' are believed to have been left behind, as it were, when the salt flowed into the diapir because of their high resistance to deformation. Similarly, less deformable salts in intergrowths with the potash are presumably left behind as the potash concentrates in the structural crest.

The situation sketched is believed to explain the features seen in Fig. 8.17, where two seismic sections intersect orthogonally above an unusual structure forming part of a low-relief Zechstein salt pillow. The base of the Zechstein is at A, whilst B marks the competent Plattendolomit band of anhydrite which separates the Z2 (Stassfurt) halite below from the Z3 (Leine) halite above. The interval B–C comprises the Z3 cycle (including the Leine halite) and the Z4 (Aller) cycle, the latter probably containing only a very thin halite development in this location, possibly with a red salt clay at its base.

The Z3 cycle is known to contain potash salts in this area, and there may also be a thin development of these in the Z4. The horizon marked C is effectively the top of the Z3/Z4 salt intervals, although the top of the Zechstein is probably an anhydrite which is producing the strong event immediately above C. The letter T indicates the event interpreted as the top of the salt (equivalent to C) over the strong antiform shape, while D is a horizon in the overburden, marked to show the degree of deformation imparted by the uplift.

A broad, gentle asymmetrical salt swell is shown in cross-section on Line A (upper illustration), being quite well defined by the Plattendolomit event B. Slightly offset from the crest of the swell, a strong antiform uplift of the post-Plattendolomit intervals is seen. Referring to Line B (lower illustration) along

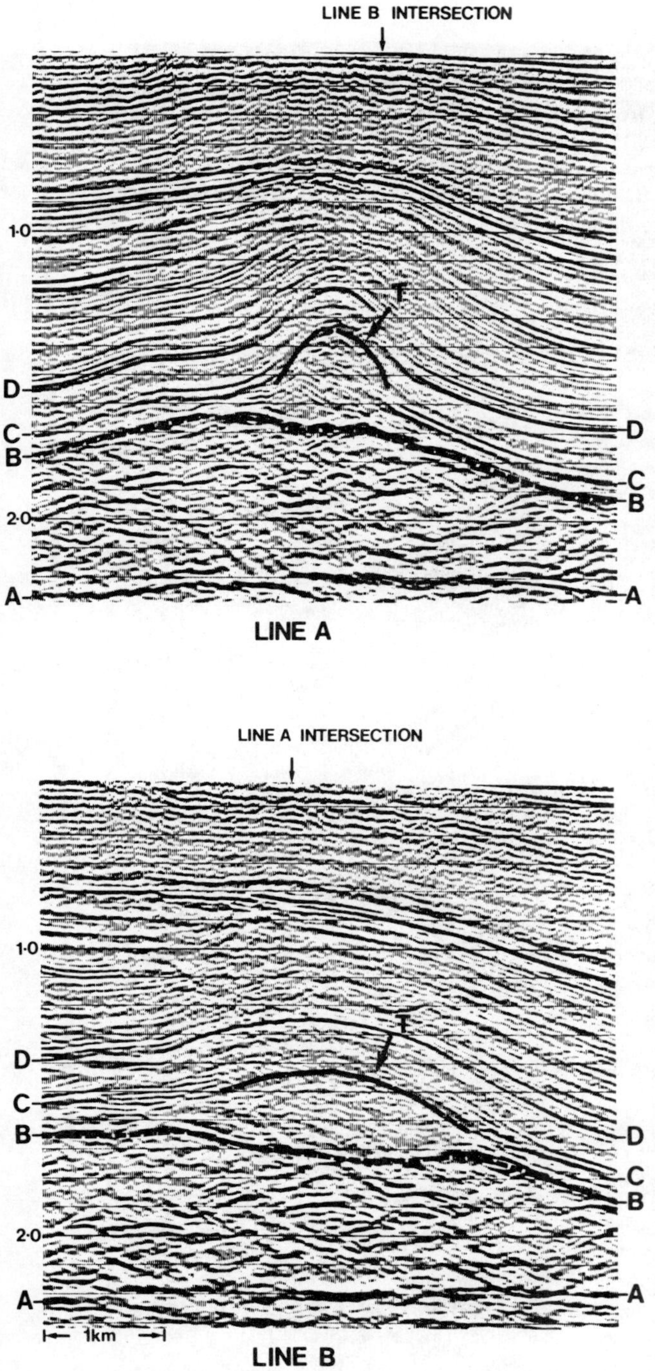

Fig. 8.17 Migrated sections of two intersecting seismic lines showing a feature interpreted as being due to flow concentration of potash salts. (Courtesy of SS(E)L., from Jenyon (1985c), with acknowledgements to the *Oil and Gas Journal.*)

the axis of the main salt swell, it is seen that the feature defined by event T is a strongly-marked hogback structure parasitic on the main swell, with the competent Plattendolomit band B showing considerable synform downward displacement beneath.

From the effects in the overburden, it is clear that this parasitic structure developed at a late stage in the rise of the main swell. The event T and the shape of event B suggest very marked thickening and flow in the upper (Z3/Z4) interval. Although there is no well information available, the circumstantial evidence is strong that it owes its origin to flow concentration in the structural crest of potash salts within the Z3/Z4. If this is so, it implies significant concentrations of potash and/or magnesium bittern salts in some locations in the British Southern Basin of the North Sea.

Apart from being of intrinsic interest as a probable example of preferential salt movement, it is also useful in that it suggests explanations for many similar features seen in the general area. Figure 8.18 shows a tight anticlinal fold in overburden, produced by a process similar to that shown in the previous figure, the lettering being the same. Note the marked variation in thickness (arrowed) of the B–C interval between flanks and crest.

It may have been wondered why this subject is being discussed under the heading of diapirism, when all the structures discussed so far have been non-piercements. In Fig. 8.19, some interesting features are to be observed, including, at location D, a piercement feature which probably involved potash salts in the first instance (note that the lettering of this figure is different from those preceding). The total Zechstein interval is Z–T, and the competent Plattendolomit band (seen only on the left-hand side now) is marked C. The crestal thickening (in the asymmetric anticline on the left), developed between horizons C and T, is clear, and the flattening of event C beneath the crest is familiar and significant. Features like this have been misidentified as 'flat-spot' gas/liquid contact events in this area.

To the right of the small salt withdrawal basin present, the small piercement D is probably an example of bittern salt diapirism. When the highly deformable salts broke through into the overburden, it is likely that some halite must have followed, and it is improbable that the bittern salts have survived; they were almost certainly removed at an early stage by dissolution in formation water.

Figure 8.20 is a further example of similar type. Base and Top Zechstein in this illustration are lettered BZ, TZ. The much-disrupted Plattendolomit is believed to be at P, with another competent band above it within the Z3/Z4 interval. Again, a marked thickening of the upper intervals in the crest of the main salt swell is evident. There is an interesting juxtaposition present here, in view of comments made on the previous figure. A real gas/liquid contact

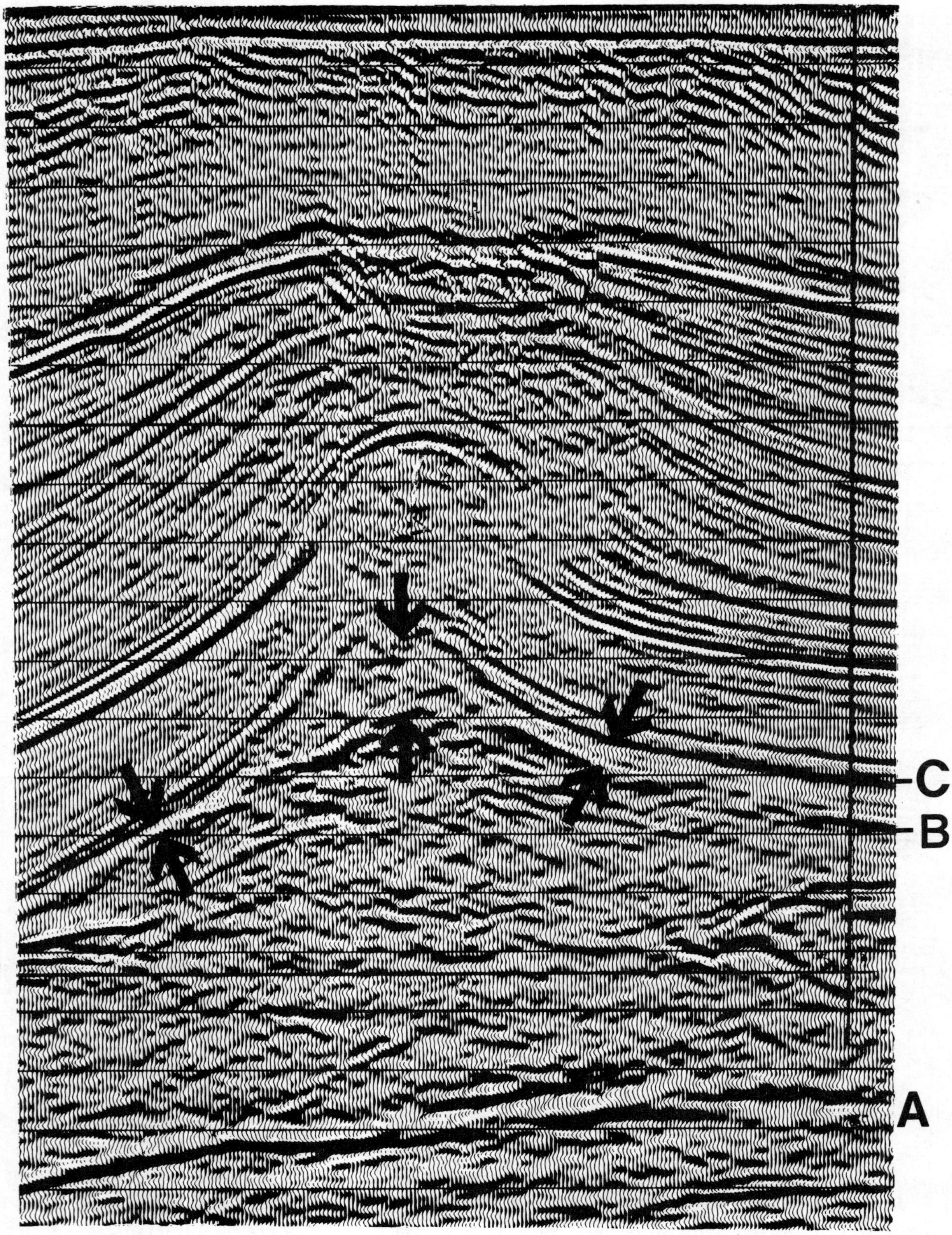

Fig. 8.18 Probable flow concentration of bittern salts in the core of a tight overburden fold. It is likely that this fold was caused by the bittern salt movement in the upper part of a Zechstein salt pillow. (Courtesy of SS(E)L.)

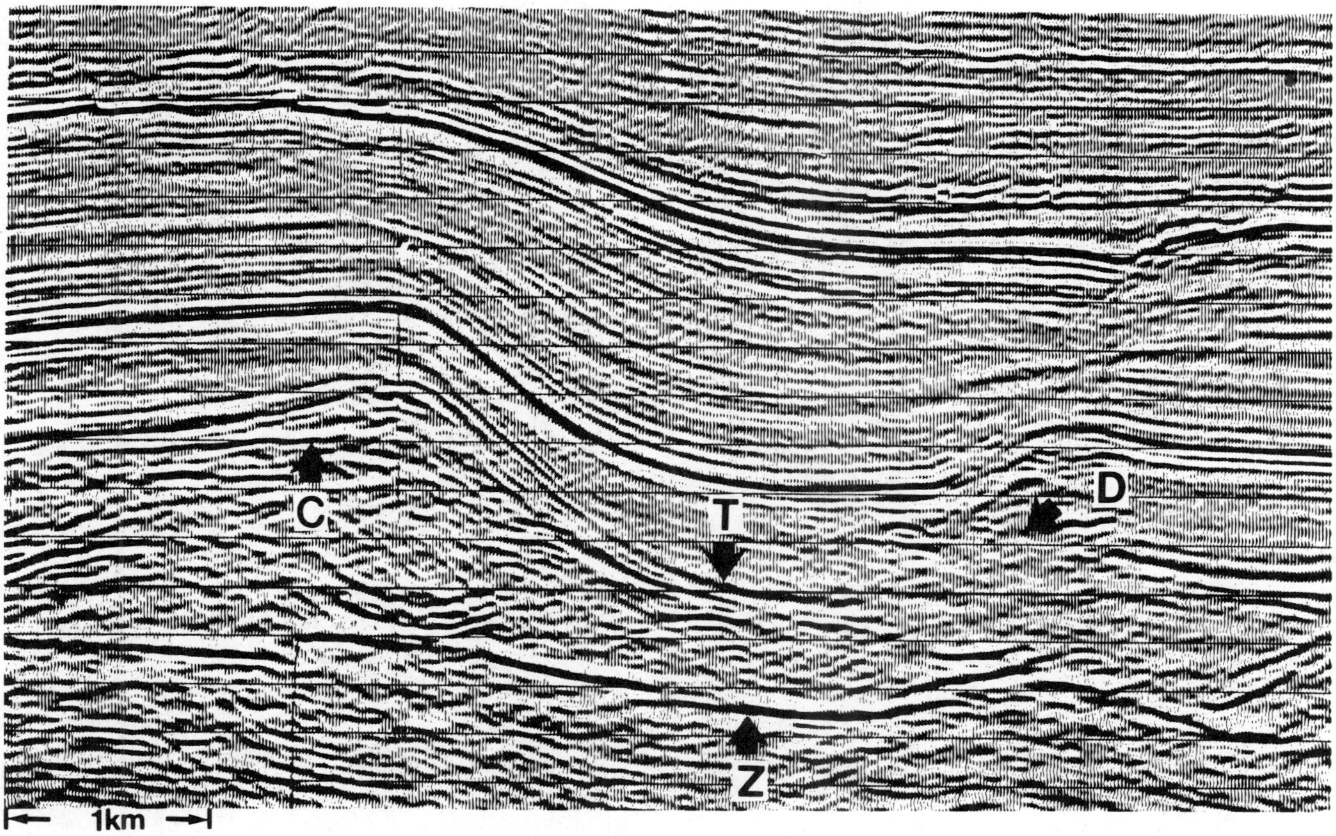

Fig. 8.19 Migrated seismic section showing what are believed to be features caused by preferential flow of potash salts. (Courtesy of SS(E)L.)

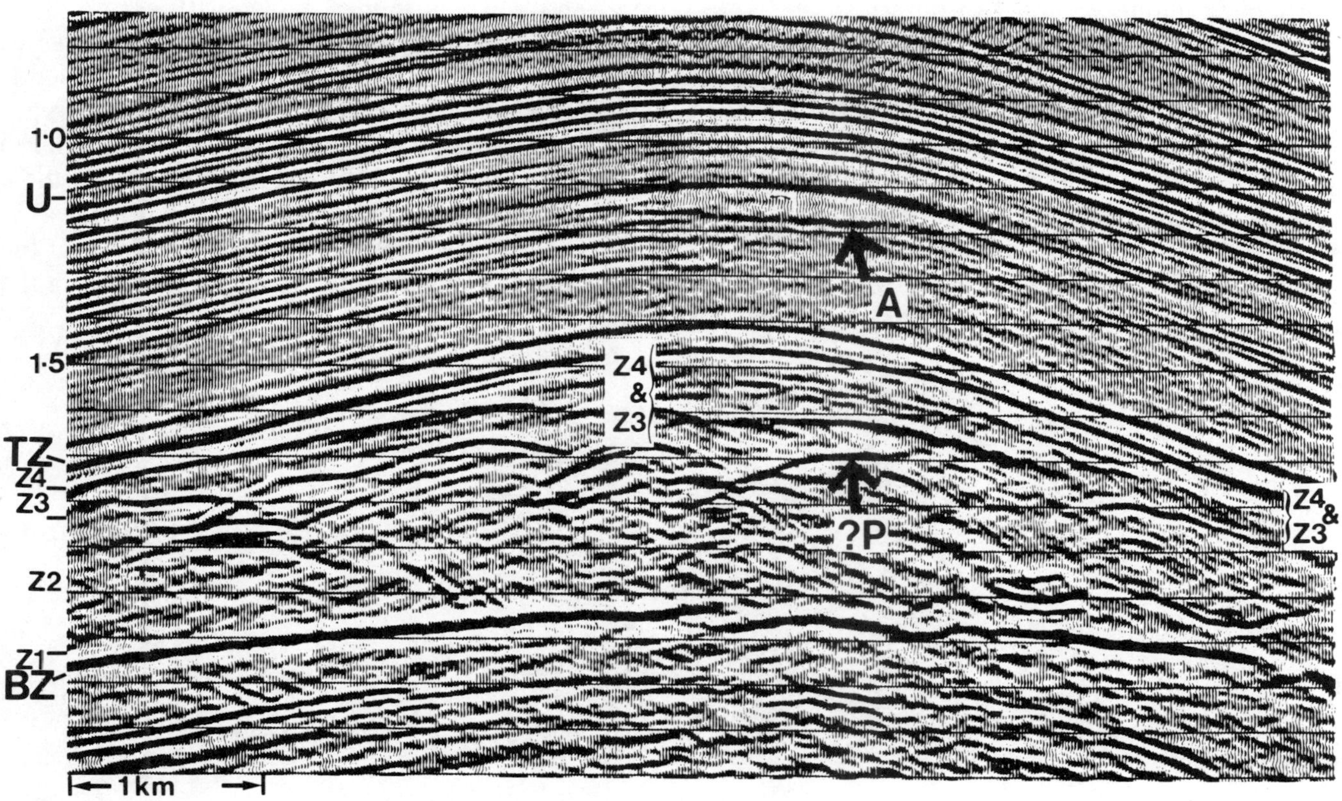

Fig. 8.20 Migrated seismic section showing thickening in a salt pillow crest by preferential salt movement. Note also the 'flat-spot' gas/liquid contact event at A; it is easy to see how this mimics effects due to preferential flow concentration—the latter are sometimes mistaken for flat-spot phenomena in this area. (Courtesy of SS(E)L.)

event—a 'flat-spot'—can be seen above at A; a drilled and proved commercial gas accumulation is present at this location.

It is noted that in pragmatic terms, the difference made between differential and preferential movement in this discussion might be regarded as rather artificial. Differential movement implies that one part of a salt interval is, under particular conditions of imposed stress, more amenable to deformation than is another part of the interval, and the same definition could be applied to preferential movement. However, the important difference lies in the fact that there is a whole order of difference between the yield stress of halite and that of some of the bittern salts, and should the latter be present in the evaporite section, unusual and striking features such as those illustrated here may be looked for under a stress regime which may have only minor effects on associated halites.

There are possible alternative explanations for the effects just discussed, since lithologies other than those of bittern salts exist which deform more readily than halite under any given stress. For example, if a clay or shale interval occurs (particularly if overpressured) near the top of an evaporite cycle, it is possible that 'flow concentration' effects similar to these might ensue. However, in the geological context here, the balance of probabilities lies with bittern salts.

8.5 'STACKED' DIAPIRISM

So-called 'stacked' or 'double' diapirism may occur in stratigraphic circumstances where two salt intervals exist, separated by a substantial thickness of non-mobile competent rock units.

In these circumstances, movement in one of the salt intervals is likely to induce movement in the other. This is shown schematically in Fig. 8.21. If both intervals are potentially mobile it is probably immaterial which moves first, although the deeper interval could normally be expected to deform first from considerations of loading and temperature as already discussed. The descriptive matter on the figure assumes that in this case the shallower interval has moved first; had the lower interval been first to move, the resulting uplift would lead to the same situation as in the sketch.

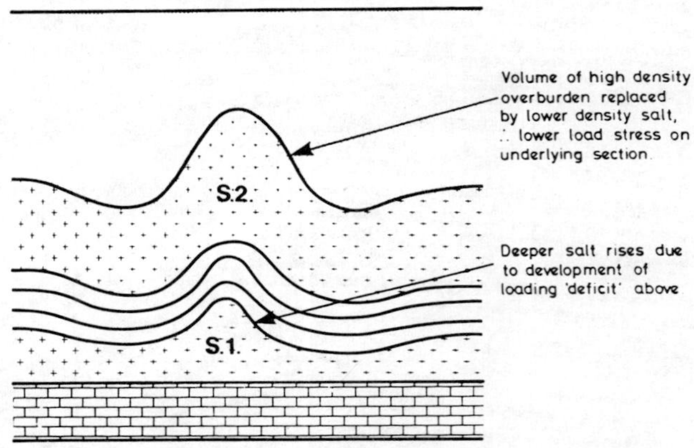

Fig. 8.21 Diagrammatic representation of the pre-piercement stage of stacked diapirism.

Both or either interval may develop into a piercement, and a structure may evolve with a core of S1 salt within a diapiric body of S2 salt. Structures of this kind are known in certain areas, one of which is Schleswig-Holstein, NW Germany. Here the (Upper Permian) Zechstein halite overlies a (Lower Permian) Rotliegendes interval comprising a substantially thick halite separated from the Zechstein by redbed clastics. Diapirism of the type mentioned has led to the development of stacked diapirs nested one within the other.

CHAPTER 9

Some Special Aspects of Salt and Salt Structures

9.1 INTRODUCTION

There are several aspects of salt rock and structures involved in salt tectonics which do not fall conveniently within the terms of reference of preceding chapters, and these will be discussed here. They include such subjects as the thermal properties of salt, matters arising from the increasing use of salt intervals and structures for the purposes of storage or disposal of materials, and certain aspects of the investigation of salt features by seismic and other means.

9.2 GEOTHERMAL EFFECTS OF SALT DIAPIRS

The thermal conductivity of samples of salt rock has been investigated by many workers, and has been found to vary by some four orders of magnitude, depending on the temperature range under consideration. According to NBS Monograph 167, Physical Properties Data for Rock Salt, the wide variation encountered in laboratory investigation is related to lattice imperfections, types of treatment of the sample before measurement, and the type and amount of any impurities present. From the same source, Table 9.1 gives 'recommended' values for the thermal conductivity, k, of salt at and above room temperatures.

Table 9.1

Temperature (K)	k ($W\,m^{-1}\,K^{-1}$)
293	6·65
300	6·57
400	4·80
500	3·67
600	2·98
700	2·47
800	2·08
900	1·85
1000	1·67

It has been found that below 100 K, the thermal conductivity is inversely proportional to the quantity of impurities present, while above this temperature, k becomes insensitive to this factor.

Taking a typical figure of the value of k for clastic sediments from Heiland (1940) as, in different units, $1{\cdot}4 \times 10^{-3}\,cal\,cm^{-1}\,s^{-1}\,°C^{-1}$, the room temperature value for k of salt rock from Table 9.1 (i.e. 6·65) converts to $15{\cdot}8 \times 10^{-3}\,cal\,cm^{-1}\,s^{-1}\,°C^{-1}$, which confirms in a general way the finding that salt rock has a relatively high thermal conductivity compared to normal clastic sediments.

As a result, a vertical salt structure of pillow or diapir form acts as a conduit for thermal flux, the heat being derived, of course, from the decay of radioactive components in the surrounding and subjacent sediments. This phenomenon, often referred to as the thermal 'chimney' effect, has been studied for a number of salt structures.

Selig and Wallick (1966) investigated mathematically the isogeothermal patterns in and around a salt intrusion from the theoretical viewpoint, and concluded that temperature anomalies depend not only on the thermal properties of salt and surrounding sediment, but also on the geometry and movement history of the system. One particularly interesting conclusion drawn from this theoretical study was that local heat generation by frictional effects resulting from the movement of the salt through overburden sediments should sensibly effect the temperature distribution in the flank zone of an intrusion.

In a recent paper on the Vejrum salt diapir in Denmark, Jensen (1983) investigated the temperature anomaly (the observed temperature field minus the undisturbed geothermal field) associated with the diapir by means of computer modelling. A simplified half-profile of the diapir, shown by the heavy line in

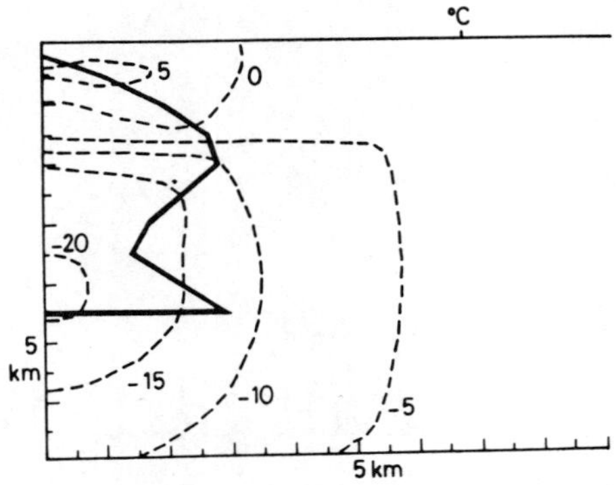

Fig. 9.1 Thermal anomaly associated with a salt diapir. This computer-derived model shows a simplified half-profile of the diapir as a solid black outline, with the anomaly isotherms (°C) as dashed lines. (After Jensen (1983); courtesy of the EAEG.)

Fig. 9.1, was constructed, and combined with borehole-derived temperature data. The anomaly reaches a value of over +5°C near the crest of the salt body, and −20°C at its root. The author suggests that (a) the effect could be used to identify salt diapirs at depth by anomalies in the temperature gradient in the groundwater of overlying formations, and (b) the possibility of geothermal energy prospecting should be considered where enhanced heat flow occurs above salt diapirs. Commenting on suggestion (a), one would say that unless an extensive system of boreholes already existed in an area, it could be an expensive method of identifying the presence of salt diapirs compared, say, to seismic survey techniques. Suggestion (b) is of considerable interest, and it may be that heat flux associated with some diapirs of large vertical dimensions should be investigated.

It is not known if the thermal flux in a salt body, and anomalous temperature effects in the contiguous sediments, has any influence on the elastic limit of the halite in different zones; it seems likely that any such effects would be quite negligible, but suggests an interesting subject for study.

9.3 MATERIAL STORAGE WITHIN SALT STRUCTURES

In recent years, an increasing amount of research and development has been effected with regard to the underground storage of liquid and gaseous hydrocarbons, and also the disposal of radioactive and other waste products. This is becoming ever more necessary because of environmental and surface storage space problems, as well as from the viewpoint of safety.

Deposits of massive halite are particularly attractive for these purposes owing to their complete impermeability with respect to the materials to be stored or disposed of, and also because of the relative ease by which large-volume caverns (necessary for projects to be economically viable) can be leached out by water circulation.

As an example of the scale of such operations, Dreyer (1982) mentions that in the Federal Republic of Germany, at the time of writing, about 140 caverns for the storage of hydrocarbons with a total volume of 22 million m^3 were either completed or in course of construction, most of these being in salt deposits. This would need to be multiplied many times to arrive at a total for only the European countries and North America. This does not include figures for storage capacity of other materials, and for strategic purposes.

A great number of factors must be investigated before the opening up of an underground storage cavern in a salt body can be undertaken. These include such matters as the stability of the bedrock beneath the salt, the tectonic framework of the area including the fault systems present, the stress patterns and movement history of the salt body, the frequency and severity of earthquakes affecting the region, the stability of the groundwater system, permeabilities of salt and country rocks, the geothermal gradient, and the internal integrity of the salt body itself, to mention but a few of the more important.

Some unusual phenomena have come to light during the investigations required. One example is provided by the behaviour in the presence of a thermal gradient of the liquid and gaseous inclusions commonly found in both primary and recrystallised halite (Holdoway, 1974; Anthony and Cline, 1974). It has been found that when a thermal gradient is applied, fluid inclusions in the salt migrate physically up the gradient (i.e. towards the higher temperature region). The rate at which this migration takes place appears to be proportional to the steepness of the gradient. Although there is still some lack of agreement as to certain aspects of the migration process, the basic mechanism seems to be as follows: the slight increase in the solubility of halite with increasing temperature means that where a temperature differential is applied across an inclusion containing brine, dissolution of halite occurs at the higher temperature end of the inclusion, and re-precipitation at the lower temperature end. This leads to the physical translation of the inclusion in the direction of higher temperature. Clearly, such observations must have considerable importance, especially in regard to the storage of radioactive waste.

The integrity of the salt body itself must be a matter

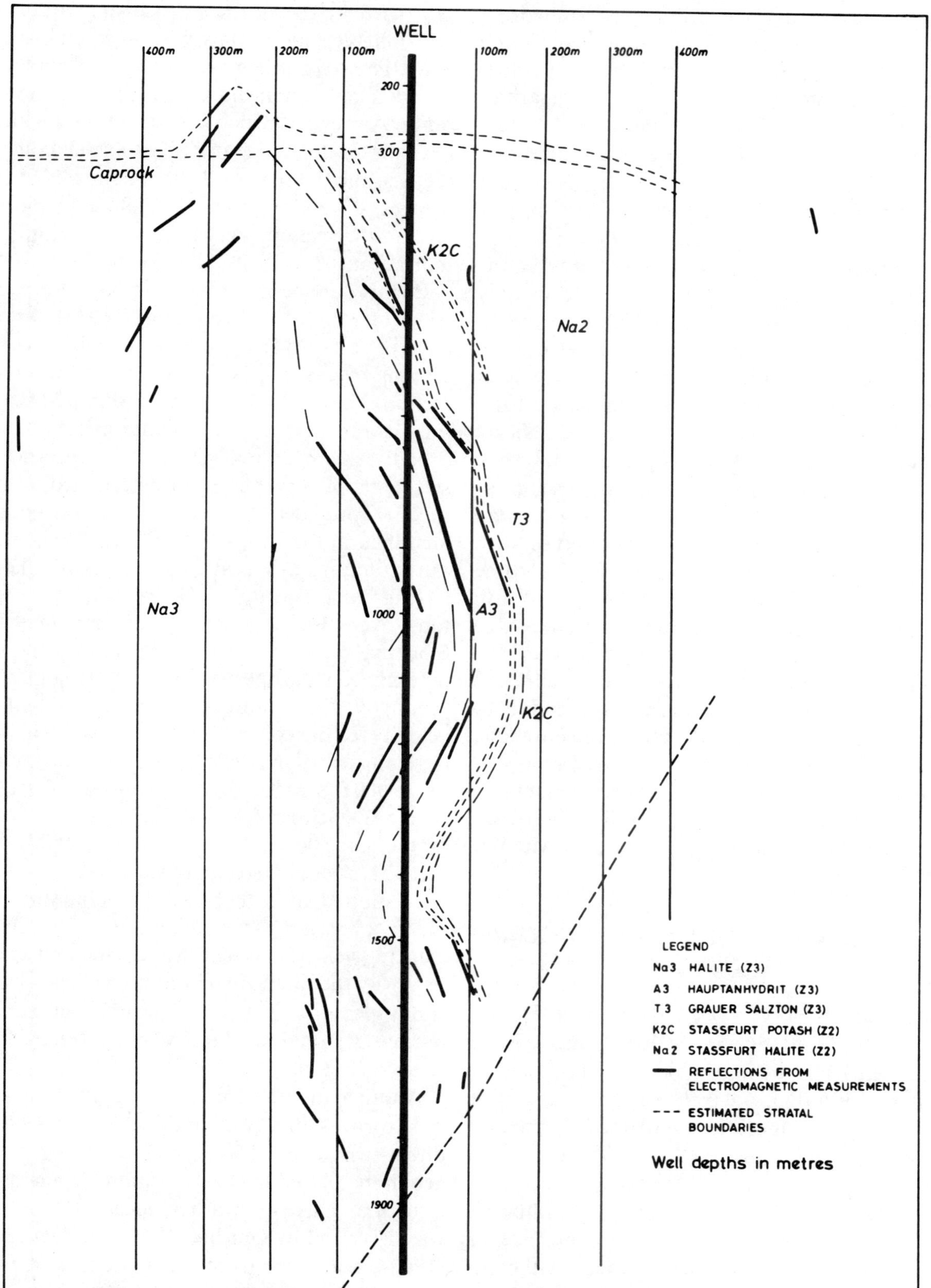

Fig. 9.2 The extension of the 'information radius' of a borehole by high-frequency electromagnetic methods. Interpretation of the observations has extended the range of useful information, in combination with the geological data from the borehole itself, to a radius of over 400 m. (After Nickel *et al.* (1983); courtesy of the EAEG.)

of careful investigation. Frequently, in the interior of a salt diapir, bands of other lithologies—anhydrite, dolomite, clays and bitterns—more soluble and more easily deformable than halite, exist together with brine-filled cavities and sometimes joints. Such irregularities and discontinuities within the salt must be detected, and their potential effects on storage assessed carefully.

Where detailed knowledge of the internal structure of a salt body is needed, the amount of test drilling required for complete reassurance would normally be prohibitively expensive. The validity of any information gained from any single borehole extends only for a limited radius—in the case of highly deformed salt structures, perhaps only a few metres—from the borehole. Recently, geophysical methods have been developed which enable the 'information radius' for a borehole to be extended to hundreds of metres. Nickel *et al.* (1983) describe high-frequency electromagnetic methods using (a) absorption—the attenuation of a radio-frequency signal propagating between two boreholes, and (b) reflection from discontinuities, with the transmitter and receiver located in the same borehole, both of which methods when combined with the geological data from the borehole itself enable detailed investigations of any irregularities in the internal salt structure to be carried out. Figure 9.2 shows such an interpretation, with results extending to distances greater than 400 m radius from the borehole, and where the interfaces/discontinuities seen represent anhydrite and clay bands within the Zechstein stages shown in the key. For further work in this context see Thierbach and Mayrhofer (1978), and also Mundry *et al.* (1983).

9.4 THE DEFINITION OF SALT STRUCTURE FLANKS

In areas of salt diapirism, one of the most difficult problems to be faced by the geologist and seismic interpreter may be the definitive location of the steeply dipping flanks of a diapir, which may show angles of dip of up to 90°—or more in the case of a structure with an overhang development.

Unfortunately, the standard seismic processing route actively discriminates against steeply dipping events which result in high angles of incidence of returning seismic signal at the surface. Although recorded (with some loss of high frequencies), interfaces dipping at angles greater than 45° will not be recoverable using conventional data processing techniques. (However, nowadays well-directed reprocessing can do much to improve the situation.)

The implication of this is that the 'stacking velocities' used in the normal processing route, whereby returning seismic signals with a common mid-point (CMP) between energy source and detector group are added together, can be kept within a quite restricted range. Any events which would require stacking velocities outside this range in order to stack coherently on the seismic section will be attenuated to varying degrees, depending upon their angle of dip and time level in the section. Very steep events dipping, say, at more than 60° will usually not be recovered at all.

In this connection, it is of interest to examine the work of May and Covey (1981, 1983), who devised an inverse ray tracing method by which very steeply dipping events can be recovered and plotted on the section. The two papers referred to should be consulted for details of the theory and methods used but, broadly, the stacking velocities appropriate to very steeply dipping events can be determined from velocity spectra (provided that the events can be positively identified in the spectra). In the 1983 paper, two real salt structures are used as examples for the method, having flank dips up to 70° in the first case, and 90° (and over, due to a small overhang) in the second. Stacking velocities for the salt flank in the first case, determined from spectra, increased to some 17 000 ft/s (about 8000 ft/s more than the equivalent stacking velocity for the adjacent sediments) while in the second case, the salt flank reflection had a stacking velocity of 25 000 ft/s at 2·8 s, which was 15 000 ft/s greater than the stacking velocity for the adjacent sediments at the same level. By using the inverse ray tracing method, clear and strong reflections for the salt flanks were plotted in both cases. The differences between the 'appropriate' stacking velocity and the normal stacking velocity, as just quoted, make it understandable why the salt/sediment interface in such structures is not seen on the standard seismic section, where the normal regional stacking velocity trends have been applied (Fig. 9.3).

Another method for the delineation of salt flanks is known as the 'proximity survey', which uses the direct arrival of refracted seismic energy at a well geophone placed in close proximity to the flank of the structure. This is not a new idea—such an approach was described by Gardner (1949) and Musgrave *et al.* (1960). With the advent of modern data processing, however, it is possible to effect more complex modelling rapidly, which has led to a renewed interest in the method.

The basic technique involves the location of a seismic energy source over the crest of a salt struc-

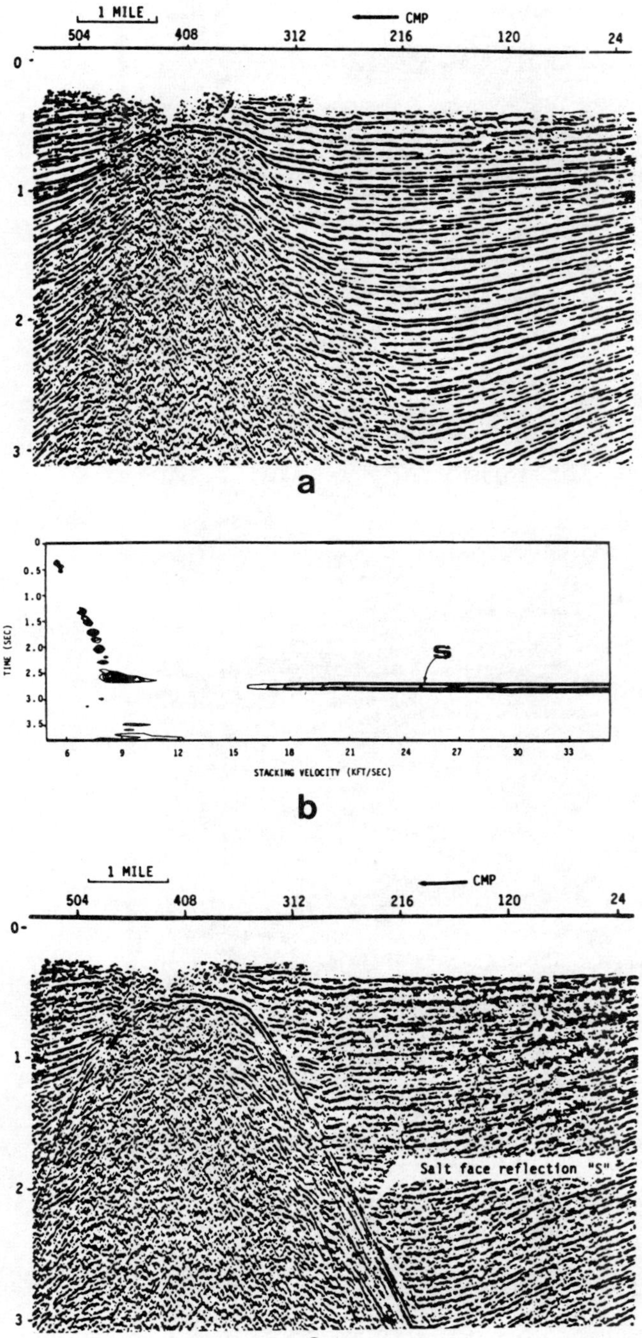

Fig. 9.3 Improvement of resolution of a salt flank by the inverse ray modelling technique of May and Covey. (a) Salt diapir in a CMP stack using off-structure regional stacking velocities; (b) stacking velocity spectra for CMP 216 on the line in (a), showing a velocity of 25 000 ft/s at 2·8 s for the salt face (S); (c) the salt diapir in a reprocessed CMP stacked section, using the higher stacking velocities for the salt face in the velocity spectra for the CMP series along the line, in the inverse ray tracing process. A dramatic improvement in the resolution of the salt/sediment interface results. (From May and Covey (1983); courtesy of the Society of Exploration Geophysicists.)

ture—usually as nearly as possible over the centre of the crest area—and travel times through the salt to a detector in a well are measured. The well is one which has been drilled close to the flank of the salt feature, and the detector (nowadays a three-component sensor is used) is positioned successively at a number of predetermined depths down the well. By virtue of the position of the well, information is available regarding the velocity and dip of the flank sediments, and also a precise measure of the velocity in the salt. It is also normal to have one tie-point on the salt–sediment interface. A model is devised for the travel path of each of the direct arrivals through the salt which is recorded at the successive depth positions of the well sensor and, with the model as control, the arrival times can be wavefront-migrated in order to define the interface (Whittle, 1985). Quite accurate results are possible by this method. A problem often encountered is steep dips in the encasing sediments, which may affect the validity of the model. Most computed raypaths may emerge at or near prominent refractors, and few points on the salt/sediment interface may be determinable.

The use of the refraction type of proximity survey presupposes a knowledge of several factors, including the salt velocity, the velocity of sediments between salt flank and borehole, and through the overburden, as well as the depth to the top of the salt body. Recently, a technique has been described by Manzur (1985) utilising reflections from the salt flank which requires no pre-knowledge of the above factors. The energy source is located in a position offset from the borehole on the other side of the well from the salt mass, so that the seismic energy is reflected from the salt flank back into the downhole sensors in the well. Iterative modelling is carried out until a match is obtained with observed VSP (vertical seismic profile) events obtained by processing the arrivals from the downhole sensors.

The refraction-type survey has the advantage that it can be used to determine the shape of a salt mass which has an overhang, whereas the reflection survey cannot. A judicious combination of the two types of survey can, as Manzur remarks, produce an accurate shape for the salt flank.

9.5 SIDE REFLECTIONS AND OTHER SEISMIC EVENTS FROM DIAPIRS

Because of the usually strong acoustic impedance contrast between a salt diapir and the clastic sediments in which it is emplaced, it is a good generator of side reflections, i.e. seismic events coming from outside the

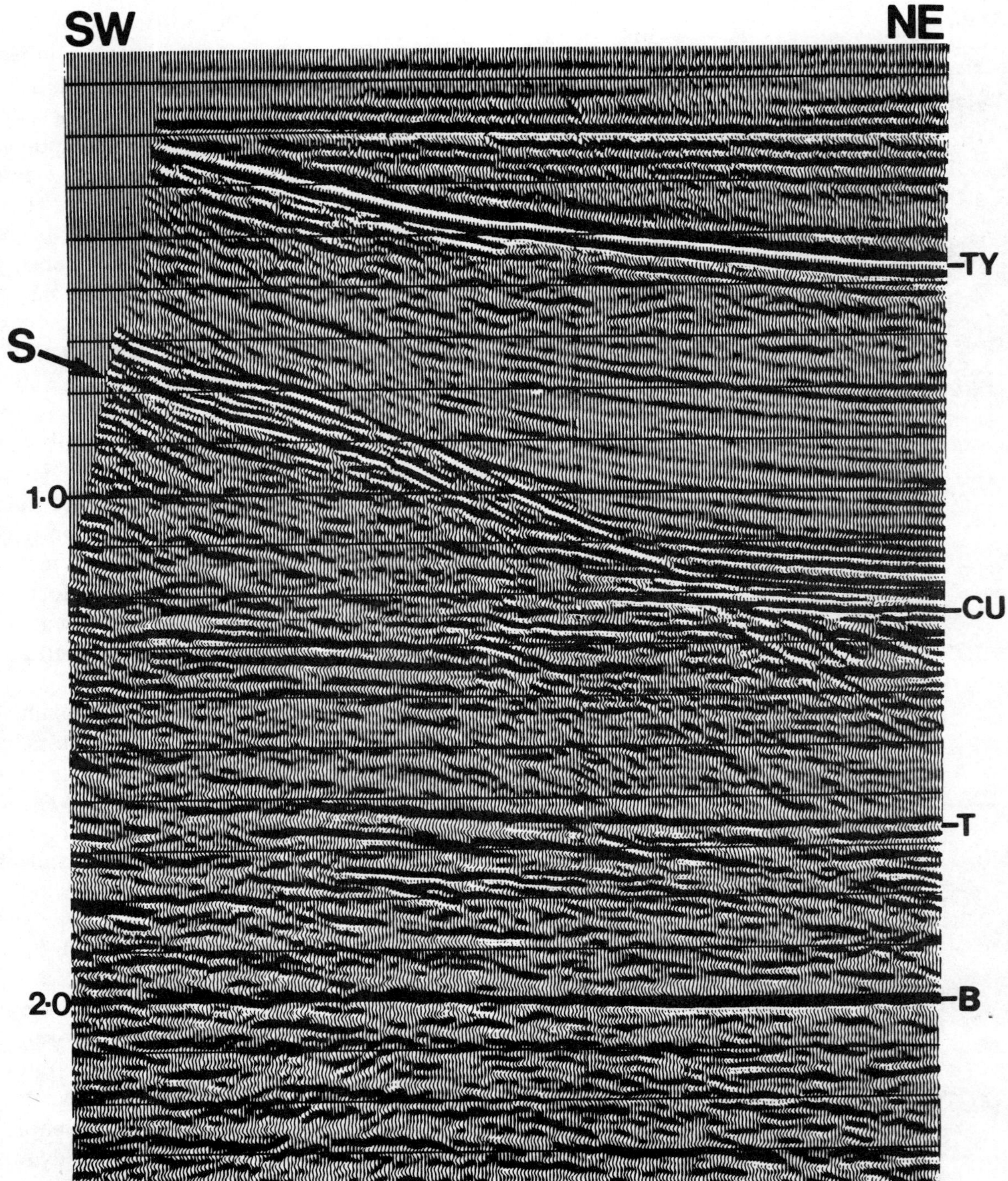

Fig. 9.4 Side events recorded from a salt diapir which is offset to one side of this seismic line. The actual salt interval on-line is between B and T. Event S is interpreted as the response to the salt/sediment interface in the diapir which lies off-line; the strong band of events above S is being produced by salt roof units, probably anhydrites and carbonates. In a gross way, the effects here are analogous to those obtained in the reflection type of proximity survey mentioned in Section 9.4 of the text. (Courtesy of SS(E)L.)

vertical plane of the section (and also, due to the steep flanks of a salt structure, coming from ahead of, and from behind, the spread of seismic detectors at any given location; the event will in this case also be plotted on the section as though arriving from vertically below the detectors). A side event of this type is seen in Fig. 9.4, returning from part of the flank of a salt diapir which lies about 2 km off line (on the side of the section remote from the reader) to the NW. The strong events parallel to and above the event, interpreted as the response to the salt/sediment interface (S), are also side events being recorded from carbonate and anhydrite units above the salt in the evaporite interval. The true salt interval on this cross-section lies between B and T, with an unconformity at CU and the Base Tertiary at TY.

Note that there is no question of being unable to 'see' the diapir flank seismically in this case, as was discussed in the earlier part of Section 9.4, since here the locus of reflecting points does not lie on a dip track down the flank, but on some near-horizontal track *along* the flank of the diapir. The convex-upward curvature shown by the event S is due to the distance between the reflecting points on the flank and the seismic detectors on the line decreasing and increasing again (although we see only half the picture here due to the line terminating) as the line passes the diapir. This will become clear if a sketch plan is drawn showing the seismic line with a circular diapir offset from it to one side at a short distance, and travel path segments drawn from equally spaced points on the line to the nearest point on the diapir.

Hospers (1985) discusses the use of events of this type in interpretation; the discussion includes the geometries involved, and also those for internal reflections which may be produced from within a diapir. The latter phenomenon originates from reflected refraction energy, whereby the seismic wave follows a normal trajectory downwards for part of its path, is then refracted along an interface in the clastic sediments adjacent to the diapir, enters the diapir, and is reflected back along the same path from the salt/sediment interface on the far (internal) side of the diapir. This is shown in the sketch of Fig. 9.5, adapted from Hospers' (1985) paper, which represents a cross-section through a salt diapir (cross symbol) and shows a ray path followed by the seismic signal. Leaving the source point A, it travels down normally to point B at the critical distance A–B for refraction along the refracting horizon, where it is refracted along B–C. It enters the salt body, and is reflected back at D, following the same route D–C–B–A back to the surface. Note that there should also be a shorter path, A–B–C–B–A producing an event which is reflected back by the first sediment/salt interface; according to Hospers, this event was not identified in his study.

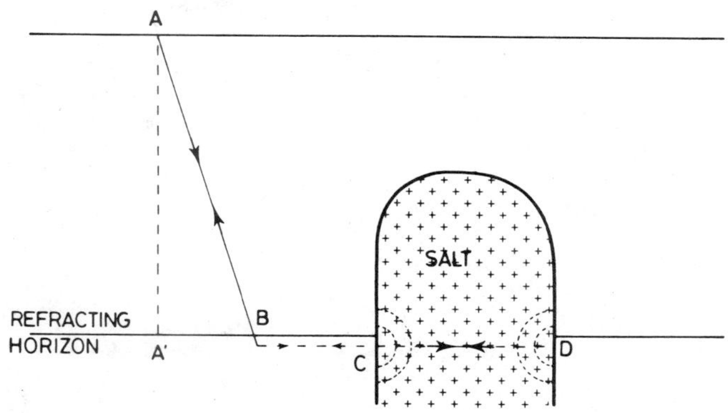

Fig. 9.5 Diagram showing in simplified form the generation of internal events from a salt body by reflected refraction. See text for explanation. The dashed semicircles within the salt body are intended to suggest the wavefronts being transmitted from the refractor into the salt (on the left), and the reflected wavefronts from the salt/sediment interface on the right. (After Hospers (1985); courtesy of the EAEG.)

Considering the complexity of geometry of a large, irregular salt feature, it may seem surprising that many more coherent noise events of the type mentioned are not seen on most seismic sections. However, the greater the path length, and the larger the number of different 'legs' in the path, the greater is the effect of attenuation by the various loss mechanisms involved in the propagation of seismic energy. Most of the more complex events which are undoubtedly generated in the circumstances described are not seen, or at most contribute their small quota to the general background 'noise' on the seismic section.

CHAPTER 10

Salt Tectonics and Hydrocarbon Traps

10.1 INTRODUCTION

There is an infinite number of possible hydrocarbon trap situations of structural, stratigraphic and combination type, and most if not all can occur related to a salt structure.

The vast literature of petroleum geology is concerned to a great extent with this matter, and many excellent general treatments exist, such as that by Halbouty (1967), or more recently a paper by Woodbury *et al.* (1980). However, considering the close association so frequently encountered between hydrocarbon occurrences and salt structures, and although this work is not intended specifically for petroleum geologists, a discussion of the subject is included here.

As a matter of consistency in a book illustrated largely with seismic examples, the device is adopted of using seismic sections of typical salt structures, and constructing 'dummy' hydrocarbon traps within them in order to show the principal classes of trap which may be encountered. It is important to remember at all times, in looking at these examples, that a hydrocarbon accumulation is three-dimensional, and the trap situations suggested will only be valid in most cases if the requisite geometry perpendicular to the cross-section is also present. The necessity to 'think three-dimensionally' is certainly not less important in hydrocarbon exploration than it is in other branches of the geosciences.

10.2 SALT AS A SEAL FORMATION

A common denominator in many of the situations discussed in this chapter is the possibility of the salt interval acting as a seal towards underlying hydrocarbon accumulations. Most of the large Lower Permian Rotliegendes gas fields in the Southern North Sea, such as West Sole and Indefatigable, are very comprehensively faulted, and owe their existence to the seal characteristics of the overlying Zechstein salt. Other possibilities may also occur, such as the injection of fault zones by salt which subsequently acts as a seal in any lateral updip migration of gas or oil that eventually encounters the fault. The range of credible situations is so great that this type of trap will not usually be mentioned in what follows, but should always be borne in mind.

10.3 SITUATIONS RELATED TO PRE-PIERCEMENT STRUCTURES

Some of the simplest trap situations are associated with doming above salt swells or pillows, and thinning above their flanks due to syndepositional movement. Where overburden sequences are pre-movement in age, and of uniform depositional thickness, trap situations may still occur due to faulting of stress-zone or other types, and also to stratigraphic porosity traps in the dipping sediments. Figure 8.20 shows a good example of an actual trap caused by updoming of Triassic sediments above a Zechstein salt pillow. Gas has found its way through the Zechstein salt interval somewhere beyond the confines of this section, and accumulated in the structural high here. The migratory route through the salt in this instance has not been determined; it may have been up a fault zone where the salt is thin, or through a zone of salt depletion (as seen in Fig. 4.6). Unfortunately, in the North Sea Southern Basin, this situation is all too rare, the Zechstein salt being a highly efficient seal. The same property which created the Lower Permian gas fields has, in general, destroyed the possibility of large numbers of traps in the overlying Mesozoic section, so that a multitude of potential structural traps

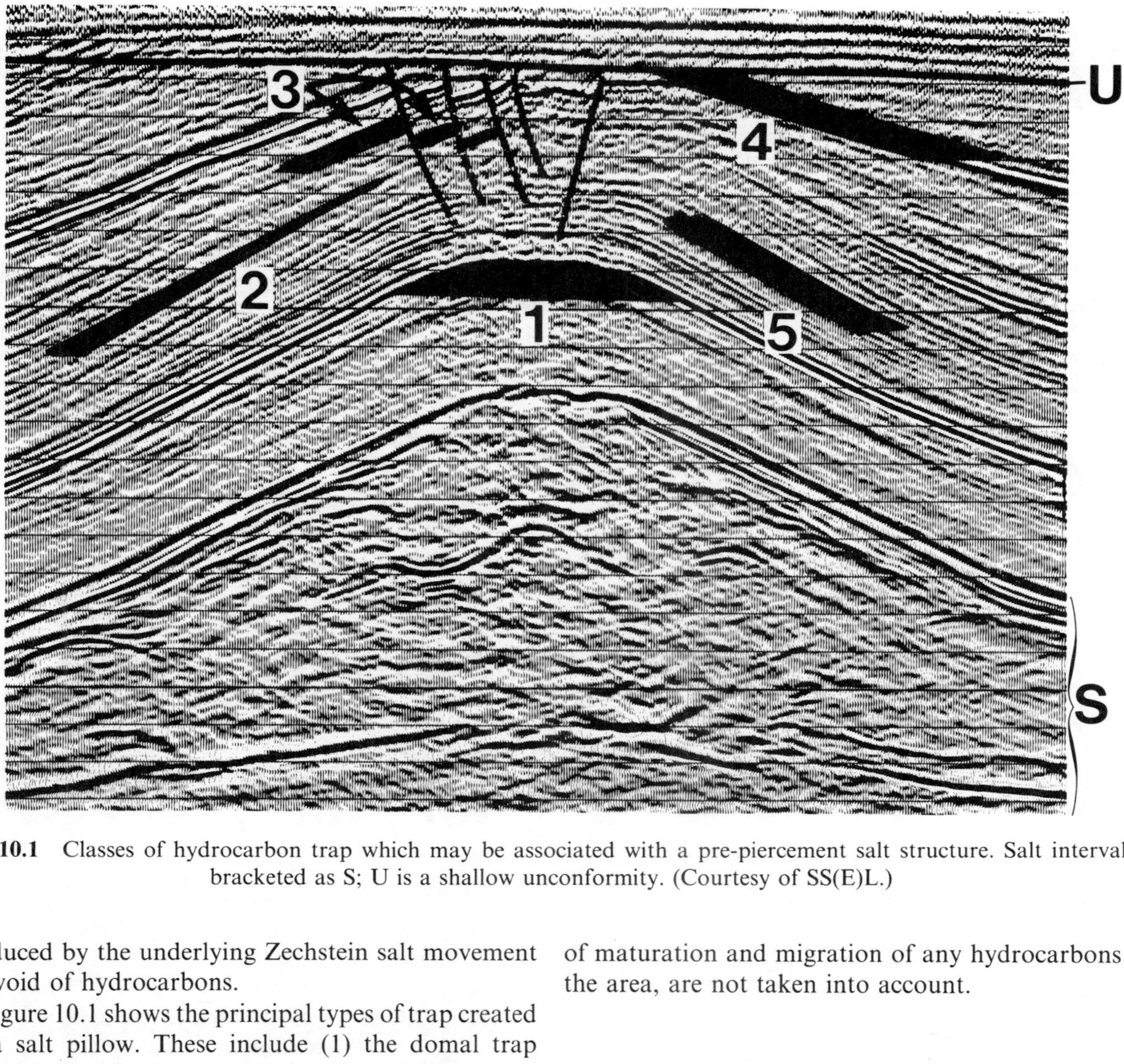

Fig. 10.1 Classes of hydrocarbon trap which may be associated with a pre-piercement salt structure. Salt interval is bracketed as S; U is a shallow unconformity. (Courtesy of SS(E)L.)

produced by the underlying Zechstein salt movement are void of hydrocarbons.

Figure 10.1 shows the principal types of trap created by a salt pillow. These include (1) the domal trap just mentioned; (2) a trap situation produced by 'pinchout'—thinning of an interval upflank, which will normally only occur when the pillow formation is syndepositional; (3) fault traps dependent on stress-zone faulting of the overburden above the pillow; (4) an unconformity trap; (5) a porosity trap—this is possible, but frequently with syndepositional movement the lithology would 'fine' downflank away from the crest of the structure. The salt interval is bracketed as S in this figure, and the unconformity is marked U.

Note that this discussion involves only structural/porosity considerations of trap formation. The many other aspects of hydrocarbon accumulation, including such factors as the presence of source, reservoir and seal formations, burial history, and timing of development of the structures relative to the periods of maturation and migration of any hydrocarbons in the area, are not taken into account.

10.4 SALT/FAULT SITUATIONS

The possibilities of trap formation in these situations are legion, and it is proposed here to discuss only two special cases. The first, illustrated in Fig. 10.2, is where salt withdrawal, perhaps combined with extrusion at the surface at some stage, had led to the development of a growth fault, F, with a rollover anticline and thickening in some overburden intervals on the downthrow side.

Situation (1) is a pinchout trap due to thickening of some stratigraphic units on the downthrow side towards the growth fault zone. Often, on the downthrow side of a growth fault, 'fines' predominate, but there may be some coarser intervals with good porosity, depending on the overall depositional pattern. At (2) there is a possible situation in the crest of

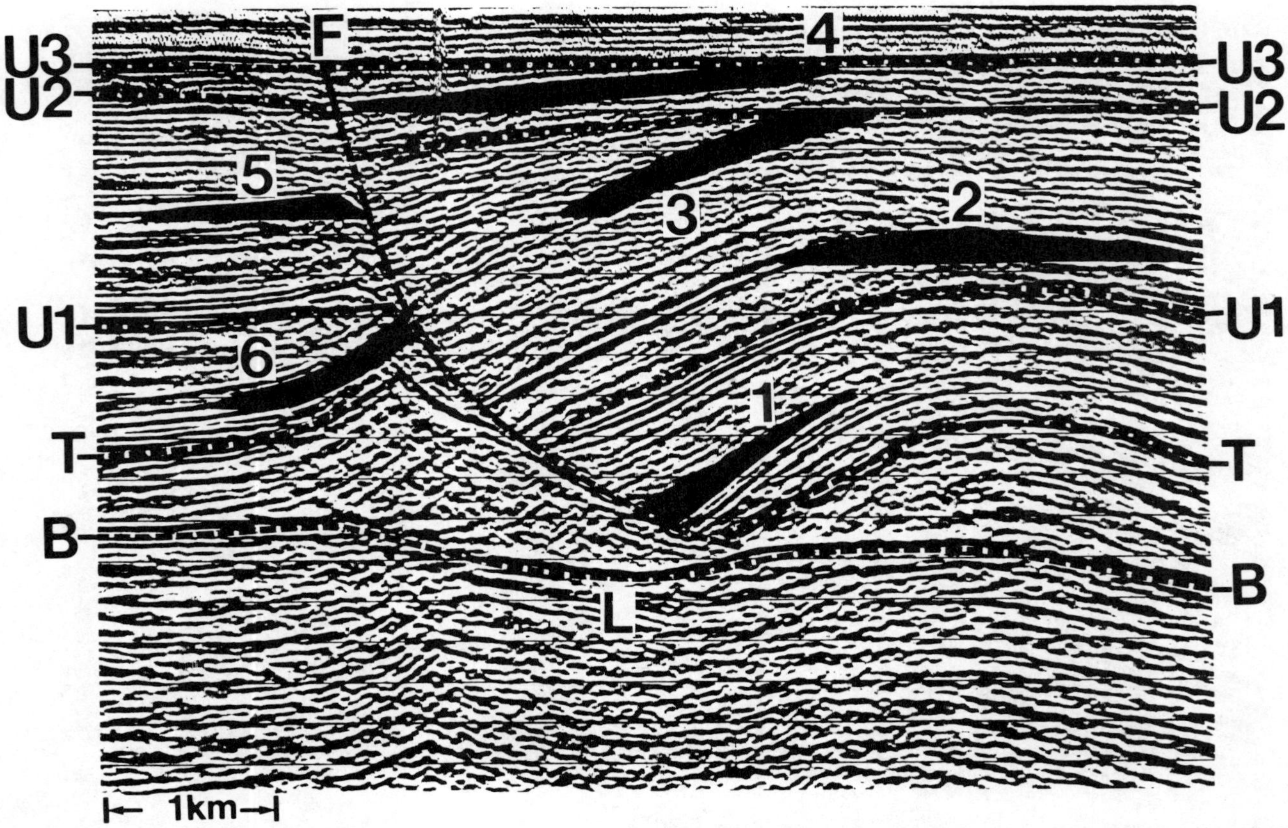

Fig. 10.2 Trap classes possibly present in a salt-withdrawal/growth fault situation. Key: B–T, salt interval; U1, U2, U3 are unconformities; L is a velocity 'pull-down' anomaly due to marked thinning of the salt interval. Note the well-defined salt uplift on the upthrow side of growth fault F. This is probably due to the flow mechanism discussed in Sections 5.2.1 and 5.2.2. (Courtesy of SS(E)L.)

the rollover anticline related to the faulting. The structure here also often includes antithetic faults (although not in this case) which can provide favourable situations. Traps related to unconformities (e.g. at U2, U3) may be present as at (3) and (4) in this example, and other possibilities of this type may exist in a section like this. At (5) a trap occurs against the main fault on the upthrow side—apart from the fault closure, in this case there may be some inherent closure present due to the marked fault-drag folds at and below this level. Beneath the unconformity U1 on the upthrow side there are possible trap situations against the main fault, as at (6), and also beneath the unconformity in toplapping units.

The marked low, L, shown at the base of the salt interval (B–T), is a velocity anomaly 'pull-down' effect resulting from the rapid thinning of the salt combined with sudden thickening of the lower velocity sediments in the downthrow subsidence. Note also the localised salt uplift on the upthrow side of the fault below (6), probably due to some flow from the downthrow side by the differential loading mechanism. Some salt may have escaped up the fault zone also, but most has probably flowed into the pillow to the right. The trap situation at (6) owes much of its potential to this local salt uplift.

The second case to be considered is that where the differential loading mechanism discussed in Chapter 5 has led to a salt uplift on the upthrow side of a fault which affected the whole salt interval (although there would be similarities with the alternative situation of a salt interval deposited across an old fault scarp). Figure 10.3 illustrates the principal trap possibilities, using an example from Chapter 5 as a model section. Trap type (1) shows two possible situations related to the upthrow side of the younger, listric fault L, whilst at (2) an anticlinal trap in the rollover against the listric fault zone is shown, together with a possible fault trap in the sliver of section isolated by fault L and a synthetic fault associated with it. At (3) a combination pinchout/unconformity trap is indicated, and at (4) a fault trap against antithetic faulting which is related to the overall growth fault configuration. As always, the possibility of stratigraphic porosity traps, and sub-salt traps sealed by the salt interval should not be overlooked.

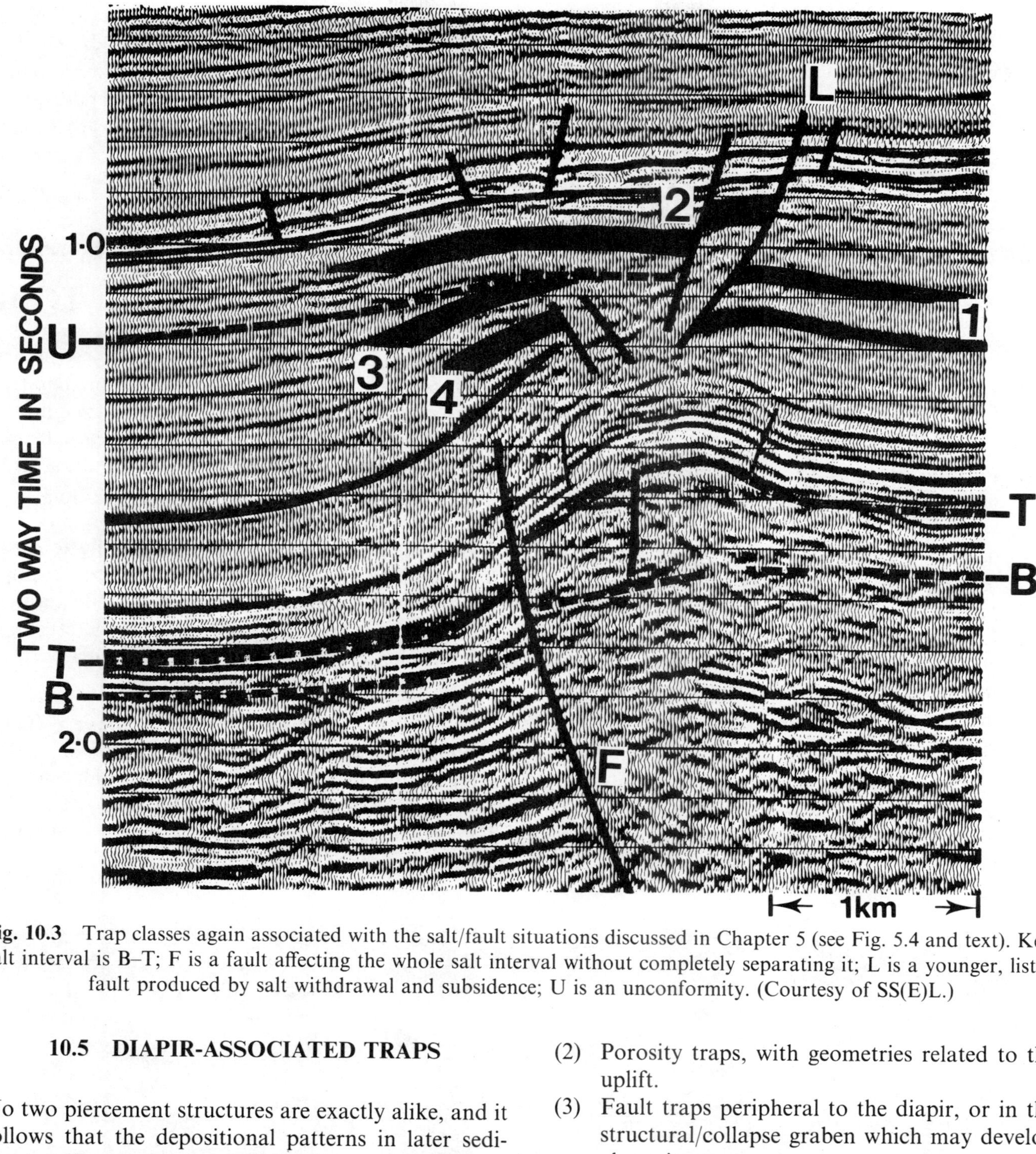

Fig. 10.3 Trap classes again associated with the salt/fault situations discussed in Chapter 5 (see Fig. 5.4 and text). Key: salt interval is B–T; F is a fault affecting the whole salt interval without completely separating it; L is a younger, listric fault produced by salt withdrawal and subsidence; U is an unconformity. (Courtesy of SS(E)L.)

10.5 DIAPIR-ASSOCIATED TRAPS

No two piercement structures are exactly alike, and it follows that the depositional patterns in later sediments will vary from one structure to another. It is possible, however, in a general way, to identify certain principal types of trap geometry associated with developed diapiric structures, not all of which will necessarily be present with any one structure. These general types of trap are represented diagrammatically in Fig. 10.4, and are:

(1) Simple anticlinal or domal traps due to uplift above the diapir.
(2) Porosity traps, with geometries related to the uplift.
(3) Fault traps peripheral to the diapir, or in the structural/collapse graben which may develop above its crest.
(4) Cap rock (usually secondary porosity) traps.
(5) Truncation traps against the salt flanks. These may be complicated by peripheral or radial faulting.
(6) Pinchout traps resulting from uplift thinning towards the diapir. They may also be due to subsidence in a salt withdrawal basin adjacent to the diapir. In either case, syndepositional movement of the salt is necessary.

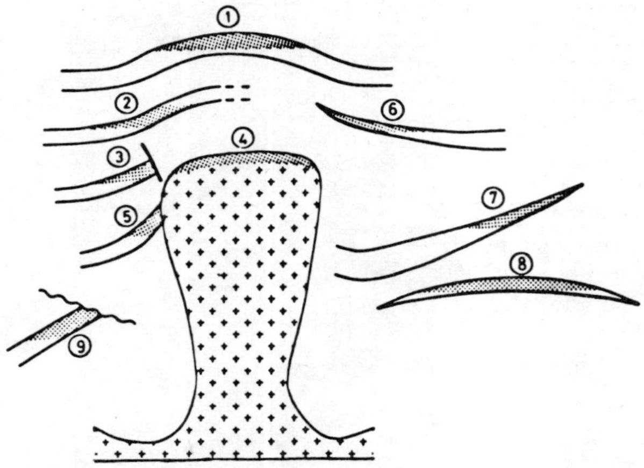

Fig. 10.4 Diagram showing principal classes of hydrocarbon trap associated with a piercement salt structure. See text for key and explanation.

(7) Pinchout traps due to secondary rim syncline sediments thinning away from the diapir.

(8) 'Turtle structure' traps in primary rim syncline deposits which have undergone structural inversion related to the secondary rim syncline subsidence.

(9) Unconformity traps; these can occur at different locations in the section, but in this instance the suggested trap is in pre-movement sediments uplifted in the pillow stage and later eroded.

The locations of some of these trap types in seismic data are illustrated in Fig. 10.5, again using a previously shown example as a model. A diapir, D, has developed from a salt interval BZ–TZ. A major unconformity, U, occurs in the section, above which there is a well-developed secondary rim syncline, S. Base Tertiary is at T, and Base Rotliegendes at R below the Zechstein salt interval. Trap types shown are:

(1) and (7)—truncation traps against the diapir flanks.

(2) —unconformity trap related to the major unconformity U.

(6) and (8)—unconformity traps related to a shallower unconformity.

(3) —pinchout trap due to thinning of secondary rim syncline sediments away from the diapir.

(4) —pinchout trap due to uplift and thinning of sediments contemporaneously with the salt rise.

(5) —fault trap related to peripheral faulting near the diapir crest.

(9) —half of a possible turtle trap in primary rim syncline sediments which are thinning towards the diapir, having subsided during the development of the secondary rim syncline. If another diapir is located adjacent to this structure, trap 9 could be the mirror image of the other portion of the trap off section to the left.

Generally speaking, anticlinal traps above a diapir are often dependent on such factors as the degree of disruption at the crest of the structure due to graben faulting of either structural or dissolution collapse origin. Such traps may often be 'empty' by virtue of

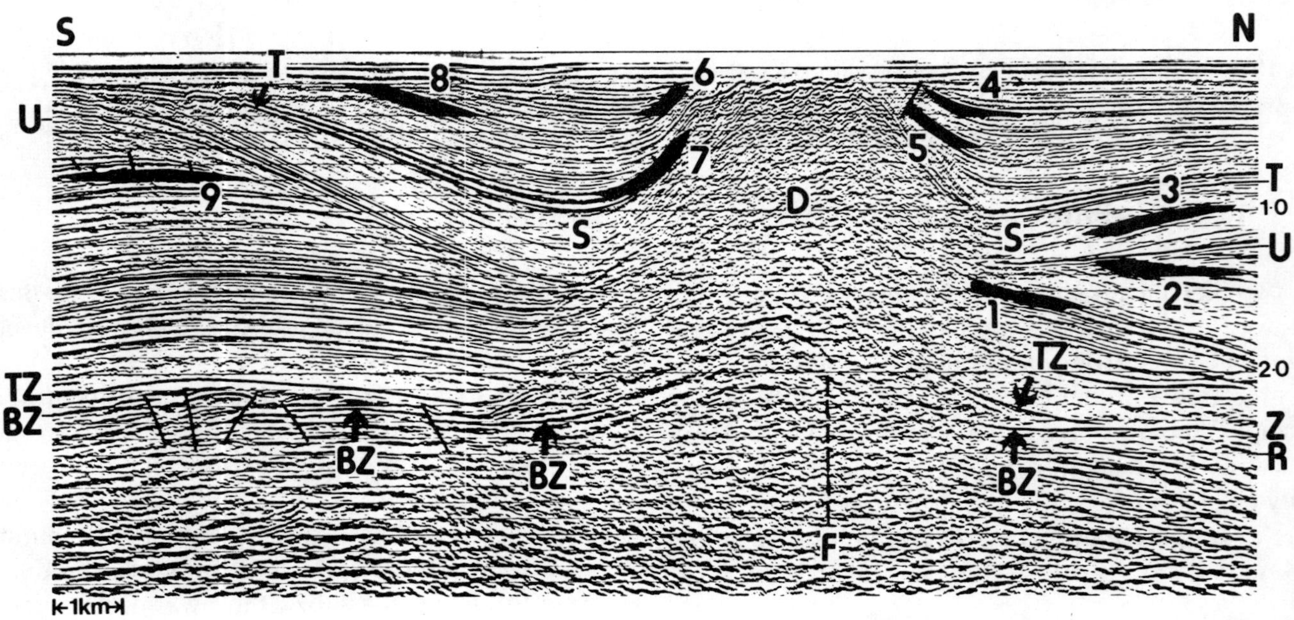

Fig. 10.5 Some of the trap classes shown in the diagram of Fig. 10.4, as modelled on a seismic section across a salt-wall diapir, D. See text for key and explanation. (Courtesy of SS(E)L.)

the very efficient sealing property of the underlying salt, as is the case with the majority of the trap types discussed. The effect this has had in the North Sea Southern Basin was mentioned earlier in the book.

Where a cap rock has retained its integrity, secondary porosity may have evolved due to dissolution of some of the minerals present. Such porosity can be very capricious in its distribution, and frequently has been recemented at a later stage by minerals exsolving from circulating formation water. In some areas, any cap rock developments which were present have undergone major disruption due to graben faulting, and are no longer recognisable in seismic data.

Unconformity traps may be of great importance. In a discussion of palaeogeomorphic unconformity traps, Halbouty (1982) points out that salt diapirs could be regarded as mobile palaeogeomorphic features—'impervious buried hills'—which, by changing shape both laterally and vertically, affect the depositional patterns in overlying and adjacent sediments, occasioning many possible trap situations of unconformity type. Where a diapir changes its direction of emplacement upwards in response to tectonism or varying rates and patterns of deposition of overburden sediments, as is the case in many Gulf Coast structures, an offset or 'shoulder' of salt body is often left at depth which may lead to various associated pinchout and unconformity traps in sediments being deposited at the time of the salt movements.

Truncation traps against the salt flanks may be ill-defined or invisible in the immediate neighbourhood of the salt in seismic data for reasons discussed previously. The beds involved may be uplifted to the vertical, and even overturned, against the salt flank. Often in this location the situation is complicated by faulting. Situations like this, if suspected, need special processing of the seismic data for their elucidation.

Where a diapir develops a spreading overhang at its crest, beds may be truncated beneath this, providing good trap potential. This situation has not been illustrated, but should be borne in mind. Dissolution and collapse of such spreading overhang features may also result in potential trap situations in the collapsed sediments which previously covered the overhang. This is a subject which is being studied at present. It seems possible that good fracture porosity may develop due to the subsidence of the overlying sediments, and it is believed that favourable trap geometries may occur. Provided overlying seal formations remain intact, features of this type could prove to be important in hydrocarbon exploration, whereas up to the present their potential does not appear to have been recognised.

To conclude this brief discussion, it is again stressed that in order to be structurally viable, a hydrocarbon trap must possess favourable geometry in three dimensions, and not just in the two dimensions of a seismic or stratigraphic section. That part of seismic stratigraphic analysis which considers the three-dimensional shape of a particular facies for diagnostic purposes is admirable as a starting point for the study of hydrocarbon traps, and is to be recommended particularly to those geologists who use seismic data in their work.

CHAPTER 11

Afterthoughts

This book might have been sub-titled 'A Perspective', since the *Concise Oxford Dictionary* defines this word in one of its senses as the 'relation in which parts of a subject are viewed by the mind'. Certainly, it has been an attempt on the author's part to express, in a rather broad-brush way, a very personal view of how various aspects of the subject relate to each other.

To have been a geologist involved in seismic interpretation over the past three decades is to have seen the progressive stripping away, as it were, of a series of veils interposed between the observer and the geological details of the subsurface. Gradually, more and more detail has become visible, and the results of the process to date are seen in the seismic examples which illustrate this work. Some veils still remain, and doubtless over the *next* three decades, more of these will be removed. Even now, however, the amount of geologically meaningful information carried by a modern seismic section would have been incredible thirty years ago.

All this means that we are now in a much better position to understand subsurface processes in general, and salt movement processes in particular, than when interpretation meant inspired correlative guesswork applied to the wide expanses between boreholes. We should therefore begin to use our new powers by asking searching questions, and testing hypotheses which were advanced before the new technology was available. Some of these hypotheses have achieved the status of dogma, and some of this dogma exists in salt tectonics. Part of the purpose of this work has been to begin to put such questions, and to suggest some possible alternative answers to those which have been accepted for many years.

It was with this in mind that, in particular, several suggestions were made in Chapter 7 as regards the initiating and controlling factors in salt movement and diapirism. It must be stressed, however, that there was no intention of 'throwing the baby out with the bath water' by underestimating the part that gravitational control plays in the development of salt structures. Once salt is in a mobile state, and load stress differentials exist in the overburden, then gravitational effects will certainly be of great importance in determining the form taken by the evolving salt structure. Also, it is dangerous to oversimplify the possible results that such effects may have. Trippet (1981) has shown that although, in depth, many individual diapirs on the outer continental shelf of the NW Gulf of Mexico may coalesce into ridges on the Bay Marchand model, as they rise through the overburden they come under the influence of lateral variations in loading. These may be due to the effects of individual depositional basins or prisms of deltaic sediments, etc., which can cause deviations in trend of elongate diapirs away from trends in the major salt ridges from which they rise. This may lead (in plan) to apparently random configurations, or even configurations orthogonal to the major underlying salt trends. This must be borne in mind when considering the simple models of basin configuration considered earlier in the book.

What can be said of possible future conceptual developments in salt tectonics?

Searches of the literature have suggested that salt movement activity—and especially peak activity involving diapirism—in many basins falls within the mid-Jurassic to mid-Cretaceous period. This seems to be especially true in the case of basins at the continental margins of the Atlantic, both North and South. If this is true, and is not due to the purely fortuitous evolution in the various basins, over the same period, of overburden loading to produce the right temperature, with differential stresses, for salt mobility to develop, then it must be of some significance. It suggests some common factor which has operated during the period in question, and which has had an initiatory effect on salt movement. The first idea which comes to mind is that of the tectonic effects attendant on lithospheric plate movements. The mid-Jurassic to mid-Cretaceous period was undoubtedly one of considerable activity in this context. Moreover there is a developing study of 'pulsation tectonics'—the effects of *changes* in the rate of plate movement. Such changes may have caused the propagation of pulses of compression through adjacent areas which could have affected any salt basin within range in which potentially mobile salt existed at the time.

A hypothetical situation of this kind is of interest from two quite separate points of view. First, it bears upon the fundamental question of what initiates salt movement in general, and possibly diapirism in particular. This could have an important influence on our thinking as regards the dynamics of the mobile salt/overburden system. Plate movement as a cause of salt movement has already been advanced hypothetically in specific areas. Asaad (1983), for instance, has invoked the relationship to explain salt structures in north-central Algeria, relating these to the subduction of the Saharan Plate below the Atlassic Plate, and proposing the term 'halokinematics' to describe the effects.

Secondly, it could give useful insights into the palaeogeography of different periods of plate movement, if patterns and ages of salt structure development in various areas are considered. If the suggestion is taken seriously, it opens up a wide field of study, and offers a new way of looking at salt tectonics on the global scale.

If the book has a general 'message', it must be to highlight the requirement to view salt structures not as occurring in isolation, but as being continuously and intimately related to other geological mechanisms and situations—to basin configuration, overburden depositional patterns, geothermal gradient, erosional phases, and tectonism in its various forms. A salt structure is a link in a chain of cause and effect.

It would be surprising indeed if all the suggestions and interpretations put forward in the book were acceptable to all geoscientists; it is hoped, at least, that they will provoke thought and discussion on an important and interesting subject which, in its wider field, has not perhaps received as much attention in the European area over the last decade as it has elsewhere.

Bibliography and References

ANDERSON, R. Y. and KIRKLAND, D. W. (1966) Intrabasin varve correlation. *Geol. Soc. Amer. Bull.*, **77**, 241–56.

ANDERSON, R. Y. and KIRKLAND, D. W. (1980) Dissolution of salt deposits by brine density flow. *Geol. Soc. Amer.* (*'Geology'*), **8**, 66–9.

ANDERSON, R. Y., DEAN, W. E., KIRKLAND, D. W. and SNIDER, H. I. (1972) Permian Castile varved evaporite sequence, West Texas and New Mexico. *Geol. Soc. Amer. Bull.*, **83**, 59–86.

ANTHONY, T. R. and CLINE, H. E. (1974) Thermomigration of liquid droplets in salt. IV Symp. *Salt*, **1**, 313–21.

ANTONOWICZ, L. and KNIESZNER, L. (1984) Zechstein reefs of the main dolomite in Poland and their seismic recognition. *Acta Geologica Polonica*, **34**, 81–94.

ARRHENIUS, Sv. (1912) Zur Physik der Salzlagerstätten. *Meddelanden fran K. Vetenskabsakademiens Nobelinstitute*, **2**, 7–25.

ARTHURTON, R. S. (1973) Experimentally produced halite compared with Triassic layered halite-rock from Cheshire, England. *Sedimentology*, **20**, 145–60.

ASAAD, F. A. (1983) An approach to 'halokinematics' and interplate tectonics (North Central Algeria). *Jour. Petr. Geol.*, **6**, 83–8.

BAAR, C. A. (1977) *Applied Salt-Rock Mechanics—1.* Elsevier, Amsterdam.

BÄCKER, H., LANGE, K. and RICHTER, H. (1975) Morphology of the Red Sea central graben between Subair Islands and Abul Kizaan. *Geol. Jb. D.*, **13**, 79–123.

BACOCCOLI, G., MORALES, R. G. and CAMPOS, O. A. J. (1980) The Namorado oil field: a major oil discovery in the Campos Basin, Brazil. In: *Giant Oil and Gas Fields of the Decade: 1968–1978.* AAPG.

BADIOZAMANI, K. (1973) The Dorag dolomitization model, application to the Middle Ordovician of Wisconsin. *Jour. Sed. Pet.*, **43**, 965–84.

BARTON, D. C. (1933) Mechanics of formation of salt domes with special reference to Gulf Coast salt domes of Texas and Louisiana. *AAPG Bull.* **17**, 1025–83.

BATHURST, R. G. C. (1971) *Carbonate sediments and their diagenesis.* Developments in sedimentology No. 12. Eslevier, Amsterdam.

BELDERSON, R. H., KENYON, N. H. and STRIDE, A. H. (1978) Local submarine salt-karst formation on the Hellenic Outer Ridge, eastern Mediterranean. *Geol. Soc. Amer.* (*'Geology'*), **6**, 716–20.

BIOT, M. A. and ODÉ, H. (1965) Theory of gravity instability with variable overburden and compaction. *Geophysics*, **30**, 213–27.

BISHOP, R. S. (1978) Mechanism for emplacement of piercement diapirs. *AAPG Bull.*, **62**, 1561–83.

BLACK, M. (1933) The precipitation of calcium carbonate on the Great Bahama Bank. *Geol. Mag.*, **LXX**, 455–66.

BOLDREEL, L. O. (1985) On the structural development of the salt dome province in North West Jutland, Denmark, based on seismic studies. *First Break*, **3**(8), 15–21.

BORCHERT, H. and MUIR, R. O. (1964) *Salt Deposits.* Van Nostrand, Wokingham.

BOUMA, H. A., MOORE, G. T. and COLEMAN, J. M. (Eds.) (1978) *Framework, Facies, and Oil-Trapping Characteristics of the Upper Continental Margin. Studies in Geology No. 7.* AAPG.

BRAITSCH, O. (1971) *Salt Deposits—Their Origin and Composition.* Springer, New York.

BRICE, S. E., COCHRAN, M. D., PARDO, G. and EDWARDS, A. D. (1982) Tectonics and sedimentation of the South Atlantic Rift Sequence: Cabinda, Angola. In: *AAPG Memoir 34*, *Studies in Continental Margin Geology*, 5–18.

BRINK, A. H. (1974) Petroleum geology of the Gabon Basin. *AAPG Bull.*, **58**, 216–35.

BRUNSTROM, R. G. W. and WALMSLEY, P. J. (1969) Permian evaporites in the North Sea. *AAPG Bull.*, **53**, 870–83.

CASHMAN, K. V. and POPENOE, P. (1985) Slumping and shallow faulting related to the presence of salt on the continental slope and rise off North Carolina. *Marine & Petr. Geol.*, **2**(3), 260–71.

CLARK, D. N. (1980) The diagenesis of Zechstein carbonate sediments. *Contr. Sedimentology*, **9**, 167–203.

CLARK, D. N. (1985) Diagenesis in evaporite basins in relation to reservoir potential. In: *The Role of Evaporites in Hydrocarbon Exploration* (*Course Notes No. 39*). Joint Association for Petroleum Exploration Courses (UK), London.

CLARK, D. N. (in press). The distribution of porosity in Zechstein carbonates. In: *Proc. Symp. 'Habitat of Palaeozoic Gas in NW Europe', Geol. Soc. Lond.*

CLARKE, F. W. (1924) Data of geochemistry. *Bull. U.S. Geol. Surv.* 770.

CLOUD, P. E. (1962) Environment of calcium carbonate deposition west of Andros Island, Bahamas. US Geol. Surv. Prof. Pap. 350.

COELEWIJ, P. A. J., HAUG, G. M. W. and VAN KUIJK, H. (1978) Magnesium-salt exploration in the northeastern Netherlands. *Geologie en Mijnbouw*, **57**(4), 487–502.

COTTON, C. A. (1956) Geomechanics of New Zealand mountain-building. *N.Z. Jour. Sci. & Technol.*, **38**, 187–200.

CRAMER, H. R. (1969) Evaporites—a selected bibliography. *AAPG Bull.*, **53**, 982–1011.

CURRIE, J. B. (1956) Role of concurrent deposition and deformation of sediments in development of salt-dome graben structures. *AAPG Bull.*, **40**, 1–16.

DABLAIN, M. A. and LINVILLE, A. F. (1985) Case study in seismic diffraction near a Gulf Coast salt dome. *O & GJ*, 14/10/85, 137–41.

DANEŠ, Z. F. (1964) Mathematical formulation of salt-dome dynamics. *Geophysics*, **29**, 414–24.

DAUZACKER, M. V., SCHALLER, H., CASTRO, A. C. M. Jr. and DE SOUZA, M. M. (1985) Geology of Brazil's Atlantic margin. *O & GJ*, 4/3/85, 142–4.

DAVIES, G. R. and LUDLAM, S. D. (1971) Origin of laminated and graded sediments, Middle Devonian of Western Canada. *Geol. Soc. Amer. Bull.*, **84**, 3527–626.

DEGOLYER, E. L. (1926) Origin of North American salt domes. In: *Geology of Salt Dome Oil Fields*. AAPG, 1–42.

DE GROOT, K. (1965) Inorganic precipitation of calcium carbonate from sea water. *Nature*, **207**, 404–5.

DE GROOT, K. (1967) Experimental dedolomitization. *J. Sed. Petrol.*, **37**, 1216–20.

DE GROOT, K. and DUYVIS, E. M. (1966) Crystal form of precipitated calcium carbonate as influenced by adsorbed magnesium ions. *Nature*, **212**, 183–4.

DE MILLE, G., SHOULDICE, J. R. and NELSON, H. W. (1964) Collapse structures of the Prairie Formation, Saskatchewan. *Geol. Soc. Amer.* (*Bulletin*), **75**, 307–16.

DICKINSON, G. (1963) Geological aspects of abnormal reservoir pressures in the Gulf Coast region of Louisiana. *AAPG Bull.*, **37**, 410–32.

DILLON, W. P., POPENOE, P., GROW, J. A., KLITGORD, K. D., SWIFT, B. A., PAULL, C. K. and CASHMAN, K. V. (1982) Growth faulting and salt diapirism: their relationship and control in the Carolina Trough, Eastern North America. *AAPG Memoir 34, Studies in Continental Margin Geology*, 21–46.

DOBRIN, M. B. (1941) Some quantitative experiments on a fluid salt-dome model, and their geological implications. *Amer. Geophys. Union Trans.*, **22**, 528–42.

DOTT, R. H. and RICE, P. (Eds.) (1971) *Origin of Evaporites*. AAPG Reprint Series No. 2.

DREYER, W. (1982) *Underground Storage of Oil and Gas in Salt Deposits and Other Non-Hard Rocks*. Halsted Press, New York.

EDMUNDS, R. H. (1980) Salt removal and oil entrapment. *Can. Soc. Petr. Geol. Memoir* 6, p. 988.

ELLIS, J. P. and MILLIMAN, J. D. (1985) Calcium carbonate suspended in Arabian Gulf and Red Sea waters: biogenic and detrital, not 'chemogenic'. *J. Sed. Petrol*, **55**, 805–8.

ESCHER, B. G. and KUENEN, Ph. H. (1929) Experiments in connection with salt domes. *Leidsche Geol. Med.* **3**(3), pt. 2, 151–82.

EWING, M. and EWING, J. (1962) Rate of salt dome growth. *AAPG Bull.*, **46**, 708–9.

EWING, T. E. (1984) Growth faults and salt tectonics. *O & GJ*, 30/1/84, 176–9.

FEELY, H. W. and KULP, J. L. (1957) Origin of Gulf Coast salt dome sulphur deposits. *AAPG Bull.*, **41**, 1802–953.

FISHER, J. H. (Ed.) (1977) *Reefs and Evaporites—Concepts and Depositional Models*. AAPG Studies in Geology No. 5.

FREY, M. G. and GRIMES, W. H. (1970) Bay Marchand–Timbalier Bay–Caillou Island salt complex, Louisiana. *AAPG Memoir 14*, 277–91.

FRIEDMAN, G. M. (1972) Significance of Red Sea in problem of evaporites and basinal limestones. *Bull. AAPG*, **56**, 1072–86.

GARDNER, L. W. (1949) Seismograph determination of salt dome boundary using well detector deep on salt dome flank. *Geophysics*, **14**, 29–38.

GARRELS, R. M., THOMPSON, M. E. and SIEVER, R. (1961) Control of carbonate solubility by carbonate complexes. *Amer. J. Sci.*, **259**, 24–45.

GOULD, D. B. and DE MILLE, G. (1968). Piercement structures in the Canadian Arctic Islands. *AAPG Memoir 8, Diapirism and Diapirs*, 183–214.

GRIPP, K. (1958) Salzspiegel und Salzhut auf dem Lande und unter dem Meere. *Abh. Naturw. Ver. Bremen*, **35**, 249–58.

GUSSOW, W. C. (1968) Salt diapirism: importance of temperature and energy source of emplacement. *AAPG Memoir 8, Diapirism and Diapirs*, 16–52.

HALBOUTY, M. T. (1967) *Salt Domes—Gulf Region, United States and Mexico*. Gulf Pub. Co., Houston, Texas.

HALBOUTY, M. T. (1982) The time is now for all

interpreters to purposefully search for the subtle trap. In: *AAPG Memoir 32, The Deliberate Search for the Subtle Trap* (Ed. M. T. Halbouty), 1–10.

HEILAND, C. A. (1940) *Geophysical Exploration*. Prentice-Hall, New York.

HILLS, E. S. (1972) *Elements of Structural Geology* (2nd edn). Chapman & Hall, London

HOBSON, G. D. (1938) Salt structures: their form, origin, and relationship to oil accumulation. In: *The Science of Petroleum, Vol. 1* (Ed. Dunstan *et al.*). Institute of Petroleum, London.

HODGMAN, C. D. (Ed.) (1948) *Handbook of Chemistry and Physics* (13th edn). Chemical Rubber Publishing Co., Cleveland, Ohio, p. 1386.

HOLDOWAY, K. A. (1974) Behaviour of fluid inclusions in salt during heating and irradiation. *IV Symp. Salt*, **1**, 303–12.

HOLSER, W. T. (1979) Trace elements and isotopes in evaporites. In: *Marine Minerals* (Ed. R. Burns). Min. Soc. Amer. Short Course Notes, 6, 295–346.

HOSPERS, J. (1985). Sideswipe reflections and other external and internal reflections from salt plugs in the Norwegian–Danish Basin. *Geoph. Prosp.*, **33**, 52–71.

HSÜ, K. J. (1972) Origin of saline giants: a critical review after the discovery of the Mediterranean evaporite. *Earth Sci. Review*, **8**, 371–96.

HUBBERT, M. K. (1937) Theory of scale models as applied to the study of geologic structures. *Geol. Soc. Amer.* (*Bulletin*), **48**, 1459–520.

HUMPHRIS, C. C. (1979) Salt movement on continental slope, Northern Gulf of Mexico. *AAPG Bull.*, **63**, 782–98.

JENSEN, P. K. (1983) Calculations on the thermal conditions around a salt diapir. *Geoph. Prosp.*, **31**, 481–9.

JENYON, M. K. (1984a) Seismic response to collapse structures in the southern North Sea. *Marine & Petr. Geol.*, **1**, 27–36.

JENYON, M. K. (1984b) Upper Carboniferous gas indications and Zechstein features in the southern North Sea. *O & GJ*, 14/5/84, 135–44.

JENYON, M. K. (1985a) Basin-edge diapirism and updip salt flow in the Zechstein of the southern North Sea. *AAPG Bull.*, **69**, 53–64.

JENYON, M. K. (1985b) Fault-associated salt flow and mass movement. *Jour. Geol. Soc. Lond.*, **142**, 547–53.

JENYON, M. K. (1985c) Differential movement in salt rock. *O & GJ*, 30/9/85, 73–5.

JENYON, M. K. (1985d) Salt movement: part 1, plastic deformation; part 2, dissolution. In: *The Role of Evaporites in Hydrocarbon Exploration. Course Notes No. 39*, Joint Assoc. for Petroleum Exploration Courses (UK), London.

JENYON, M. K. (1986) Some consequences of faulting in the presence of a salt interval. *Jour. Petr. Geol.*, **9**, 29–52.

JENYON, M. K., CRESSWELL, P. M. and TAYLOR, J. C. M. (1984) The nature of the connection between the Northern and Southern Zechstein Basins across the Mid North Sea High. *Marine & Petr. Geol.*, **1**, 355–63.

JENYON, M. K. and FITCH, A. A. (1985) *Seismic Reflection Interpretation*. Gebrüder Borntraeger, Stuttgart.

JENYON, M. K. and TAYLOR, J. C. M. (1983) Hydrocarbon indications associated with North Sea Zechstein shelf features. *O & GJ*, 5/12/83, 155–60.

JOFFÉ, A. F. (1928) In: *The Physics of Crystals* (Ed. L. B. Loeb). McGraw-Hill, New York.

JUKES-BROWN, A. J. (1892) *The Building of the British Isles*. George Bell & Sons, London.

KENDALL, A. C. (1978a) Facies models 11: Continental and supratidal (Sabkha) evaporites. *Geosci. Canada*, **5**, 66–78.

KENDALL, A. C. (1978b) Facies models 12: Subaqueous evaporites. *Geosci. Canada*, **5**, 124–39.

KENT, P. E. (1979) The emergent Hormuz salt plugs of Southern Iran. *Jour. Petr. Geol.*, **2**, 117–44.

KING, R. H. (1947) Sedimentation in the Permian Castile Sea. *AAPG Bull.*, **31**, 470–7.

KINGMA, J. T. (1958) Possible origin of piercement structures, local unconformities, and secondary basins in the Eastern Geosyncline, New Zealand. *N.Z. Jour. Geol. Geophys.*, **1**, 269–74.

KINSMAN, D. J. J. (1966) Gypsum and anhydrite of Recent age, Trucial Coast, Persian Gulf. *Proc. 2nd. Salt Symp.* (Ed. J. L. Rau), **1**, 302–26. Northern Ohio Geol. Soc.

KINSMAN, D. J. J. (1969) Modes of formation, sedimentary associations, and diagenetic features of shallow water and supratidal evaporites. *AAPG Bull.*, **53**, 830–40.

KINSMAN, D. J. J. (1974) Calcium sulphate minerals of evaporite deposits: their primary mineralogy. *4th Symposium on Salt, Vol. 1* (Ed. A. H. Coogan). Northern Ohio Geo. Soc. Inc., Cleveland, Ohio, pp. 343–8.

KINSMAN, D. J. J. (1976) Evaporites: relative humidity control of primary mineral facies. *Jour. Sed. Pet.*, **46**, 273–9.

KINSMAN, D. J. J. (1985) Tectonic and climatic constraints on evaporite formation and examples from the Recent. In: *The Role of Evaporites in Hydrocarbon Exploration*. Course Notes No. 39, Joint Assoc. for Petroleum Exploration Courses (UK), London.

KIRKLAND, D. W. and EVANS, R. (1981) Source-rock potential of evaporitic environment. *AAPG Bull.*, **65**, 181–90.

KNIPE, R. J. and WHITE, S. H. (1978) On the deformation of quartz containing bubbles. *Jour. geol. Soc. Lond.*, **135**, 313–16.

KUPFER, D. H. (1968) Relationship of internal to external structure of salt domes. *AAPG Memoir 8, Diapirism and Diapirs*, 79–89.

LANGSTROTH, W. T. (1971) Seismic study along a portion of the Devonian salt front in North Dakota. *Geophysics*, **36**, 330–8.

LECOMTE, P. (1965) Creep in rock salt. *J. Geol.*, **73**, 469–84.

LEHNER, P. and DE RUITER, P. A. C. (1977) Structural history of the Atlantic margin of Africa. *AAPG Bull.*, **61**, 961–81.

LOHMANN, H. H. (1972) Salt dissolution in subsurface of British North Sea as interpreted from seismograms. *AAPG Bull.*, **56**, 472–9.

LOHMANN, H. H. (1980) Seismic differentiation between salt diapirs, shale diapirs, and similar bodies. Symp. Interpretational Information from Seismic Data, Norsk Petroleumsforening, Kristiansand, Norway.

LOTZE, F. (1957) *Steinsalz und Kalisalz, erste Teil.* (2nd edn). Gebrüder Borntraeger, Berlin.

LUCIA, F. J. (1972) Recognition of evaporite-carbonate shoreline sedimentation. In: *Recognition of Ancient Sedimentary Environments* (Ed. J. K. Rigby and W. K. Hamblin). S.E.P.M. Spec. Publ. **16**, pp. 160–91.

MAIKLEM, W. R. (1971) Evaporative drawdown—a mechanism for water-level lowering and diagenesis in the Elk Point Basin. *Can. Soc. Petr. Geol. Bull.*, **19**, 487–501.

MANZUR, A. (1985) Delineation of salt masses using borehole seismics. *O & GJ*, 7/10/85, 147–9.

MAY, B. T. and COVEY, J. D. (1981) An inverse ray method for computing geologic structures from seismic reflections—zero offset case. *Geophysics*, **46**, 268–87.

MAY, B. T. and COVEY, J. D. (1983) Structural inversion of salt dome flanks. *Geophysics*, **48**, 1039–50.

McCLAY, K. R. (1977) Pressure solution and Coble creep in rocks and minerals: a review. *Jour. geol. Soc. Lond.*, **134**, 57–70.

MORGAN, J. P., COLEMAN, J. M. and GAGLIANO, S. M. (1968) Mudlumps: diapiric structures in Mississippi Delta sediments. *AAPG Memoir 8, Diapirism and Diapirs*, 145–61.

MRAZEC, M. L. (1915) Les plis diapirs et le diapirisme en général. *Rumania Inst. Geol. Comptes Rendues* **4**, 226–70.

MUEHLBERGER, W. R. and CLABAUGH, P. S. (1968) Internal structure and petrofabrics of Gulf Coast salt domes. *AAPG Memoir 8, Diapirism and Diapirs*, 90–98.

MUNDRY, E., THIERBACH, R., SENDER, F. and WEICHERT, H. (1983) Borehole radar probing in salt deposits. *Proc. 6th. Int. Symp. on Salt*, v.1, Toronto, 589–99.

MURRAY, R. C. (1964) Origin and diagenesis of gypsum and anhydrite. *J. Sedim. Petrol.*, **34**, 512–23.

MURRAY, G. E. (1968) Salt structures of Gulf of Mexico Basin—a review. *AAPG Memoir 8, Diapirism and Diapirs*, 99–121.

MUSGRAVE, A. W. and HICKS, W. G. (1966) Outlining of shale masses by geophysical methods. *Geophysics*, **31**, 711–25.

MUSGRAVE, A. W., WOOLLEY, W. C. and GRAY, H. (1960) Outlining of salt masses by refraction methods. *Geophysics*, **25**(1), 141–67.

NBS MONOGRAPH 167 (1981) *Physical Properties Data for Rock Salt.* National Bureau of Standards, US Department of Commerce.

NEEV, D. and EMERY (1967) The Dead Sea; depositional processes and environments of evaporites. *Bull. Israel geol. Surv.*, **41**.

NETTLETON, L. L. (1934) Fluid mechanics of salt domes. *AAPG Bull.*, **18**, 1175–204.

NETTLETON, L. L. (1943) Recent experimental and geophysical evidence of mechanics of salt dome formation. *AAPG Bull.*, **27**, 51–63.

NETTLETON, L. L. (1955) History of concepts of Gulf Coast salt dome formation. *AAPG Bull.*, **39**, 2373–83.

NICKEL, H., SENDER, F., THIERBACH, R. and WEICHART, H. (1983) Exploring the interior of salt domes from boreholes. *Geoph. Prosp.*, **31**, 131–48.

ODÉ, H. (1968) Review of mechanical properties of salt relating to salt-dome genesis. *AAPG Memoir 8, Diapirism and Diapirs*, 53–78.

OJEDA, H. A. O. (1982) Structural framework, stratigraphy, and evolution of Brazilian marginal basins. *AAPG Bull.*, **66**, 732–49.

OLIVER, T. A. and COWPER, N. W. (1983) Wabamun salt removal and shale compaction effects, Rumsey area, Alberta. *Bull. Canadian Petroleum Geology*, **31**(3), 161–8.

PARKER, J. M. (1967) Salt solution and subsidence structures, Wyoming, North Dakota, and Montana. *AAPG Bull.*, **51**, 1929–47.

PARKER, T. J. and MCDOWELL, A. N. (1951) Scale models as a guide to interpretation of salt dome faulting. *AAPG Bull.*, **35**, 2076–86.

PARKER, T. J. and MCDOWELL, A. N. (1955) Model studies of salt-dome tectonics. *AAPG Bull.*, **39**, 2384–470.

PETTIJOHN, F. J. (1957) *Sedimentary Rocks.* Harper Bros., New York.

POSNJAK, E. (1940) Deposition of calcium sulphate from seawater. *Am. Jour. Sci.*, **35A**, 245–72.

PROSPERO, J. M. (1985) Records of past continental climates in deep-sea sediments. *Nature*, **315**, 279–80.

RAJ, R. (1982) Creep in polycrystalline aggregates by matter transport through a liquid phase. *Jour. Geoph. Res.*, **87**, 4731–939.

RAMBERG, H. (1981) *Gravity, Deformation, and the Earth's Crust* (2nd edn). Academic Press, London.

RAUP, O. B. (1970) Brine mixing: an additional mechanism for formation of basin evaporites. *AAPG Bull.*, **54**, 2246–59.

RAUP, O. B. (1982) Gypsum precipitation by mixing seawater brines. *AAPG Bull.*, **66**, 363–7.

REUSCH, E. (1867) Über eine besondere Galtung von Druckgangen im Steinsalz und Kalkspath: Pogg. *Annales Phys. u. Chemie*, **132**, 441–51.

RICHTER-BERNBURG, G. (1959) Zur Palaegeographie der Zechsteins. In: *I giacimenti gassiferi dell Europa Occidentale 1*, Accad. Nazionale dei Lincei, Rome, pp. 88–99.

RICHTER-BERNBURG, G. (1960) Zeitmessung geologischer Vorgänge nach Warven-Korrelationen im Zechstein. *Geol. Rundschen*, **49/1**, 132–48.

RICHTER-BERNBURG, G. (1970) Post-depositional structures in salt. *Proc. Symp. Geol. Technol. Gulf Coast Salt, Louisiana State Univ., Baton Rouge*, pp. 145, 146, 161–6.

RUTTER, E. H. (1983) Pressure solution in nature, theory and experiment. *Jour. geol. Soc. Lond.*, **140**, 725–40.

SANNEMAN, D. (1960) Zur Entwicklung von Salzstock-Familien. (Original German version of Sanneman, 1968.)

SANNEMAN, D. (1968) Salt-stock families in Northwestern Germany. *AAPG Memoir 8, Diapirism and Diapirs*, 261–70.

SCHMALZ, R. F. (1969) Deep-water evaporite deposition: a genetic model. *AAPG Bull.*, **53**, 798–823.

SCHOELL, M., BÄCKER, H. and BAUMANN, Q. (1974) The Red Sea geothermal systems. New aspects on their brines and associated sediments. In: *Problems of Ore Deposition. 4th IAGOD Symp., 1 Varna*, **1**, 303–14.

SCHULZE, G. (1960) Stratigraphische u. genetische Deutung der Bromverteilung in den mitteldeutschen Steinsalzlagern des Zechsteins. *Freiburger Forschungsh. (C)*, **83**, 1–114.

SCIENTIFIC STAFF, CRUISE *BANNOCK* 1984–12 (1985) Gypsum precipitation from cold brines in an anoxic basin in the eastern Mediterranean. *Nature*, **314**, 152–4.

SCRUTON, P. C. (1953) Deposition of evaporites. *AAPG Bull.*, **37**, 2498–512.

SELIG, F. (1965) A theoretical prediction of salt dome patterns. *Geophysics*, **30**, 633–43.

SELIG, F. and WALLICK, G. C. (1966) Temperature distribution in salt domes and surrounding sediments. *Geophysics*, **31**, 346–61.

SELIG, F. and WERMUND, E. G. (1966) Families of salt domes in the Gulf Coastal province. *Geophysics*, **31**, 726–40.

SENGBUSH, R. L. (1983) *Seismic Exploration Methods*. IHRDC, Boston, Massachusetts.

SENI, S. J. and JACKSON, M. P. A. (1983) Evolution of salt structures, East Texas diapir province—pt. 1, pp. 1219–44; pt. 2, pp. 1245–74. *AAPG Bull.*, **67**.

SENI, S. J. and JACKSON, M. P. A. (1984) Sedimentary record of Cretaceous and Tertiary salt movement, East Texas Basin. Bureau of Economic Geology, Univ. of Texas at Austin.

SHEARMAN, D. J. (1966) Origin of marine evaporites by diagenesis. *Trans. Inst. Min. Metall. Sect. B*, **75**, 208–15.

SHEARMAN, D. J., KHOURI, J. and TAHA, S. (1961) On the replacement of dolomite by calcite in some Mesozoic limestones from the French Jura. *Proc. Geol. Assoc.*, **72**, 1–12.

SHERIFF, R. E. and GELDART, L. P. (1982) *Exploration Seismology, Vols. 1 and 2*. Cambridge University Press, Cambridge.

SHLICHTA, P. J. (1968) Growth, deformation, and defect structures of salt crystals. In: *Saline Deposits*. Geol. Soc. Amer. Special Report 88, 597–617.

SMITH, D. A. and REEVE, F. A. E. (1970) Salt piercement in shallow Gulf Coast salt structures. *AAPG Bull.*, **54**, 1271–89.

SMITH, D. B. (1972) Foundered strata and collapse-breccias in the English Zechstein. In: *Geology of Saline Deposits* (Ed. G. Richter-Bernburg). UNESCO, Paris, pp. 255–69.

SMITH, D. B. (1981) The evolution of the English Zechstein Basin. In: *Proc. Int. Symp. Central European Permian, Jablonna, Poland*, 1978.

SMITH, D. B. (1985) Gravitational movement in Zechstein carbonate rocks in Northeast England. In: *The Role of Evaporites in Hydrocarbon Exploration* (*Course Notes No. 39*), Joint Assoc. for Petroleum Exploration Courses (UK), London.

SMITH, D. G. and PULLEN, J. B. (1967) Hummingbird structure of south-east Saskatchewan. *Can. Soc. Petr. Geol. Bull.*, **15**, 468–82.

SONNENFELD, P. (1984) *Brines and Evaporites*. Academic Press, New York.

SØRENSEN, K. (1986) Rim syncline volume estimation and salt diapirism. *Nature*, **319**, 23–7.

SPIERS, C. J., LISTER, G. S. and ZWART, H. J. (1982) The influence of fluid-rock interaction on the rheology of salt rock, and on ionic transport in the salt: First results. Pub. Eur. Atomic Energy Community, WAS-153-80-7N, 268–80.

SPIERS, C. J., URAI, J. L. and LISTER, G. S. (1986) The effect of brine (inherent or added) on rheology and deformation mechanisms in salt rock. *Proc. 2nd. Conference on the Mechanical Behaviour of Salt, Hannover*, 1984 (Ed. Hardy and Langer). In press.

SZATMARI, P., MILANI, E., LANA, M., CONCEIÇÃO, J. and LOBO, A. (1985) How South Atlantic rifting affects Brazilian oil reserves distribution. *O & GJ*, 14/1/85, 107–13.

TALBOT, C. J. and ROGERS, E. A. (1980) Seasonal movements in a salt glacier in Iran. *Science*, 208, 395–7.

TALBOT, C. J., TULLY, C. P. and WOODS, P. J. E. (1982) The structural geology of Boulby (Potash) Mine, Cleveland, United Kingdom. *Tectonophysics*, **85**, 167–204.

TAVIANI, M. (1984) Submarine 'sinkholes': A review.

Proc. 1st. Multidisc. Conf. on Sinkholes, Orlando, Fla., 15–17 Oct. 1984 (Ed. B. F. Beck), Univ. of Central Florida.

TAYLOR, R. E. (1938) Origin of cap rock of Louisiana salt domes. *Louisiana Dept. Conserv. Geol. Bull.*, **11**.

TAYLOR, J. C. M. (1980) Origin of the Werraanhydrit in the UK Southern North Sea—a reappraisal. In: *The Zechstein Basin. Contr. Sedimentology 9* (Ed. Füchtbauer, H. and Peryt, T. M.). Schweitzerbart'sche Verlag, Stuttgart, pp. 91–113.

TAYLOR, J. C. M. (1981) Zechstein facies and petroleum prospects in the Central and Northern North Sea. In: *Petroleum Geology and the Continental Shelf of North-West Europe* (Ed. Illing, L. V., and Hobson, G. D.). Heyden & Son Ltd, London, for the Institute of Petroleum.

TAYLOR, J. C. M. (1984) Late Permian–Zechstein. In: *Introduction to the Petroleum Geology of the North Sea* (Ed. K. W. Glennie). Blackwell Scientific Publications, Oxford.

TAYLOR, J. C. M. (1985) Principal basin models. In: *The Role of Evaporites in Hydrocarbon Exploration* (*Course Notes No. 39*) Joint Assoc. for Petroleum Exploration Courses (UK), London.

TAYLOR, J. C. M. and COLTER, V. S. (1975) Zechstein of the English sector of the Southern North Sea Basin. In: *Petroleum and Continental Shelf of North-West Europe* (Ed. A. W. Woodland). Elsevier Applied Science Publishers, London.

THIERBACH, R. and MAYRHOFER, H. (1978) Elektromagnetische Reflexionsmessungen in Salzlagerstätten. *Proc. 5th. Int. Symp. on Salt*, Hamburg, 393–403.

THOMPSON, E. G. (1965) An experimental technique for the investigation of the flow of halite and sylvinite. Univ. of Texas at Austin, Ph.D. Thesis.

TOLMACHOFF, I. P. (1926) A salt dome, Solenaya Sopca, in northern Siberia. *Econ. Geol.*, **31**, 774–81.

TRABANT, P. K. and PRESLEY, B. J. (1977) Orca Basin, an anoxic depression on the continental slope, Northwest Gulf of Mexico. *AAPG Studies in Geol.*, **7**, 303–11.

TRIPPET, A. R. (1981) Characteristics of diapirs on the outer continental shelf—upper continental slope boundary, Northwest Gulf of Mexico. *Transactions, Gulf Coast Assoc. Geol. Socs.*, **31**, 391–7.

TRUSHEIM, F. (1960) Mechanism of salt migration in northern Germany. *AAPG Bull.*, **44**, 1519–40.

TUCKER, M. E. (1981) *Sedimentary Petrology*. Blackwell Scientific Publications, Oxford.

TWENHOFEL, W. H. (1950) *Principles of Sedimentation*. McGraw-Hill, New York.

URAI, J. L. (1983) Water assisted dynamic recrystallisation and weakening in polycrystalline bischofite. *Tectonophysics*, **96**, 125–57.

USIGLIO, J. (1849) Analyse de l'eau de la Méditerranée sur les côtes de France. *Ann. Chim. Phys.* (3) 27 (2) 92–107; (3), 177–91.

VAN T'HOFF, J. H. and WEIGERT, F. (1901) Untersuchungen über die Bildungsverhältnisse der ozeanischen Salzablagerungen, insbesondere des stassfurter Salzlagers. *Sitzungsber. k. preuss. Akad. Wiss.*, **23**, 1140–8.

VAN TUYL, F. M. (1930) Contribution to salt-dome problem. *AAPG Bull.*, **14**, 1041–7.

WALTON, A. W. (1978) Evaporites: relative humidity control of primary mineral facies: a discussion. *Jour. Sed. Petr.*, **48**, 1357–9.

WARDLAW, N. C. and SCHWERDTNER, W. M. (1966) Halite-anhydrite seasonal layers in the Middle Devonian Prairie Evaporite Formation, Saskatchewan, Canada. *Geol. Soc. Amer. Bull.*, **77**, 331–42.

WELLS, A. J. and ILLING, L. V. (1964) Present-day precipitation of calcium carbonate in the Persian Gulf. In: *Deltaic and Shallow Marine Deposits* (Ed. L. M. J. U. van Straaten). Elsevier, Amsterdam, pp. 429–435.

WEYL, P. K. (1967) The solution behaviour of carbonate materials in sea water. *Stud. Trop. Ocean, Miami*, 178–278.

WHITTLE, A. S. (1985) Salt flank delineation using the proximity survey—a technique revisited. (A presentation at the 47th. EAEG Meeting, Budapest, Hungary.)

WILGUS, C. K. and HOLSER, W. T. (1984) Marine and non-marine salts of western interior, United States. *AAPG Bull.*, **68**, 765–7.

WILSON, J. L. (1975) *Carbonate Facies in Geologic History*. Springer Verlag, New York.

WOODBURY, H. O., MURRAY, I. B. and OSBORNE, R. E. (1980) Diapirs and their relation to hydrocarbon accumulations. In: *Facts and Principles of World Petroleum Occurrence. Can. Soc. Petr. Geol. Memoir 6* (Ed. Miall), pp. 119–142.

YANAT'EVA, O. K. (1955) Effect of aqueous solution of gypsum on dolomite in the presence of carbon dioxide. *Akad. Nauk. SSSR., Doklady*, **101**, 911–12.

Index